ENVIRONMENTALISM GONE MAD:

HOW A FORMER SIERRA CLUB ACTIVIST AND SENIOR EPA ANALYST DISCOVERED A RADICAL GREEN ENERGY FANTASY

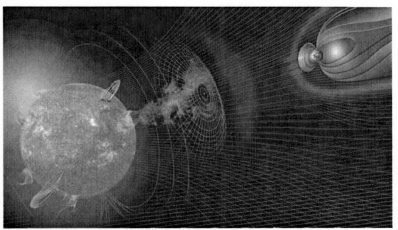

Alan Carlin

Stairway Press, Mount Vernon, Washington

Grayscale ISBN: 978-1-941071-13-7

Color ISBN: 978-1-941071-92-2

eBook ISBN: 978-1-941071-14-4

STAIRWAY⹀PRESS

www.StairwayPress.com

1500A East College Way #554

Mount Vernon, WA 98273

Cover photo: Lenticular cloud over Fortuna Bay on the North Shore of South Georgia Island in the South Atlantic Ocean; Credit: Alan Carlin.

Title page illustration: Storms from the Sun—artist illustration of events on the sun changing the conditions in Near-Earth space. Credit: NASA at: http://www.nasa.gov/mission_pages/sunearth/news/storms-on-sun.html

Chapter icons: Satellite image of Earth's interrelated systems and climate. Credit: NASA at: http://grin.hq.nasa.gov/ABSTRACTS/GPN-2002-000121.html

Contents

Appendices (on separate disk from publisher)

A   2009 Comments to EPA on Draft EPA GHG TSD

B   John Broder's Spin Job on Alan Carlin by Marlo Lewis

C   A More Technical Discussion of CAGW Science, Alternatives to It, and Economic Cost-Benefit Analysis of EPA Regulations

D   The Politics of EPA's Endangerment Finding, a Congressional Staff Report

List of Acronyms

AFD	anti-fracking doctrine as defined in Chapter 1
AGW	anthropogenic global warming
AID	(US) Agency for International Development
AR1...5	Assessment Report 1...5 of UN IPCC
BSER	Best System of Emission Reduction
^{14}C	carbon 14, an isotope of carbon used as a proxy for cosmic ray intensity
°C	degrees Centigrade
CAGW	catastrophic anthropogenic global warming
CAP	Central Arizona Project
CEI	Competitive Enterprise Institute
CCSP	(US) Climate Change Science Program
CFC	chlorofluorocarbon
CIC	climate-industrial complex defined in Chapter 1
CO_2	carbon dioxide
COP	Conference of the Parties under UNFCCC
CR	cosmic rays
CRU	Climatic Research Unit of the University of East Anglia
CSF	climate sensitivity factor
CWP	current warm period
DDT	dichlorodiphenyltrichloroethane
ENSO	El Nino southern oscillation
EPA	(US) Environmental Protection Agency
EKC	environmental Kuznets curve
EWD	extreme weather doctrine as defined Chapter 1

GCCD	global climate change doctrine as defined in chapter 1
GCR	galactic cosmic rays
GDP	gross domestic product
GEF	green energy fantasy as defined in Chapter 1
GHG	greenhouse gas
GISP	Greenland Ice Sheet Project
GWD	global warming doctrine as defined in Chapter 1
HadCRUT	dataset of monthly instrumental temperature records formed by combining the sea surface temperature records compiled by the Hadley Centre of the UK Meteorological Office and the land surface air temperature records compiled by the Climatic Research Unit (CRU) of the University of East Anglia
IPCC	(UN) Intergovernmental Panel on Climate Change
LIA	Little Ice Age
LNG	liquefied natural gas
LNT	linear no threshold
MACT	maximum achievable control technology
MIT	Massachusetts Institute of Technology
NAO	North Atlantic oscillation
NASA	National Aeronautics and Space Administration
NCA	(US) National Climate Assessment
NCEE	(USEPA) National Center for Environmental Economics
NGO	non-government organization

List of Acronyms

NIPCC	Nongovernmental International Panel on Climate Change
NOAA	National Oceanic and Atmospheric Administration
NO_x	nitrogen oxides
NRC	National Research Council
NRDC	Natural Resources Defense Council
NSF	National Science Foundation
^{18}O	oxygen 18, an isotope of oxygen used as a proxy for temperature
OAP	Office of Air Programs, USEPA
OECD	Organization for Economic Cooperation and Development
OMB	(US) Office of Management and Budget
ORD	Office of Research and Development, USEPA
PDO	Pacific decadal oscillation
PM	particulate matter
$PM_{2.5}$	particles less than 2.5 micrometers in diameter; sometimes called "fine" particles
RFS	renewable fuel standard
RIA	regulatory impact assessment
RPS	renewable portfolio standard
RSS	Remote Sensing Systems
SCC	social cost of carbon
SM	scientific method
SO_2	sulfur dioxide
SOE	state-owned enterprises
SRM	solar radiation management
TSD	technical support document
TSI	total solar irradiance

UAH	University of Alabama at Huntsville
UHI	urban heat island
UN	United Nations
UNEP	United Nations Environment Programme
UNFCCC	United Nations Framework Convention on Climate Change
US	United States
US$	United States Dollar
USEPA	United States Environmental Protection Agency
USG	United States Government
USGCRP	US Global Change Research Program
WG1...3	Working Group 1...3 of the UN IPCC

Preface

THE PURPOSE OF this book is to explain why I changed from my lifelong support of the environmental movement to extreme skepticism concerning their current primary objective of reducing emissions of carbon dioxide. My change was based on a personal journey of discovery which initially looked at the "solutions" that climate alarmists proposed to avoid alleged catastrophic changes in the climate they prophesized and later examined the science they used to argue for the alleged need for these solutions. Along the way I learned that there are many others, including a number of prominent researchers in the field, who had reached similar conclusions concerning alarmist climate science. My review of this science, in turn, led to an exploration of the principal groups supporting climate alarmism and an analysis of the climate alarmism and radical environmentalism they advocate from a multidisciplinary perspective. This book outlines what I learned in this voyage of discovery.

Although I and the many other climate skeptics are now referred to as "deniers" by the climate alarmists, that does not change the science—that there is no valid scientific basis for the alarmists' catastrophic climate predictions—or justify their fantastically expensive and useless "solution."

The discussion draws on my long involvement with environmental issues as a 38 year senior EPA employee and on my experiences as a former environmental activist and Sierra Club leader.

Unfortunately, it is increasingly difficult to carry on a meaningful discussion of the climate issue because of its intense politicization and the continuing deliberate distortions and misrepresentations made by climate alarmists. In the United States this politicization has long involved the media and scientific worlds but has now reached the highest levels of government, which proclaim that the alarmists' views are supported by science, that the science should be decided on the basis of a "consensus," and that it is "settled." Furthermore, contrary views are not tolerated by the Federal Government and those who hold them are not welcome to be employed by the Government. It is appropriate, supporters believe, to describe those who hold contrary views by using denigrating epithets and *ad hominem* attacks. The suppression of unpopular views may be common in religion and politics, but it is not a good route to scientific knowledge or free speech or meaningful discussion of public policy, and those government officials who advocate such suppression should be the first to understand this. In this case their advocacy has increasingly authoritarian overtones.

Unlike religion and politics there is a correct answer to a scientific question, although it may take some time and considerable effort to discover what it is. And it is never "settled" or based on "consensus." It is based on applying the scientific method to determining whether a hypothesis conforms to available real world observations relevant to the topic.

My background is somewhat unique among those interested in the topic since I have an extensive background in both science and economics. I consider myself a strong environmentalist. I had and have nothing to gain from taking any particular stand on the climate issue, so hope that my views may provide a new perspective.

The book is multidisciplinary, which is appropriate considering the nature of environmental policy in general and climate in particular. It is part autobiography, part environmental policy analysis, part scientific analysis, part economic analysis, and part political and legal analysis. It is also likely to be controversial since it reaches very different conclusions from those of climate alarmists, much of the mainstream media, many liberal politicians, President Obama, many Western European governments, the European Union, and the United Nations. These conclusions are of considerable importance, however, from all these perspectives.

Although more sophisticated approaches using science and economics will be presented in Chapters 9 and 11 and Appendix C, a simple way for readers to make up their own minds on climate proposals is to ask whether previous alarmist predictions have proved accurate. The climate alarmists have now been making their apocalyptic predictions for almost 30 years and it is now possible to compare their predictions with actual physical observations. This book provides the information needed to judge these predictions.

This book argues that environmental policy has been hijacked by radicals intent on imposing their ideology by government fiat on the rest of us whether we like it or not. This would not be of much concern except that they are being supported by many Western European countries and the Obama Administration.

I argue that environmental policy should be based on good science and economics and respect for the laws and Constitution of the United States rather than ideology or invalid scientific and economic findings often based on vague policy slogans. I do not know of any useful alternative. If environmental policy is based on governmental fiat or "green" policy prescriptions the results have been and are very likely to continue to be disastrous. The most likely consequence is that policy will be based on the shrill emotions of those who feel most strongly and have the largest public relations budgets; policy will likely be much too stringent, inconsistent, and likely to cause unintended and even deadly consequences. Overly strict environmental regulations can and have resulted in very bad outcomes. Alarmist climate policy has already had significant adverse effects.

I have a strong preference for data rather than ideology and arguments from authority. A good data set is worth far more than all the opinions that have been expressed on a scientific subject. Science should be based on good and reproducible data and the scientific method. Environmental policy should be based on the resulting good science together with sound economics and law.

Generally speaking, people appear to reach conclusions concerning science-based issues either by examining the data themselves or by trusting the views of alleged experts on the subject. The UN Intergovernmental Panel on Climate Change appears to have been set up primarily for the purpose of providing a body of science expertise that people would be likely to trust. It was carefully structured so as to keep the ultimate policy conclusions within a small group sympathetic to the UN's policy conclusions and by the governments most involved. Their approach to scientific assessment, which was formulated before the days of widespread use of the internet, has been remarkably successful in influencing global thinking on the subject, particularly in Western Europe; it might even have been adopted by more of the developed world except for the efforts of a few brave politicians and a dedicated, largely volunteer, group of concerned citizens and scientists, mainly in the US and Australia, whose views would have been drowned out by the mass media if it had still held its former sway.

Fortunately, the skeptics have raised enough doubts about the UN science that this will be more difficult. Despite unremitting and disgraceful attacks by climate alarmists, the doubts raised by the skeptics have been sufficient so that many Congressional Republicans have adopted a strongly negative (and I believe appropriate) view on the UN climate science.

I would like to thank Larry Bell and my daughter, Nancy Bundics, for their ideas, help, and encouragement.

My website, http://carlineconomics.com, will provide selected updated information on related topics.

1
Introduction

ALTHOUGH I DID not plan it that way, my career has been framed by the rise of the environmental movement. I was a Sierra Club activist and Chapter leader during one of the pivotal environmental battles of the 1960s which brought the environmental movement to public consciousness and the Sierra Club to prominence. I also played a role in the great global warming/climate scare as a senior analyst at the US Environmental Protection Agency on what has become the signature issue of the environmental movement in recent decades—global warming, later changed to climate change, and most recently to extreme weather.

This scare can only be described in superlative terms. It was and is audacious, deceptive, bold, mad, scandalous, and has come closer to achieving its purposes than it should have given its flimsy and invalid scientific basis. The proposed "solution" advanced by the environmental movement is even worse. This book attempts to explain how I reached my conclusions as to its scientific validity, or rather the lack thereof, how I came to play a role in the issue, how the Obama Administration's climate/energy policy is wasting very large sums on non-solutions to minor or non-problems, why the science strongly suggests following just the opposite policy, why the "solution" is just as much a fantasy as the science, and what I learned along the way about American journalism, science, and government, the attempts by the Obama Administration to undermine the economic criteria long used to judge regulatory policy, and finally the very important role that fossil fuel development can play in improving our economy and national security if we ignore or reject the radical environmental ideology so opposed to such development. It also examines the evolution of the environmental movement from its bipartisan wilderness preservation and real pollution control days of my first involvement with it to its current far left radical ideology concerning restricting energy generation and use through dubious and even illegal means.

How the Environmental Movement Lost Its Way

I actively supported the environmental movement in the 1960s and early 1970s during its early rise to prominence in the United States and on the basis of this experience decided to devote most of my career to supporting its goals by joining the US Environmental Protection Agency in 1971 and working there for over 38 years. However, it is now clear that between then and now the movement lost its way and was taken over by radicals advocating fantasies that would actually harm the environment, the economy, national security, and human welfare. Broadly speaking, the movement became preoccupied with the effects of energy generation and use rather than meaningful environmental problems. This is the story of how I went from supporting a movement to fighting it. What has changed is the movement rather than me. The absurdity of what they now advocate with regard to energy generation and use will be discussed in many other parts of this book.

A Few Important Definitions

Because of the often contradictory phraseologies used in the climate debate I propose to introduce several acronyms. The first is Green Energy Fantasy or GEF. GEF is an expansion on the term Global Warming Doctrine (GWD), as formulated by Vaclav Klaus, then President of the Czech Republic, in a speech in London in early

Introduction

2011. He describes GWD as an ideology if not a religion that has several loosely related claims and beliefs:[1]

> *1. It starts with the claim that there is an undisputed and undisputable, empirically confirmed, statistically significant, global, not regional or local, warming;*
>
> *2. It continues with the argument that the time series of global temperature exhibits a growing, non-linear, perhaps exponential trend which dominates over its cyclical and random components;*
>
> *3. This development is considered dangerous for the people (in the eyes of soft environmentalists) or for the planet (among "deep" environmentalists);*
>
> *4. The temperature growth is interpreted as a man-made phenomenon which is caused by the growing emissions of CO_2. These are considered the consequence of industrial activity and of the use of fossil fuels. The sensitivity of global temperature to even small variations in CO_2 concentration is supposed to be high and growing;*
>
> *5. The GWD exponents promise us, however, that there is a hope: the ongoing temperature increase can be reversed by the reduction of CO_2 emissions...;*
>
> *6. They also know how to do it. They want to organize the CO_2 emissions reduction by means of directives (or commands) issued by the institutions of "global governance". They forget to tell us that this is not possible without undermining democracy, independence of individual countries, human freedom, economic prosperity and a chance to eliminate poverty in the world. They pretend that the CO_2 emissions reduction will bring benefits which will exceed its costs.*

---

[1] Klaus, 2011.

3

The first five points are claims potentially subject to scientific verification. Point 6, on the other hand, is a belief as to what society should do. It is now necessary to expand the GWD term used by Klaus because the leaders of the alarmist movement have tried to broaden the appeal of their ideology and/or solve the problems posed by nature's stubborn refusal to exhibit any global warming trend for at least the past 17 by first claiming that it is now necessary to reduce CO_2 emissions in order to prevent climate change and in 2012 to prevent extreme weather events. They have not changed their "solution," however, so my new term, Green Energy Fantasy or GEF, includes all of GWD plus the two additional phenomena, climate change and extreme weather events, that the movement now claims that government can influence by government-dictated reductions of greenhouse gas (GHG) emissions and believes that it should do so. So it is necessary to expand Klaus' six points to include several additional ones to reflect these more recent developments:

> 7. *The belief that electric power should be increasingly generated by "renewable" sources including wind, solar, and biomass, but excluding hydro, and that electric motors should even be used to power transportation vehicles.*
> 8. *The claim that human-caused GHG emissions will result in climate change, and the beliefs that preventing climate change should be the new objective and that climate change should be reduced, if not prevented, by reducing these emissions.*

Point 7 is an amplification of point 6 but needs to be made more clearly since it is such an important part of the alarmist ideology. Point 8 results from the redefinition of the objectives of the movement from preventing alleged global warming to preventing climate change in the early 2000s, so that the new acronym might better be Global Climate Change Doctrine or GCCD.

Introduction

In 2012 the definition of the movement's goals was expanded still further by including:

9. The claim that unusual or extreme weather events are being made worse because of changes in human-caused emissions of CO_2 and the belief that such events should be reduced or even prevented by reducing these emissions.

10. The claim that CO_2 levels and emissions can and should be substantially reduced by using regulations to require CO_2 emissions reductions.

Point 9 arose apparently in response to several unusual weather events in the US during 2012 and elsewhere in 2013 and possibly because there has been no global warming since 1998. Presumably there will always be extreme weather even though such events have been decreasing in recent years by most if not all measures (as discussed at the beginning of Chapter 9). This claim does not fit within the GWD or GCCD frameworks since it alleges a different threat—but exactly the same "solution."

Point 10 arose in response to defeat of the Obama Administration's cap and trade bill in the 111th Congress and is unique to the US. The Obama Administration claims that they can and will accomplish the same goals by having the EPA use the Clean Air Act to write regulations reducing CO_2 emissions. Such regulations were proposed by EPA in 2014.

For the purposes of this book I will include the first ten points as part of GEF or GWD/GEF. Point 8 alone will be referred to as GCCD. Point 9 alone will be referred to as the Extreme Weather Doctrine or EWD.

Many alarmists, including the Sierra Club, go even further by advocating what I will call the Anti-Fracking Doctrine or AFD, namely:

11. The claim that hydraulic fracturing is dangerous and the belief that it should not be used to gain access to natural gas and oil resources contained in shale formations with low permeability.

Point 11 has been added to the ideology only in the last few years as fracking has become more common and immensely valuable as a technology for increasing oil and gas supplies, so far primarily in the US. Fracking along with horizontal drilling has proved useful to free oil and natural gas tightly contained within shale formations. In the case of shale gas, the new technologies may be feared by environmentalists because they make renewables even more expensive relative to fossil fuel sources for the indefinite future. AFD is currently of concern primarily in a few blue states such as New York and California and in many parts of Western Europe, but is not publicly supported by the Obama Administration,[2] the Cameron Government in Great Britain, or the Brown Administration in California. These three administrations publicly support only points 1 through 10, or what I call GEF.

The attempt to impose GEF is best understood as a widespread, organized conspiracy primarily in Western countries involving a very wide variety of groups which have come together to promote their common interests in bringing it about. I believe these efforts to be harmful and often underhanded, but generally not illegal until very recently in the US. I propose to use the term climate-industrial complex or CIC to describe these groups as a whole. This term is patterned after the military-industrial complex that President Eisenhower warned against in his 1961 Farewell Address,[3] although it also relates to the second, but less well known, threat to democracy that he posed—the scientific-

---

[2] A recent minority report from the Environment and Public Works Committee of the US Senate (2014c) argues that the Obama Administration is quietly seeking to regulate fracking and expand Federal regulation of the oil and gas industry, so the Administration's public opposition to AFD may not be sincere.

[3] Eisenhower, 1961.

technological elite. With regard to this latter threat he stated:

> *In the same fashion, the free university, historically the fountainhead of free ideas and scientific discovery, has experienced a revolution in the conduct of research. Partly because of the huge costs involved, a government contract becomes virtually a substitute for intellectual curiosity. For every old blackboard there are now hundreds of new electronic computers. The prospect of domination of the nation's scholars by Federal employment, project allocations, and the power of money is ever present—and is gravely to be regarded. Yet, in holding scientific research and discovery in respect, as we should, we must also be alert to the equal and opposite danger that public policy could itself become the captive of a scientific-technological elite.*

With the substitution of GEF-related industries for military industries and narrowing the scientific-technological elite to the climate science elite, the CIC poses both dangers that Eisenhower warned of in his Farewell Address and for the reasons that he stated, including the danger that Federal funding of scientific research could warp the research findings. GWD/GEF/EWD is the ideology used to persuade the public and governments at many levels to mandate certain energy-related policies that will benefit the CIC. CIC is intended to be very broad by including all individuals and institutions that advocate or benefit from GWD/GEF/EWD. It thus includes those organizations supporting GWD/GEF/EWD in scientific organizations who have made official statements on climate science, such as the National Academy of Sciences, environmental organizations such as the Sierra Club, their funders such as the Sea Change and Energy Foundations, the sympathetic mainstream press, such as the *New York Times*, the *Guardian*, and the British Broadcasting Corporation; the alarmist blogosphere, such as *Wikipedia,* commercial groups that benefit from Federal and state GEF-related subsidies such as corn ethanol producers, corn farmers, wind turbine producers and installers, solar energy equipment makers and installers, electric car companies, light bulb manufacturers who benefit from outlawing incandescent bulbs, international organizations, such as the UN and the EU, various national and state governments such as the Obama Administration, California, Britain, and Germany, and some academic researchers and journal editors, particularly climate model builders. I will often refer to people who advocate GWD/GEF/EWD as alarmists or CIC supporters.

All these components—scientists, propaganda, government supporters at the national and international levels, environmental organizations, and funding groups—have worked together to promote GEF. Such a combined effort is much more effective than individual efforts by each of these groups. In fact the CIC is fairly tightly organized at many levels, particularly in its funding functions.

Thus the mass media and alarmist blogosphere provide the propaganda that the CIC hopes will keep the public receptive to sacrifices required by GWD/GEF, and keeps supporters informed as to the current CIC agreed point of view. National governments provide the regulations and subsidies needed to make it possible for private windmill/solar/ethanol producers/non-incandescent light bulb manufacturers to make a profit selling their products that would otherwise be unsalable due to their inferior performance, provide grants/contracts/payments to researchers to provide research supporting GWD/GEF, and provide funding for the UN to produce their five year assessments and never ending conferences in many parts of the world promoting GEF/CAGW and international agreements in support of GEF. Researchers provide the "science" used to support the UN Intergovernmental Panel on Climate Change (IPCC) assessments and the subject matter used in mass media propaganda in return for government funding. CIC industries provide campaign contributions to politicians in return for favorable regulations and subsidies. Environmental organizations attract members and contributions to lobby government on behalf of GWD/GEF and encourage people to vote for CIC-supporting politicians. Environmentally oriented foundations provide substantial funding to environmental organizations to support their GWD/GEF efforts, and are reportedly in turn funded by wealthy left-wing contributors.[4]

So, this is just what Eisenhower was talking about, only much more so since it involves many additional components. If all these components work together with the aim of achieving an end that is harmful or not in the public interest, it can reasonably be described as a conspiracy.

At times I will also use the acronym CAGW, or catastrophic anthropogenic global warming, one of the major components of the GWD and GEF ideologies, when I am referring only to that scientific hypothesis rather than the broader GWD/GEF/AFD ideologies, which include non-scientific components. CAGW is the principal scientific viewpoint underlying these ideologies. GWD/GCCD/GEF/AFD is a system of beliefs concerning both science and government policy that is primarily but not exclusively based on the CAGW hypothesis. The C in CAGW is important. There is indeed some anthropogenic global warming (AGW), particularly in large urban areas, which I would argue is good, not bad, but it is the claim that there will be *catastrophic* AGW which is central to the alarmist ideology and will be the subject of much of the scientific discussion in this book. There is no reason to be concerned about mild AGW; there is for CAGW if it should ever occur, but for which no scientific case has been made. The difference is whether the AGW results in great or significant effects and damage. EWD is a policy but is allegedly also based on "science," which will be referred to as EWD science.

Another important definition concerns the phrase "climate change." The United Nations under the United Nations Framework Convention for Climate Change (UNFCCC) defines climate change as "change of climate which is attributed directly or indirectly to

---

[4] *Ibid*.

human activity that alters the composition of the global atmosphere and which is in addition to natural climate variability observed over comparable time periods."[5] In this book, I will do the opposite, so that climate change means the more normal definition of changes in longer-term weather trends without regard to how they may be caused.

Principles for Deciding Environmental and Natural Resource Development Issues

Despite my seeming change of course on environmental policy as advocated by the environmental movement, I actually have had and still have a set of fundamental principles in reaching my conclusions as to an appropriate response to environmental problems and how they are best dealt with in general.

The short version is that I have long believed in both economic development and growth as well as environmental protection and the rule of law. I think I am not unusual in these beliefs among most Americans, but I certainly am at odds with many modern radical environmentalists, who too often believe that environmental protection and development are incompatible and even that pursuit of their environmental ideology should go beyond what is permitted by current law and the Constitution. I, on the other hand, believe that while there can be some conflicts between development and protection, development makes possible environmental protection in the longer run.

The longer version is that my approach follows closely my background in economics in that I believe that environmental problems and resource development issues should generally be left to the private market unless there is a demonstrable market failure in the economic sense[6] which can partially or wholly be corrected

---

[5] Article 1 (2).

[6] A market failure occurs when a market is not efficient. Technically, a market failure occurs when there exists another outcome in which one

10

Introduction

by governmental action whose incremental social economic benefits exceed the incremental social costs. The benefit-cost requirement is included because if society as a whole would be made worse off economically by government intervention in the market to correct a market failure the intervention is not worthwhile in my view.

In addition, this general principle assumes that the science used in determining the economic benefits is scientifically valid in the sense that it can stand up to review using the scientific method and available observational evidence. This statement could be abbreviated by omitting the "in addition" sentence because only benefits that have been demonstrated to have objective reality can reasonably be described as benefits. But given the current state of the climate science debate, it is probably better to make the scientific validity qualification explicit. There are obviously a number of non-economic considerations that should also be considered in deciding on regulations. One is that regulations should also take account of distributional effects (who gains and who loses), national security concerns, and be in accordance with current law. In most cases the national security concerns will be satisfied if the economic ones are, but in the unusual case of conflict, the security aspects also need to be taken into account.

Market failures can arise from a variety of causes, but in the case of environmental and natural resource issues the most common is as the result of what economists call "externalities," where the effects of purchase, production, or use of a product or service are imposed on others who did not have a choice and whose interests were not taken into account. As a result of negative externalities private costs of production are sometimes lower than their societal or social cost. An example is the effects of noise pollution from airplanes on those not using the airplanes causing the noise.

Most pollution problems arise from externalities. In the case of air pollution in Los Angeles in the mid-Twentieth Century, those operating motor vehicles did not take into account the adverse

---

participant can be made better off without making anyone else worse off.

effects their vehicles were having on citizens who were breathing the air. Thanks to the work of Arie Haagen-Smit of Caltech and other scientists, we learned what was causing these effects, and government action to reduce motor vehicle emissions has greatly alleviated the problem. These government interventions had substantial costs, but they were more than counter-balanced in my view by the benefits to humans who breathed the improved quality air.

Although the use of the market failure criteria may seem like commonsense to many people, it is surprising how many government environmental interventions do not satisfy these minimal standards despite many years of efforts to develop and use such benefit-cost standards. Many radical environmentalists believe, for example, that human use of natural resources should be directly regulated by government to the point of deciding exactly how much energy from what source should be used. They apparently do not realize that such detailed government intervention is normally counter-productive where it has been tried, as in Eastern Europe during the Soviet era.

The East German government decided that ordinary mortals should be happy with the Trabant, a small, underpowered vehicle with a skin made of plastic reinforced with recycled fibers (that horses were said to eat at times) and a waiting list of decades. Similarly, but closer to home, the Obama Administration decided that people would have to buy new cars that averaged no more than 54.5 miles per gallon starting in model year 2025. It apparently never occurred to either government that everyone would be better off to leave such decisions to the market place so that people could buy what best meets their needs and they can afford, not what government in its wisdom decided they should have. In both cases, there were no significant market failures in my view and the government cure was far worse than the problems they claimed they were trying to solve.

In the case of increasing the US mileage standard, the Obama Administration decided that reducing pollution and fuel use were more important than the negative effects, particularly greater highway deaths from driving lighter, less crash worthy vehicles. These increased deaths are less visible and difficult to identify but are nevertheless a very real cost of their environmentally-inspired decision. As a result those who think that their lives are more important will have fewer means to meet their needs.

It is often difficult for environmentalists to understand why all pollution of any kind should not be eliminated by government action. They recently have even taken to calling an essential trace gas, carbon dioxide, a pollutant in order to argue for a drastic reduction in emissions of it. The reason all actual pollution cannot be eliminated is that we live in a world with unlimited wants and limited means. Expenditures for unlimited pollution control would mean that there would be reduced resources available for other needs such as education or health care. Economics provides a conceptual basis for understanding these trade-offs and making rational choices between them. This is often anathema to radical environmentalists, who do not realize that either directly or indirectly these trade-offs will and must be made. The important choice is not whether they will be made but whether they will be made in a rational manner.

Radical environmentalists sometimes go beyond what would be suggested by economics by supporting a "precautionary principle" that says that we should assume that environmental problems need to be addressed before we know enough to do so according to the above principles in order to avoid environmental damages that would occur before the science can be established. This approach has and will lead to governmental action when it is not necessary, damaging, and unwise, and increases the chances that the resulting actions will actually decrease social welfare.

Other, even more radical, environmentalists believe that environmental concerns should be given preference over implementing environmental laws as written. I regard such challenges to the rule of law as dangerous and counterproductive since abandoning the rule of law endangers the democratic basis for our government. Similarly, compromising national security endangers the existence of government and ordered markets, and thus the rule of law.

As will be discussed in Chapter 11, government actions to reduce CO_2 emissions in order to reduce alleged global warming/climate change/extreme weather do not meet my conditions for interfering with market-determined solutions because the science has not been proven valid using the scientific method, and there are few if any (and probably negative) benefits and very high costs of such government interference. This is a major instance where radical environmentalists want to pursue imaginary or even negative benefits at very high costs. This is irrational and damaging to society and human welfare.

Outlook Towards the Environment

I have long believed myself to be an environmentalist. I have always believed as strongly in what used to be called conservation as anyone I knew. I believe in preserving the natural world whenever such preservation can be justified according to the scientific and economic principles just outlined, including taking into account the damages caused by development. Use of these principles has put me on the side of both the environmentalists and the developers at different times and on different issues. But I believe the principles are fundamental to good environmental and natural resource policy. The object of such policies should be to minimize government involvement while maximizing economic welfare. In those very rare cases where there is real doubt after careful analysis, and a decision simply must be made before the doubt can be resolved, I believe that preservation should usually be given the benefit of the doubt. Normally, however, there is no reason to rush such decisions and every reason to make them carefully.

I came to these environmental views in part as a result of a continuing interest in natural history beginning in high school, from working summers in high school at a natural history museum, and a lifetime spent exploring wilderness areas in the United States and over 60 other countries. I have sought out opportunities to explore, observe, and photograph wildlife and landscapes and the few human habitations wherever I have gone and recorded what I saw in a large collection of photographs. I have participated in private and organized trips to natural areas on every continent, almost all at my own expense. This may be a little unusual for economists but nevertheless reflects my background and interests.

I have spent an unusual amount of time in the Arctic and Antarctic, where I have cumulatively spent several months in each. I have learned first-hand about the fragile environment of these polar areas as well as the harsh realities of their climate and often total isolation from the services provided by civilization. I have camped and hiked in Greenland and observed polar bears eating seals from a hundred feet away. I have watched albatrosses learning to fly and visited million-strong penguin rookeries on South Georgia Island in the extreme South Atlantic. I have visited an Emperor penguin colony and the unearthly dry valleys in Antarctica. Two of the ships I sailed on to these areas later sank. One sank as a result of hitting sea ice; the other sank because it hit an uncharted submerged rock. These sinkings are not surprising given the ice flows and less than comprehensive navigation charts of the areas that "adventure travel" ships visit. Happily I never suffered more than a little seasickness on the roughest passages to Antarctica and nearby islands.

I do not, however, wish to live in the polar areas on a more permanent basis. They are beautiful but inhospitable to humans, particularly during the winter, and best visited with ships specially designed for sailing in ice-clogged/covered seas. I do not comprehend why anyone would be particularly concerned about somewhat warmer temperatures in these areas as long as the major icecaps do not substantially melt, which they have not and seem unlikely to do so at this stage of the ice age cycle. Since the icecaps did not significantly melt during the temperature highs of the current interglacial period more than 3,000 years ago, there need be no concern that this will happen as a result of the very minor warming experienced in the late Twentieth Century. Many of the humans that prefer the Arctic are Inuit, whom I visited in Greenland and Northeastern Canada, who have developed unique survival strategies to cope with their very harsh environment since their arrival after the end of the last ice age. The Greenland Viking settlers, who had a more difficult time adapting to the Little Ice Age, eventually left.

Reconciling Environmental Protection with Economic Development

I have never regarded my environmental protection and economic development viewpoints as conflicting in the longer run. If the price mechanism does not take into account the full social costs and benefits of production and consumption (creating externalities in economic jargon, as just referred to) but has been corrected by governmental intervention, economic development and resource use benefit human health and welfare by providing services that humans desire without adversely affecting others not involved, including ecological systems of value to humans. Despite great effort by those holding opposing views, there is simply no evidence thus far that the normal activities of man have or will result in catastrophic outcomes for either man or nature. Obviously some non-normal activities such as nuclear war are an exception.

Since humans are clearly better off economically as a result of development as a whole and have repeatedly demonstrated their preferences by engaging in development activities, there is no economic reason to halt or even slow economic development as a whole. Obviously some development activities can and do result in aesthetically offensive and even environmentally damaging outcomes if not tastefully and carefully carried out with environmental protection in mind, but that is why we need judicious environmental protection agencies and environmental advocacy groups to avoid such outcomes. But since environmentally catastrophic outcomes have never occurred as a result of development and no one has been able to show that they will, further development is an easy choice.

Introduction

From a larger perspective a strong economy is the best prescription for enhanced environmental protection because it is usually only when people feel economically secure that they will concern themselves with it. Environmental protection activities should be undertaken only when they are supported by sound economics and sound science because otherwise they may well harm the economy (and thus the opportunity for future environmental improvements), not produce the promised environmental improvements promised, and may harm humans and the environment. Changes to natural systems from economic development and resource use are inevitable. But it is far from clear that all such changes are adverse, as many radical environmentalists often now claim. For example, if increased use of energy resulted in avoiding a new ice age (unlikely)—which would end up wiping out the natural world as we know it at higher northern latitudes—it is clear to me that such increased use would be good for humans and good for all the other organisms now living in the higher latitudes. The claim that the natural world must be left exactly as it currently is, as some environmentalists believe, has no reasonable basis in a world where natural climate change will bring about important and at times very damaging environmental changes anyway. And surely not all such natural climate changes will be favorable for either humans or other organisms.

This is not to say that humans have not had adverse effects on the environment that government should be used to control. They have, and government has played an important and useful role in reducing many of these adverse effects. The examples are many. As mentioned, smog in the Los Angeles Basin had been a continuing and increasing problem before strict emission limits were imposed. Industrial and household heating emissions in parts of England many decades ago and parts of China now endanger health. Oil and other water pollution needs to be better controlled. Although significant advances have been made as a result of efforts at many levels of government and elsewhere, even more needs to be done in carefully selected areas even in developed countries. Urban

17

development has too often resulted in increased runoff of water, oil, chemicals, and soil. We know how to control these problems, but too often this knowledge has not been used. Much has been done to preserve unique and wilderness areas, and the US has one of the best national parks and wilderness protection systems I have experienced anywhere I have traveled in the world.

But stopping or even significantly slowing economic development as a whole will have very little impact on these usually more local problems; it will rather reduce the ability of society to afford the resources to make further progress on them. This is where I part company with the radical environmentalists who now control most environmental organizations and enjoy strong support from the Obama Administration and the European Union. Whether they realize it or not, what they advocate would significantly slow development as a whole and do little or nothing towards improving the environment. There is no justification for this and it would result in decreased environmental protection in the long run, not increased protection.

My professional efforts to promote both economic and environmental improvements over many years have included trying to better understand economic development, improving economic analysis techniques for determining the economic benefits of environmental protection, assessing the science available to determine when environmental protection measures might be useful to control particular pollutants, and applying sound economics to determining what environmental protection measures might be useful. US EPA was a good place for me to pursue these interests and objectives until the arrival of the Obama Administration, with its radical environmentalist ideology. Based on the Administration's actions, their view appears to be that the policies advocated by radical environmental groups (with the possible exception of AFD) should be implemented at all costs whether or not they represent sound economics or science or even law.

The Administration has generally not attempted to

independently assess climate science but rather accepted the science as portrayed by the United Nations, environmental organizations, and US Government review groups, and used whatever economic assumptions necessary to justify their preconceived "green" policies. This abandonment of sound science and economics is one of the major environmental dangers we face since the result is all too likely to be slower growth, less meaningful environmental protection, and monumental waste of resources badly needed for useful purposes.

Organization of This Book

This book is organized into three general sections and a summary, with four appendices available containing more detailed analyses. The first section, constituting Chapters 2 through 6, is a chronological history of my career relevant to my approach towards environmental issues in general and climate in particular. The second section (Chapters 7 through 10) describes what I learned about four of the major components of the CIC—journalism, government, climate science, and the environmentalists—during this odyssey. The third section (Chapters 11 through 13) is an analytical overview of climate alarmism from a number of other perspectives.

The first section is broken down into my early interest in the environment (Chapter 2), my renewed interest when working at the RAND Corporation (Chapter 3), my first 30 years at EPA (Chapter 4), and my last ten years (Chapter 5) at EPA, after I became actively engaged on the global warming/climate change issue, and what I have learned after leaving EPA (Chapter 6). The second section is divided into Chapters 7 and 8, which discuss what I learned about the state of American journalism and government, respectively, Chapter 9, which summarizes in non-technical terms what I learned about global warming/climate change science, and Chapter 10 which discusses environmentalism and the future of environmental regulation. The first half of Chapter 9 deals with what is wrong with CAGW science and ends with a detailed summary of what actually is known about alternative scientific explanations for climate change.

The third section starts with Chapter 11, which takes an economic and national security perspective of climate alarmism. It suggests how fossil fuel development could contribute to improving the country's economy and national security if only the government and the radical environmentalists would allow it to happen. Chapter 12 discusses how we might get out of the mess created by the GEF caper and how we can avoid similar attempts to game the system in the future. Chapter 13 presents an overall summary from all these perspectives.

The book and Appendix C contain two levels of scientific discourse. The intention has been to make the main text readable by the non-specialist. I have also written an Appendix C, however, which presents a more detailed and technical discussion of the science and economics. Appendix C.1 supplements the first major section in Chapter 9. on why the scientific basis for CAGW is invalid). Appendix C.2 supplements the discussion in Chapter 9 concerning alternatives to CAGW science. Appendix C.3 supplements the discussion of geoengineering in Chapter 9. Finally Appendix C.4 supplements the discussion of the application of benefit-cost analysis to EPA's 2014 proposed regulations to limit CO_2 emissions from existing power plants in Chapter 11. These appendices allow a more technical format which facilitates the use of more charts, tables, and technical terminology, which makes it much easier to make the more technical arguments for those interested.

Appendix C as well as three others of more historical interest are not included in the main text in order to reduce its length and complexity. The four appendices are rather included on a separate computer disk which will be shipped with this book if purchased directly from the Publisher. References in the book to the figures and tables in Appendix C will be in the form Figure/Table C-X where X represents the figure/table number.

2
Early Interest in Environmental and Natural Resource Issues

School Years

I WAS FORTUNATE to have attended a private school, Cranbrook (now more widely known as a result of Mitt Romney's attendance and reported activities while a boarding student there), during my junior high and high school years. Their excellent science curriculum particularly emphasized the importance of the scientific method and allowed me to undertake some unusual research projects. I should have realized my interest in natural resource issues much earlier than I did since when I was forced to pick subjects for major year-long term papers I selected the Tennessee Valley Authority in the eighth grade and conservation in the twelfth grade. Conservation was one of the important predecessors of environmentalism; it was more narrow in its concerns but included some of the same general issues. But I was also interested in public affairs and science, and like many in that period selected physics as my major interest in college while still in high school. This did not preclude an interest in natural resources, but the emphasis in physics at the time was on nuclear physics, which I thought had little to say about natural resources, or so I thought until recently.

But climate is very multidisciplinary and it may actually have some interesting things to say about climate science.

Conservation, as it was known in the mid-1950s, involved essentially non-wasteful use of natural resources. The most widely cited example was the prevention of soil erosion, which had been a major problem during the Great Depression both because of water and particularly wind erosion. The Soil Conservation Service in the US Department of Agriculture exists to the present day, but is not often regarded as a major part of the modern environmental movement. I believed strongly in non-wasteful use of natural resources and preservation of scenically, biologically, and historically significant areas.

Physics at Caltech

When I had originally decided to become a physicist while in high school, I applied to what I considered to be the then pre-eminent educational institution in that area, the California Institute of Technology located in Pasadena, California, where I majored in physics as an undergraduate. I attended during what may prove to have been one of its best known periods, the late 1950s, when a number of luminaries such as Linus Pauling, Richard Feynman, and Murray Gell-Mann, were all on the faculty. I worked for Frank Press (later President of the National Academy of Sciences) and the US Geological Survey one summer on a geophysics research project in the Owens Valley of California. I took a small class instructed by Pauling himself and got to know Feynman through informal interactions with him over several years.

Feynman made a particularly lasting impression on me with his friendly approachability, colorful lectures, amusing contempt for government security regulations during his wartime work on the Manhattan Project, and willingness to take an interest in students (he once even inquired about my high school science project). I was particularly impressed by his overwhelming belief in the importance of the scientific method, which strongly reinforced what I learned in high school on this subject. In his view, there was no such thing as

22

permanent or settled science or even consensus science. Only an open mind and continuing research can insure that science represents the real world. Richard Feynman[7] has since expressed this as follows:

> *In general, we look for a new law by the following process. First, we guess it. Then we compute the consequences of the guess to see what would be implied if this law that we guessed is right. Then we compare the result of the computation to nature, with experiment or experience; compare it directly with observation to see if it works. If it disagrees with experiment it is wrong. It's that simple statement that is the key to science. It does not make any difference how beautiful your guess is. It does not make any difference how smart you are, who made the guess, or what his name is—if it disagrees with experiment (observation) it is wrong.*

But I also learned that most of my fellow students had very little interest in public affairs, which I had always followed avidly. Fortunately, Caltech also had invested substantially in building a teaching faculty in what it called the "humanities," which included everything which was not a physical or biological science. Although it did not offer degrees at that time in these areas, it offered courses in a number of areas, including economics. So by my senior year I decided to make a jump to economics instead of continuing with physics at the graduate level since I thought it would provide a much greater opportunity to pursue my interest in public affairs. At that time it was not possible to major in any of what Caltech called the humanities (which it now is), so I completed the requirements for a physics major and graduated with honor.

My years at Caltech had several other completely unforeseen effects on my future interests. One had to do with my observations

---

[7] Feynman, 1965, p. 156.

concerning air quality in Pasadena. At the time I arrived there the air quality left much to be desired. Caltech is located a little over five miles south of the foothills of the San Gabriel Mountains, which help to trap the ozone pre-cursers emitted into the atmosphere in the Los Angeles Basin along with other surrounding mountains. I rated good air quality as days when I could clearly see the San Gabriels from the campus. There were very few of them. The odor caused by high ozone levels was clearly evident on those days. I (and most everyone else except Professor Arie Haagen-Smit) did not know much about air quality at the time but I strongly suspected that the then current low air quality was very unhealthful, especially since I noticed that I was much more likely to cough on low visibility days. I strongly felt that something needed to be done about this problem, and wondered whether scientific and economic analysis could play a useful role.

Because Caltech is located in the Far West and because a number of my friends were also greatly interested in the outdoors, I spent several summers and school vacation periods visiting a number of the Western wilderness and scenic areas that could easily be reached from Los Angeles. This included a number of private trips into the Sierra Nevada of California, where I hiked most of both of the major trails, the Arizona, California, and Baja California deserts, and importantly, as it turned out, Havasu Canyon in Northern Arizona.

At the time I was largely unaware of the Sierra Club's role in wilderness preservation or the role of other environmental/conservation groups. What I did know was that I found the wilderness outdoors fascinating, magnificent, interesting, and photogenic, and like the Sierra Club wanted at least the more unique parts of it preserved in as natural a state as possible. I later came to support most organizations, private or public, that worked to make this possible. I was also unaware of free market environmentalism and knew very little about economics in general, but thought that the Park Service was performing a very valuable service by preserving those areas they had responsibility for. I also found all too many examples of what too often happens to unprotected areas as a result of human carelessness and thoughtless private development. This was particularly evident in unprotected desert areas with their very fragile ecology and lengthy regeneration periods. There was a clear difference between the condition of the protected and the unprotected areas I visited, and I much preferred the conditions I found in the protected areas. The 1950s were also prior to the mass visitation of these areas. Bears had not learned to follow human campers over the Sierra Crest into the Eastern Sierra so no precautions were necessary to avoid bruin predation in some of the most scenic areas of the Sierra. No rationed permits were required to visit the Sierra high country controlled by the Park and Forest Services. A hike up Mount Whitney did not have to be completed in one day to avoid the necessity of a scarce permit. The only limits to what could be done were a result of individual limitations, the usually good weather, and interest. This was a new experience for me since I had grown up on the much less scenic East Coast where there are comparatively few protected areas and much less benign weather.

Havasu Canyon was formed by a tributary of the Colorado River and offers breathtaking desert and canyon vistas as well as exquisite waterfalls and pools with blue-green travertine near its lower end. The Canyon is inhabited by the Havasupai, an Indian tribe that has lived there for over eight hundred years. The Canyon is reached from the Canyon rim on the west side of Grand Canyon National Park through the village of Supai, past the beautiful waterfalls and pools, and offering the possibility of even walking all the way down to the Colorado River if desired. Later, in 1959, I took a sightseeing flight over the Grand Canyon which included a flight over both the Grand and Havasu Canyons, which further reinforced my view concerning the uniqueness and beauty of both the Grand and Havasu Canyons and the need to protect these national and world heritages. Later, in the mid-1960s, near the start of the Sierra Club's campaign against the proposed Grand Canyon dams, they sponsored flights from the South Rim to Toroweap

Overlook for the news media just north of the River which I was fortunate enough to participate in. This excursion was at a much lower level than the previous Canyon-wide flight I had taken and further cemented my views concerning the need to preserve the Grand Canyon for future generations in its natural state.

In subsequent years I initially regarded my undergraduate physics degree as a digression from what I then expected would be my life's work in economics in general and economic development in particular. But such did not prove to be the case. I now realize that I learned a number of really important things while at Caltech, including a basic background in a number of the physical sciences and the vital role played by the scientific method, both of which I later used at the Environmental Protection Agency, as well as not to be afraid of scientific terminology and complexity.

Summers at RAND

My introduction to the working world came very rapidly in the summer of 1959. I was fortunate to work in the Economics Department of the RAND Corporation in nearby Santa Monica, California, the first two summers after graduating from Caltech. RAND is a multidisciplinary "think tank" created by the US Air Force after World War II "to conduct an integrated program of objective analysis on issues of enduring concern to Air Force leaders." Many people credit it with being the first such "think tank," and providing the template for many others that have been created since then. RAND endeavors to improve policy and decision making through research and analysis, primarily of the paper and pencil variety (at least before personal computers arrived). It was an early leader in rational analysis of problems involving many uncertainties and several disciplines through systems analysis and other techniques which it helped to develop. As an offshoot of the Douglas Aircraft Company, the main office is located near the former Douglas plant and close to the Santa Monica Pier. In more recent years it has expanded its list of clients first to other Federal departments, then to states and cities, and finally to (friendly) foreign governments, particularly in the Middle East.

In the summer of 1959 RAND undertook a broad study of US strategic nuclear capabilities and vulnerabilities for the Air Force, its then primary sponsor. My role was to assist in making the computations involved. By the end of the summer I had had a fascinating overview of military systems analysis and its interaction with public affairs.

The background[8] is that at the end of the Eisenhower Administration, there was a major disagreement within the US intelligence community concerning how many Intercontinental Ballistic Missiles (ICBMs) the Russians had. The Air Force was arguing that there could be hundreds and the CIA for no more than a dozen. (The number later turned out to be four.[9]) The Air Force had a self-interest in large numbers perhaps because it could lead to a larger budget and role, but was the source of much of the intelligence information. There had been an earlier controversy about the number of heavy bombers the Russians had. Eisenhower had taken the advice of a general he had asked to examine the question that the number was relatively small and sided against the Air Force, but Senator Symington, a key Democrat, and later Senator John Kennedy, had sided with the Air Force on the missile question. RAND had played an important role in this controversy not because it had better intelligence (it did not—it had to depend primarily on the Air Force for that) but because it took a more analytical approach to determining whether a possible "missile gap" made any difference. The analytical approach used (summarizing greatly) was to argue that what made a difference was not the difference in the total number of launch-ready ICBMs (the previous measure used), but rather whether the US could mount an effective retaliatory strike after an initial hypothesized strike by the Soviet Union on the US. This approach required some analytical effort

---

[8] See Bird and Bird, 2011.
[9] Day, 2006.

since estimates had to be made both about the number and capabilities of the assumed Soviet first strike but also about the location and "hardness" of the US nuclear forces and thus their ability to survive such a first strike.

The RAND employee generally credited with this approach was Albert Wohlstetter. An unclassified version of his approach, entitled "The Delicate Balance of Power"[10] had just been published a few months before my arrival. Many others were involved, however, since in the late 1950s a large group of "strategic analysts" had coalesced at RAND including Henry Rowen (later in the US Budget Bureau during the Johnson Administration, RAND President, the Defense Department, and Stanford University), Alain Enthoven (later in the Defense Department and Stanford University), Herman Kahn (later founder of the Manhattan Institute), Andrew Marshall (later of the Defense Department), Thomas Schelling (later at Harvard and the University of Maryland), and Daniel Ellsberg (later of Pentagon Papers "fame"). Many of these strategic analysts took an active role in the 1959 study, so I interacted with many of them that summer.

The alleged missile gap was an issue in the 1960 election campaign but although the Kennedy Administration ultimately decided that there was no significant missile gap, the RAND/Wohlstetter approach to determining whether a missile gap was of any significance was widely adopted by the Defense community. So although the Air Force and its contractor, RAND, ended up on the losing side of the argument as to the number of Soviet missiles, its methodology was widely used and many of those involved at RAND took major roles in the Defense Department during the Kennedy-Johnson Administrations, where they were sometimes referred to as the "Whiz Kids."

I learned several things by participating in this study. The first was the importance of understanding the big picture where a few obscure technical or intelligence assumptions in widely diverse

---

[10] Wohlstetter, 1959.

fields could change the conclusions substantially. The second was to carefully consider and question each major assumption, particularly intelligence estimates. The third was that researchers who work on a problem for an extended period of time start to believe that their analyses and assumptions are actually valid representations of reality, even in cases such as this where the conclusions were highly sensitive to multiple poorly understood assumptions.

As the world now knows, the strategic nuclear missile face-off came to a head not in the way RAND had assumed but rather three years later in 1962 during the Cuban Missile Crisis, when Kennedy actually had to deal with related issues when the Soviets did actually deploy a much lower-technology, lower-cost, and more practical system with some of the same capabilities by placing medium range nuclear missiles in Cuba. Hindsight suggests that the US strategic posture of the time would have met the RAND/Wohlstetter second-strike test if the Soviet Union had actually used its strategic nuclear forces in Cuba against the US since the missile/bomber sites in the US Northwest could not have been reached from Cuba with medium range missiles or been seriously damaged by the minor number of ICBMs they had at the time. But then the Soviet Union apparently had different objectives from those assumed by RAND/Wohlstetter in 1959. These objectives apparently included getting the US to remove its missiles from Turkey and safeguarding their ally, Cuba, both of which they achieved at the cost of terrifying the world as a result of the Cuban Missile Crisis.

Economics at the Massachusetts Institute of Technology

Near the end of my undergraduate work at Caltech, I decided to transfer from physics to economics for graduate work so as to maintain my interest in public affairs. I was particularly interested in a comparatively new field called economic development concerned with how and why countries develop, and I discovered that the Massachusetts Institute of Technology had a substantial group working on it.

So I undertook graduate work in economics at the

Massachusetts Institute of Technology. Although the Institute as a whole is known mainly for its science and engineering programs, its economics program is also one of the best known in the world, primarily because of a number of its early faculty members such as Paul Samuelson and Robert Solow. At the time I attended in the early 1960s they were active in teaching graduate courses. Although the primary reason I wanted to attend MIT for graduate work was because of their substantial area of expertise in economic development, the Economics Department was much better known for other areas of economics. The MIT Center for International Development had an office in New Delhi, had several staff members from the Economics Department, and hosted visiting Indian scholars at MIT so was quite involved with Indian economic problems, which contributed to my interest in India. My interest in economic development continued through my years at MIT, through a year spent in India doing research evaluating US aid to India, and my very early professional career.

At MIT, besides learning the tools of the economic trade, I came to appreciate the extent to which society is driven by economic forces and the importance of looking even for hidden economic incentives in people's behavior. Economics provides a systematic approach towards analyzing economic efficiency and distributional issues. Cost-benefit analysis provides a basis for judging how much of an environmental good is worthwhile and when it is no longer worthwhile to do more.

So by the time I graduated from MIT with my PhD in economics I had developed a strong interest and background in the physical sciences, economics, economic development, and national security, as well as continuing my longstanding interest in public affairs. I did not realize it at the time, but there are very few people with such a background and interests. Very few economists have a background in the physical sciences or vice-versa. And little did I realize how useful such a background would later prove to be.

Observations on the Role of Energy and Government Regulation on Economic Development

Economics has many specialties. During graduate school and my early professional career, mine was economic development, which concerns how and why countries develop economically. This reflected my interest in understanding what causes economic development, which in turn reflected my enthusiasm for improving the standard of living of people living in the less developed countries. I regarded such development as much to be desired by everyone in both the developed and less developed worlds, and still do despite the efforts of modern day radical environmentalists. It did not occur to me at the time that there would be people in the developed world who would not want the less developed countries to grow and develop economically. I still find it unbelievable and reprehensible.

As part of my graduate work in economics at MIT I spent a year in India on a Ford Foundation Foreign Area Fellowship where I gathered data for my PhD thesis on an evaluation of US aid to India. In addition to my economic development research, I continued my interest in military strategy largely as a result of the border conflict which erupted in 1962 between China and India while I was in India, and had some interesting interactions with Albert Wohlstetter on the subject.[11]

My early area of concentration on economic development may not seem to have much to do with climate change, the principal subject of this book, but it actually does since I learned two important things concerning growth and development very relevant to climate—the crucial roles played by fossil fuel energy and the absence of extensive government regulations.

The reason that CO_2 emissions reductions have not been substantially implemented in the less developed countries is that they are strongly opposed to reducing their CO_2 emissions, and

---

[11] Carlin, 1962.

rightly so. Decreasing CO_2 emissions means that less fossil fuel energy will be available to assist humans in their daily tasks and thereby increase their productivity and income. Renewable sources of energy are not restricted under GWD and GEF but are much more expensive and unreliable, which makes it that much harder for poor countries to afford the reliable sources of energy they need to develop. It is the availability of such energy which made possible the Industrial Revolution and the escape of humanity from the Malthusian trap,[12] which the present day less developed countries would like to achieve as well. In other words such energy use has a direct influence on economic growth and development. GWD/GEF will, if successful in reducing CO_2 emissions, decrease both compared to what would otherwise be achieved.

The less developed countries realize all this, and have been fighting to avoid decreasing their emissions because of the crucial role they can play in their efforts to become more developed countries. I see no reason they should be denied this opportunity or have it made more difficult for them to achieve.

The CIC's first response to this problem was to agree that emissions reductions would only be required by developed countries. This made possible the Kyoto Protocol in 1992 since the less developed countries no longer had any such concerns with respect to the Protocol.

The problem, however, was that the Protocol had very little if any impact on world CO_2 emissions, in part because of the huge increases in emissions in rapidly developing less developed countries, particularly China and India. So if they were free to continue increasing emissions, their rapidly increasing emissions would and did soon more than offset any decreases in the developed world. The developed countries in recent years have accordingly pushed to include the less developed countries in emissions reductions. This has resulted in the current impasse. In 2009 at a failed summit in Copenhagen the effort to create a new agreement

---

[12] Goklany, 2012.

on implementing GWD/GEF, the less developed countries rejected the GWD/GEF-inspired proposed treaty, and as of the end of 2014 no real resolution has been found despite many meetings and much talk.

Thus an understanding of the underlying factors making economic development possible is crucial to understanding why no effective worldwide controls on CO_2 emissions have occurred or are likely to occur. The results have been dramatic in terms of lack of agreement on worldwide CO_2 reductions and the unusual nature of the continuing negotiations. What has evolved has been a three stage ritual[13] at each of the many UN climate meetings called Conferences of the Parties (or COPs) in recent years. First the world is treated to absurd, unscientific hype concerning how the threat of GWD/GEF is greater than ever and action must be taken immediately if the world is to be saved. Then when everyone is gathered in some agreeable venue, the less developed countries say that they are opposed to mandated reductions in emissions, and certainly not unless the developed countries pay them very large sums, usually stated to be $100 billion per year since 2009, to pay the cost of their CO_2 mitigation or, more recently, "reparations" for alleged damages from past developed country CO_2 emissions, or most recently, compensation for damages from severe weather allegedly caused by higher CO_2 levels, or better, all of these. Finally, in the early morning hours after the COP has exceeded its scheduled closing time, when it has once again become clear that these sums have not and probably will not be made available, a "breakthrough" is announced which amounts to a largely meaningless document that avoids committing anyone to anything of significance, but can be used for PR purposes to justify still more such meetings.

Among other things, my research on Indian economic development taught me that government regulations can also play a critical role in economic development. India came into the post-

---

[13] Booker, 2014.

World War II era with many advantages compared to its Asian neighbors. English was and is widely understood among educated Indians and allowed most of the important economic players to easily communicate with each other and the outside world despite the substantial differences among the many ethnic and language groups in the country. (This has recently led to the outsourcing of many customer service operations from the US to India.) The railway network was and is one of the largest in the world and similarly connected the country for economic purposes. The civil service was well trained and able to accomplish its objectives in administering such a vast, heterogeneous land despite the fact that decisions at that time were often made by bureaucratic consensus using folders literally tied with actual red tape. But somehow the country never seemed to grow as rapidly as many of its neighbors in Southeast Asia despite receiving significant development aid from the US and others. This anomaly troubled me and others interested in economic development. How could this be?

I concluded that there were two principal reasons: The first was that the government bureaucracy was too effective in enforcing its commandments on the private sector, a problem that many much more rapidly growing Southeast Asian countries did not have. The other was that Jawaharlal Nehru and others at the head of the government were overly enamored with the Soviet model of central planning and government control of much of heavy industry. The result was myriad regulations and permit requirements with usually vague but high-sounding purposes that most companies had to follow and that could be enforced by the relatively effective bureaucracy. The term often used for this state of affairs in India was Permit Raj. This was compounded by a public sector that too often was ineffective in carrying out its assumed responsibilities for vital sectors such as power supply. Electric supply was perpetually less than demand in many parts of the country with the result that rotating power blackouts were the norm. There was a permit required for everything and much of new heavy industry was reserved for Government ownership.

I am not alone in my views concerning the causes of the disappointing rate of Indian economic growth, particularly prior to the economic reforms of the early 1990s. Although I had the opportunity to get to know Amartya Sen[14] while attending MIT, my economic sympathies lie with his professional nemesis, Jagdish Bhagwati, who has offered similar views to mine over many years.

Many of the countries in Southeast Asia did not have such an effective bureaucracy and did not pursue government ownership of heavy industry; instead, they encouraged a simple market- and export-led approach to growth, with the spectacular results which are now so widely understood. India, on the other hand, actually discouraged exports by keeping the rupee overvalued for many years and by its Permit Raj. The idea was that the Government through its five-year plans could more effectively stimulate development than private markets could independently do through export-led growth. One key difference between China and India is that China has striven to keep its currency undervalued compared to the US dollar, which encourages exports and decreases imports. India, on the other hand, kept its rupee overvalued until the 1990s by maintaining low fixed exchange rates.

I mention this because of its implications for the Permit Raj, which is all too rapidly developing in the United States and is now far advanced in California to satisfy the alleged need to implement

---

[14] In Sen (2014) he has recently expressed his views on the general energy use/climate issue as well. I am, of course, highly sympathetic with his views concerning the importance of encouraging the less developed world to access cheap electric power in order to meet their needs for energy and development. I am doubtful, however, that there is much practical need for a more complicated "normative framework" for making energy decisions or that solar power offers a useful alternative to fossil fuels (except in special cases). If as I propose governments should stay out of most energy use and development decisions except when there are clear and proven externalities, there is little need for extending the "normative network" beyond conventional economic analysis. that can be corrected economically efficiently.

GWD/GEF/EWD measures such as decreasing carbon dioxide emissions. Whatever effect these regulations may or may not have on carbon emissions, they will certainly discourage economic growth and development just as the Permit Raj did in India after independence. India has done better in terms of economic growth in recent years after a loosening of government controls in the early 1990s and after the Information Technology revolution, an area which the Government had not gotten around to regulating or delegating to its own control. But while Southeast Asia was booming, India grew only slowly in the mid-twentieth century despite its substantial infrastructure, widespread public education, and other public investments during British rule. The 2014 general election in India, which brought a Hindu nationalist party, the Bharatiya Janata Party (BJP), to power on an economic development platform may or may not result in changes that will finally enable India to break free of its government-imposed regulatory chains.

The implications are obvious. Centralized government control over the means of production is generally neither effective nor efficient in elevating a country's standard of living or its economic well-being. Unfortunately that is exactly what the Obama Administration's climate/energy policies will lead to for the US economy. Despite all the lofty words about the need for energy efficiency and alternative energy and even "green jobs" that it will bring, the reality is that GHG emission controls will necessitate very extensive and intrusive government regulations and their enforcement, with all the unintended consequences so easily seen in mid-Twentieth Century India.

The arguments are different—environmental preservation rather than alleged social justice and planned development—but the results are very likely to be exactly the same. In one case the economy was haphazardly regulated by a central economic planning organization and government ministries; in the US case, it would be regulated by a central environmental bureaucracy with probably even less detailed knowledge of the adverse effects of what it proposes to do or how to avoid them. It is abundantly evident that neither approach has or ever would have the detailed knowledge concerning the economy to achieve its supposed goals in an efficient way. In the Indian case, the approach is called socialism. In the US case it is probably best characterized as state industrial policy moving towards centralized government control of the economy. Each had/have lofty-sounding goals; each has been/will be an economic failure and for similar reasons.

The most successful economies are those that decentralize economic decisions to those with the most knowledge of the alternatives and the most to lose or gain from the decisions made. Bureaucracies live by different rules and goals which are not particularly attuned to either the demands for different products or the resources required to produce them. The US Government does not use file folders actually wrapped in red tape or operate by consensus, but the results are similar. California is currently leading the way with its ever more complicated and expensive environmental and energy regulations.

3
Renewed Interest in Environmental and Natural Resource Issues after Returning to RAND

Background

AT THE END of my graduate work in the early 1960s, RAND offered me a position as an economist in their Economics Department in their Santa Monica office. Most of the strategic analysts who I had worked with earlier had left except for Albert Wohlstetter, who departed soon after I returned. RAND had taken on much more non-defense related research. Most of my early work as a permanent employee was on economic development, my major at MIT, including research on India, Turkey, and what is now the West Papua Province of Indonesia but was then known as West Irian Jaya or West New Guinea, each of which required a trip to these countries. These overseas missions were of some interest from a development viewpoint since the countries involved were at widely different levels of development and highlighted some of the crucial aspects of economic development.

Although it was much less clear at that time, economic

39

development has been most successful where the countries pursued an "export-led" approach and minimized the involvement of government. Many Western economists and left-of-center governments tended to believe that foreign aid was the best route towards encouraging development in the less developed world. Generally speaking, the US has tended to provide development assistance to economic sectors where the US Government provides considerable assistance within the US, such as infrastructure (roads, railways, irrigation), and tended to avoid investments in areas where private enterprise plays the most prominent role. This was probably done for US domestic political reasons, but avoided the larger issue of how useful foreign economic aid is as a tool to promote development. Although the MIT economic development faculty generally supported the foreign aid approach at least during the Kennedy-Johnson Administrations, my experiences in India resulted in a more skeptical approach on my part, which influenced my attitude toward Indian and Turkish development.

A region of continuing interest to me during my early years at RAND was India. In 1966, during the Johnson Administration, the US Agency for International Development mission in India was considering whether a large push of US aid might generate "self-sustaining" growth, perhaps as hypothesized by one of my MIT professors, Walt Rostow (who served in both the Kennedy and Administrations). I was asked to consider the railway portion of such a possible push since I had developed some expertise on the railways in my thesis research. I argued that the railways, which were government owned, were more than able to generate improvement projects to use almost unlimited US aid if offered, but that this was quite unlikely to result in the hypothesized "self-sustaining" growth. I argued that India's economic problems were rooted in the continuing counterproductive interference and control by the Government in the economy. Until this problem was solved, a greatly increased US aid program for the railways or anything else was not likely to achieve the desired "self-sustaining" growth.

Renewed Interest in Environmental and Resource Issues

My second but much less intensive effort was on Turkey. The US AID mission to Turkey was to be commended in my view for being interested in the economic aspects of Turkish development. My primary concern, however, was that like many less developed countries, including India, Turkey followed a fixed exchange rate regime which as elsewhere tended to result in an overvalued local currency since domestic inflation was not immediately reflected in the prices paid by those wishing to import goods from Turkey. The result was lower export demand than would have existed under a freely floating exchange rate and slower growth. Turkey ultimately adopted floating exchange rates in the 1980s but was handicapped by sticking with fixed exchange rates with occasional devaluations prior to that. The most successful developing country, China, has wisely (from their viewpoint, not ours) pursued exchange rates that undervalue their currency so as to stimulate exports. Unfortunately, many other less developed countries, such as India and Turkey, stuck with fixed exchange rates far too long. This encouraged imports but not the exports to pay for them, resulting in balance of payment problems and counter-productive foreign exchange controls where used.

A third and very different region of interest during my RAND years was Western New Guinea, which occupies the western half of the island of New Guinea in Southeast Asia. It had been ruled by the Dutch until 1962, when it was turned over to the United Nations briefly and then to Indonesia in 1963. In 1967 the Dutch Government decided to give the UN significant funds for the development of Indonesian Western New Guinea, and the UN commissioned a study to advise how it might best be spent.[15] I was fortunate to be a member of this group and spent six weeks there in mid-1967. West New Guinea was and is about as undeveloped as any area in the world. At that time the roads petered out very close to the few major towns and may still do. Hundreds of different languages were spoken by relatively isolated tribes scattered over

---

[15] United Nations Fund for the Development of West Irian, 1968.

41

the large western portion of the island, which was better known for head hunting than for economic development. The primary means of communication were ships operating around the coast and limited air service provided by Indonesian airlines and particularly some Christian missionary groups. At the time there had been no airline flights from the capital at Jayapura on the northeastern coast to the southern portion of the island for at least six months, and it was doubtful whether any fuel could be obtained there for a return flight. The Indonesian airline operations within Western New Guinea used primarily DC3s.

Our task was to recommend how the Dutch development aid could best be spent. Improving transportation and communications were obvious possibilities. But since there was little commerce between various parts of Western New Guinea, almost all of which depended on subsistence hunting-gathering and primitive agriculture, it was not clear whether improved roads would help much or would be used or maintained if built. Eastern New Guinea, at that time administered by Australia, had similar communication problems but had built an all-weather road from a major coastal town into the highlands, which was used to haul crops such as coffee to the coast for later export. Western New Guinea, however, did not have the plantations that had developed in the East, so there were no major crops to haul since most agricultural activity was much more primitive.

What was needed were export industries that would bring in substantial income that would pay for its own infrastructure. Agriculture, forestry, fishing, and possibly tourism were possibilities but probably could not support much infrastructure development or maintenance. The best possibility was natural resource development such as mining carried out by Western companies. And that depended on finding ore deposits rich enough to pay for either air support or extensive road development through some of the most forbidding terrain on Earth. In the 1970s, such a deposit was developed in the central mountains containing both copper and gold, and nearby deposits are still being exploited today by a private, non-Indonesian mining company. This was about as remote from any means of transportation as it was possible to get, but has proved rich enough to be exploited nevertheless. But it was representative of the type of activity that would be likely to result in economic development and hopefully increasing economic welfare for the population.

Subsistence hunting/gathering and agriculture may appear romantic to some in the developed world, but do not lead to improved economic well-being such as the Dutch Government wanted to promote. Subsistence living brings many problems, including lack of health care and sometimes hunger. A major continuing problem in terms of economic development was and is the continuing conflicts between the Indonesian government and the native peoples of Western New Guinea, which further raises the cost of any development efforts. Some may believe that the indigenous people should be left to the subsistence living they have evolved over long periods prior to the arrival of Europeans. I disagree. I believe that the primitive life is far from idyllic and it is they rather than Westerners that should decide such questions.

I developed some expertise in transportation economics, where some of my later work for RAND was concentrated. This included an interesting project[16] for the Port of New York Authority on how they might be able to increase the capacity of their airports; one of the recommendations was that the Port Authority change the basis for its fee structure for use of their airports from aircraft weight, the traditional approach, to actual time spent using the critical scarce resource in the heavily used New York airports, the runways. The weight of aircraft using runways has almost no effect on the operations of the airports they use since the main requirement of handling heavier aircraft is a little longer and heavier runways, but does result in most of the costs being paid by airline passengers flying in large, heavy airliners. The really crucial resource in a period when some airport resources are

---

[16] Carlin and Park, 1970.

strained by more demand and static capacity is rather the time taken by each airplane landing and taking off at a crowded airport since that time cannot be used by any other aircraft on the same runway. Our study was followed by a long series of related studies by others, and various attempts have been made to introduce related policies. But the basic idea has not met with much support from the business community, which may not have wanted to pay the added fees that their business aircraft would have had to pay. Obviously this was the point since they were and still are using the critical scarce resource at the airports without paying what I regard as their fair share based on their use of the critical resource. This may be good for business, particularly for the executives who most often use business aircraft, but pricing based on the use of the critical resources would certainly be good for improving the performance of the New York airports and other runway-constrained airports for the traveling public as a whole. Since most other airports in the US use similar weight-based pricing mechanisms, those with a shortage of runway capacity would be much more efficient if they charged on the basis of the scarce resource—use of scarce runway capacity.

The Federal Government has attempted to cope with the delays caused by overuse at such airports by restricting the number of slots available for airline use. Until recently these "slots" have effectively become the property of selected airlines rather than being auctioned off. This reduces congestion but does not insure that the most economically valuable (often those with the highest passenger loads) airline flights get to use these congested airports or that business aviation fully pays for the scarce resources they use. In 2011 auctions were held for some slots at La Guardia and Reagan National Airports, but the proceeds went to the existing holders rather than for airport expenses or improvements to increase capacity.[17]

So far the organizations representing business aircraft owners

---

[17] Papagianis, 2012.

have managed to prevent our proposed change both in New York and elsewhere in the US to the detriment of airline passengers. Most airports find it easier to charge much less organized airline passengers than the highly organized business aircraft lobby. Very few airline passengers, who pay most of the bills at busy airports, even realize that they are subsidizing the use of business aircraft to their own detriment. If business aircraft paid more for landing rights, fewer would use busy airports at peak hours, which would mean more capacity for airlines and lower prices for airline passengers both because of the lower landing fees that airlines would have to pay and the scarcity value that airlines are able to get passengers to pay at busy airports with hard to find airline seats and scarce airline "slots."

Opportunities to increase economic efficiency and improve income distribution are sometimes found in the most unlikely and unusual places, of which this is one.

Increasing Involvement with Environmental Issues

Shortly after I joined RAND as a permanent employee, I resumed my wilderness outings to the Sierra and many other parts of the Southwest, but this time I often went on trips organized and led by the local chapter of the Sierra Club—the Angeles Chapter—because they sponsored most of the nature-oriented outdoor activity events in the Los Angeles area, Southern California, and the Eastern Sierra. I soon became a member of the Sierra Club when I found it much more convenient to be a member than to buy individual outings schedules.

The first issue of the Club's national magazine I received after joining, however, primarily concerned a major conservation project they were undertaking to prevent the US Bureau of Reclamation from building two hydroelectric dams in the Grand Canyon of Arizona, which I had come to enjoy and love. The lower of the two dams, Hualapai, was to be located on the Indian Reservation of the same name but would back up water into Grand Canyon National Park. The upper dam, Bridge Canyon, was to be constructed on the

45

Navajo Indian Reservation just above the Park and would flood the Canyon from the proposed dam upstream to the Glen Canyon Dam.

I did not hesitate in agreeing with the Club that such dams would have a serious adverse effect on the Grand Canyon as I had known it and would destroy one of the few remaining free flowing sections of the Lower Colorado River, by far the easiest way to enjoy the wonders of the Canyon. The Club's longstanding philosophy of encouraging wilderness use to involve people with natural areas had worked in my case even though my initial trip to Havasu Canyon had been a private one.

I wondered whether the dams were economically justified despite the Bureau of Reclamation's favorable benefit-cost analyses. I had read several papers/books suggesting that the Bureau's analyses were sometimes more self-serving than consistent with the best practices. I agreed that there would be a continuing need for power in the Southwest, but the actual uses for the power to be generated and the reasons for building the dams in the first place were far from clear at that time and needed investigation, I felt. I wondered if there might be alternative means of generating power that would avoid flooding one of the world's premier national parks as well as the Colorado River east of the Park, which I knew had many park-like properties. I was aware that the Bureau had already built other dams higher up on the Colorado, including Glen Canyon, just above the proposed site for the proposed Bridge Canyon Dam, and wondered whether it was really necessary to do so in the last free-flowing section of the Lower Colorado, which was beginning to attract increasing numbers of river runners in a stretch known for some of the most formidable rapids in the world.

The Sierra Club has had a long-standing adversarial relationship with dams, at least partly as a result of its first major conservation effort, to prevent the City of San Francisco from building a dam within Yosemite National Park to supply reliable drinking water for the City. Despite the best efforts of John Muir and others, they lost that fight. In more modern times, they won a battle against Echo Park Dam in Dinosaur National Monument on the Upper Colorado but agreed to allow Glen Canyon Dam to be built in the process. They were determined not to lose again, no matter what it took. The Bureau, Stuart Udall, then Secretary of the Interior, and the Arizona congressional delegation were equally determined to authorize the dams as part of the Central Arizona Project. Exactly what the real purpose of the dams was and why they were so essential was a continuing mystery, however. They were presented as an integral part of the Project, but would supply "peaking" power rather than the less expensive baseload power that the Project would need.

The Campaign Against the Proposed Grand Canyon Dams

As a result of that first issue of the *Sierra Club Bulletin* I received, I discussed the economics of nuclear power with a RAND colleague, William Hoehn, who had been working on nuclear power economics, and contacted the Club's Executive Director, David Brower, in San Francisco to see if he might be interested in an economic study of the feasibility of the proposed dams. Hoehn speculated that nuclear power was likely to be very competitive with hydropower at that time because the nuclear industry was offering attractive low-cost package deals to utilities that eliminated most of the economic risks of cost inflation in hopes of promoting enough interest to get their industry underway.

Shortly afterward I had a rather prophetic meeting with David Brower alone in San Francisco. Although I had frequent contact with him during the Grand Canyon campaign, this was the only time that I discussed more general energy policy issues with him. At this first meeting he was rather concerned, I thought, that I was an economist and inquired whether I thought that fossil fuel energy use was sustainable (to use a much more modern phrase) and desirable. He obviously thought that it was not. I said that I thought that energy use was the basis for modern civilization and that there were likely to be adequate supplies for the indefinite future because of human ingenuity and technical innovation and that the workings of

47

the markets for energy would see to it that alternative energy supplies would be developed when there was a market for them. I further pointed out that energy prices were continuing to fall in most energy markets and that was hardly a sign of shortages. (This was prior to the oil price increases of the early 1970s!)

I mention this because I have often thought that Brower's concerns were rather prophetic in terms of the concerns of modern environmentalists. Somehow he held the views they would hold and felt that they were important. But in my case, it provoked a strong negative response. I did not like his apparent view that energy use was undesirable and that availability was headed for a peak any more than I agree with the modern formulation of this idea as advocated by many present day radical environmentalists. He was definitely ahead of his time in this respect in the environmental community, and right to be concerned that an economist such as myself might take strong exception to his views on energy. I did and still do. His perception that economists might not accept his ideology was entirely correct. I believed strongly in economic development (my early interest in economics) as well as environmental protection— and still do—and believe that while environmental concerns need to be carefully considered in development decisions, these two objectives need not be in opposition to each other in the broader scheme of things. Yes, there will be trade-offs between environmental and development effects of particular projects, and these need to be carefully considered, but the longer term consequences of development are good for the environment, not bad.

What I regarded as his misplaced concern—indeed preoccupation—about the undesirability of increased energy use and the inevitability of energy scarcity did not influence the Grand Canyon dam campaign, but did raise questions in my mind as to the longer-term compatibility of his views and mine. For the near term, however, I found his prosecution of the anti-Grand Canyon dam campaign to be brilliant and well executed based on my own observations of him in Club, social, and media settings. I might have placed a little more emphasis on some of the non-environmental arguments which I and others developed since most of the politicians involved probably did not share his environmental views, but it is difficult to argue with success and his greater ability to express an environmental rather than an economic or scientific viewpoint. The Club's newspaper ads concerning the folly of flooding the Sistine Chapel to get a better view of the ceiling can only be described as inspired and undoubtedly widely recognized as such by his intended audience. Whenever approached by the media he tried to leave them with a memorable quote they might be able to use, which they often did. I had nothing but praise for his work opposing the dams, as did every other Sierra Club member and activist I knew.

In response to my original question to him concerning the usefulness of an economic feasibility study of building hydroelectric dams in the Grand Canyon, Brower said that he was interested (I was offering my work and that of Hoehn for free, after all). Subsequent experience, however, suggested that he wanted to keep some distance from it. His stated reason was to make the study appear independent rather than coming from the Sierra Club, but I strongly suspected that he also did not want the Club or himself to be associated with even a hypothetical alternative nuclear plant which we used for comparative purposes, perhaps as a result of concerns about nuclear safety or the Club's long and successful fight to avoid construction of a nuclear plant at Bodega Head north of San Francisco or perhaps a personal dislike of nuclear energy.

Nevertheless, he did offer to reimburse us for incidental expenses involved in the study and did promote our effort within the Club. So although I would have preferred more direct support of our dam economics studies, I later understood why he found it better to make the support indirect rather than direct. Not only did he have a history of anti-nuclear activism with regard to the Bodega Bay controversy but there were many in and outside the Sierra Club who opposed it on safety concerns. As the campaign developed it became evident that he was open to new ideas to counter the views

49

of dam advocates and collected a small group of advisors, three of whom had graduated from MIT, to feed him ideas and arguments, including Larry Moss, Jeff Ingram, and myself. Unfortunately, there was never again an opportunity to discuss the larger relationships between development, energy, and the environment given the constant need to deal with current issues concerning the campaign.

With the vital partnership of William Hoehn several versions of our economic analysis of the proposed dams were prepared and presented at Congressional hearings under our names. The changes in the analysis were necessitated by rapid changes in the prices and conditions offered by the nuclear industry for building plants. In the first analyses[18] the case was fairly ironclad because the industry was offering turn-key contracts to build plants for a low specified price in order to stimulate interest and demand from utilities.

By the time of the second set of hearings in 1967, however, these turn-key prices were no longer being offered by the industry, so we substituted informed estimates.[19] In both cases, however, the analyses concluded that the dams were not economically justified because nuclear plants could be built and operated at a lower cost. There were two hearings on the proposed dams in 1967, in March 1967 before the House Interior and Insular Affairs Committee, and in May 1967 before the Senate Committee on Interior and Insular Affairs. In 1968 I published an article summarizing our experiences.[20]

It is hard to say exactly how much influence particular efforts by particular organizations and individuals had on the outcome. The first major action against the dams was that Secretary of the Interior Stewart Udall decided that they would no longer be supported by the Department of the Interior. Then the Navajo Nation (on whose land the proposed Marble Canyon Dam would have been built) came out against them. Finally the Arizona politicians backing them

---

[18] Carlin and Hoehn, 1966 and 1966a.
[19] Carlin and Hoehn, 1967.
[20] Carlin, 1968.

also abandoned them, and they were dropped from the Central Arizona Project authorization bill and a power plant was added to use Navajo coal and supply base load power for the Project. But the media often attributed the outcome to the Sierra Club, perhaps in part because of the eye-catching full page ads the Club ran in the *New York Times* and the *Washington Post*.

The question of who was responsible for the outcome and even what the purpose of the proposed dams was has long been controversial. Pearson[21] argues that the primary factors in the outcome were the opposition of Senator Henry Jackson of Washington State, the Chairman of the Senate Interior and Insular Affairs Committee, to any diversion of water from the Pacific Northwest to the Southwest (many involved suspected that the real purpose of the dams was to generate the funds needed to build the facilities that would be required for this interbasin transport for such a diversion, but it has been impossible to prove this), the opposition of the Navajo tribe to the construction of the proposed Marble Canyon Dam, and the immediate need to authorize the Colorado River Basin Project including the Central Arizona Project (CAP) before its chief Congressional sponsors retired. California had defeated a previous attempt to authorize the CAP on the basis that some of the Colorado River water it had long enjoyed would instead be used by Arizona. But if additional water could be brought from the Columbia River this objection might be overcome. But if no inter-basin transfer was to take place because of Jackson's opposition, one of the major possible reasons for the dams would no longer exist. Further, a coal-based power plant to be located near Page, Arizona, could supply the actual power needed by the CAP, which would provide jobs and income for the Navajo nation and would provide base load power, which is what the CAP needed rather than the more valuable "peaking" power which can be supplied by hydroelectric plants. The peaking power supplied by the proposed dams would presumably be sold for a higher price and

---

[21] 2002.

baseload power purchased at a lower price. The profit may have been intended as a source of funds to build or operate inter-basin transfer facilities. But construction of the dams proved in the end not crucial to the viability of the Central Arizona Project, the main purpose of the legislation. After they were removed and some maneuvering concerning Upper Colorado Basin water development, the Colorado River Basin Project was approved by Congress and signed by President Johnson on September 30, 1968.

My comment is that even if our economic analysis and the Sierra Club's campaign did not directly play a major role in the outcome as Pearson argues, the Club used its contacts to facilitate the opposition of the Navajo tribe at a key point of the campaign, whose opposition even Pearson thinks was important. At least one key Sierra Club official at the time, William Siri, has expressed his belief that our economic efforts were crucial to the outcome. [22] And in 1972, after I had left Los Angeles, the Angeles Chapter awarded me its Weldon Heald Conservation Award for my work on the Grand Canyon dams. It is awarded to a Sierra Club member who has provided long-term and outstanding service to the Angeles Chapter in the field of conservation. In many ways the death of the Grand Canyon dams proposal marked the end of major Western dam projects and the beginning of the Club's major role in national environmental policy.

Although there will probably be continuing debate as to whether the Sierra Club or our economic analyses or Henry Jackson or some combination thereof killed the two Grand Canyon dams, there is little doubt as to the influence that it had on my thinking. I concluded that there were useful applications of economics to environmental issues, which had hitherto played little role in the environmental movement. The water resource development industry had long used cost-benefit analysis to value their projects. Because the Grand Canyon dams were a water development project, their proposal included such an analysis. We were

---

[22] Siri, 1979.

essentially disagreeing with the analysis done by the Bureau of Reclamation. In doing so we were primarily using our interpretation of a reasonable water-development-based approach to benefit-cost analysis, which was supposed to use market-based data on the benefits and costs of projects.

It was already becoming evident, however, that we were omitting important benefits of environmental protection—in this case the benefits of not building the dams which could not be captured in market prices. Economic theory argues that the appropriate value to use in an economic analysis is what people would be willing to pay for a particular good—in this case environmental protection. Clearly most of that willingness-to-pay for this and other environmental protection efforts was not being captured in the market-based prices used in cost-benefit analyses of the day. We tried to include the benefits from the continued possibility of river running without the dams, but so few people would be doing such river running that the economic benefits were quite small relative to the value of the electricity that would have been generated by the dams. But based on the outpouring of support for not building the dams, we were clearly not capturing some important benefits. This was later to be an important insight for my future work at the US Environmental Protection Agency (EPA), where I tried to find a way to value such benefits.

In 1970 I went on a Sierra Club rafting trip down the Colorado River through Grand Canyon, one of the most exciting such trips in the world. The trip more than fulfilled my expectations, and I was again pleased by the outcome of the Grand Canyon dam debate since such a trip would have been impossible if the dams had been built. And I was very happy that I had contributed to this outcome.

Although I was unsure as to the influence of our economic study, I was quite sure that I had found my calling. This was exactly the type of public policy issue that interested me. It involved economics, energy, the environment, and engineering. And it foreshadowed the later rise of a new specialty in economics,

53

environmental economics, which would occupy most of my efforts in future years.

Involvement with the Diablo Canyon Nuclear Plant Controversy

At the successful conclusion of the Grand Canyon fight, when Congress dropped the two proposed dams from the Central Arizona Project, a new energy issue was developing in California which had a direct bearing on the internal politics of the Sierra Club as well as being an important energy resource decision in California, and greatly influenced my thinking concerning radical environmentalists including David Brower. The Pacific Gas and Electric Company, a major California utility, proposed to build a nuclear plant at the Nipoma Dunes on the California Coast near San Luis Obispo. The Club took an early position against this location, in part because of the active opposition of the local Chapter. Given the prior Bodega Head debacle concerning an earlier proposal to build a nuclear plant north of San Francisco, this time the Company took the precaution of consulting with a number of the more "conservative" members of the Sierra Club Board of Directors, who constituted a majority at that time, before proceeding with authorization and construction.

The Company suggested that they would be willing to consider several alternative sites on the Coast. The Board members involved agreed that Diablo Canyon, a remote site on picturesque private land on the Central California Coast in San Luis Obispo County might be a suitable location, perhaps because there was no prospect that it could become parkland and thus preserved for use by the public. When this became known in the Angeles Chapter a number of the Chapter leaders, including me, visited the site to see for ourselves, and concluded that the more "conservative" board members were pursuing a good policy under the circumstances. If a plant were to be built this looked like as good a place to put it as there was along the central and northern California Coast, with good access to plentiful cooling water in the Ocean and remote from any urban area. Brower, who was strongly opposed to the proposal, argued that it offered considerable promise as parkland and that the Club should not be in the business of approving proposed energy developments. I particularly objected to this last argument.

Environmental progress, in my view, is not usually achieved by total opposition to a particular type of project or viewpoint but rather by negotiation with the opposition to achieve an environmentally superior outcome, preferably without major economic losses, and if possible with economic gains. There are important trade-offs between environmental and economic objectives in the case of particular projects that need to be carefully weighed rather than be decided on a general principle concerning whether one is for or against energy development, or in this case for or against all nuclear plants on the California Coast. I and others in the Angeles Chapter believed that the Club should not oppose all power plants—only the ones that would have a strong negative impact on what are now called major environmental resources. The Grand Canyon dams, while not actually located in Grand Canyon National Park within its boundaries at the time, would have flooded the Canyon through most of the Park and upstream from it, including into the exquisite Havasu Canyon where I had hiked while at Caltech, and would have made river running impossible. Further, the land was already federally owned and could easily be added to the Park, which did in part occur as a result of the dam controversy.

Brower had led the successful fight against the proposed Echo Park Dam in Dinosaur National Monument but the Board of Directors had compromised by agreeing that the Club would not oppose other dams as part of the Colorado River Storage Project if Echo Park and Split Mountain were deleted from the proposed Act. Brower believed that he should have more actively opposed the compromise and tried to prevent construction of Glen Canyon Dam despite the Board's prior agreement; and he was determined that additional dams not be built on the Lower Colorado. Given the existence of public land adjacent to Grand Canyon National Park on the Lower Colorado, one of the premiere holdings of the National

Park Service, I could only agree. I even sympathized with Brower's views on the pre-dam beauties of Glen Canyon, which is also on public land, but also believed that defeating Glen Canyon was unachievable at the time and that the dam's destruction (which Brower advocated in later years) was out of the question once it was built. But there was no obvious source of funds or public interest in buying the private land to create a park out of the Diablo Canyon site so I could see no useful recreational purpose in opposing the site given that PG&E was willing to give up the earlier proposed Nipoma Dunes site.

The Sierra Club Board of Directors decided to ask the three Southern Californians who had worked on the Grand Canyon dams issue and had some knowledge of nuclear power to prepare a report on the Diablo Canyon issue. Brower was hoping that we would favor his viewpoint, and the majority on the Board was hoping the opposite. We were given very little time to prepare the report and had no time or resources to do background research on obvious potential issues such as earthquake faults that might be located near the proposed site. There was no internet at that time to quickly marshal information; rather, information had to be gradually accumulated from library resources, which took considerable time, which we did not have. The outcome was two reports, one favorable and one unfavorable to the proposed site. I was one of the two authors of the report favoring the Diablo Canyon site. So from the Board's view there was no change in the internal situation as a result of our efforts. Brower was unhappy with the views of Hoehn and me, and the more "conservative" Board members were unhappy with the report by Larry Moss unfavorable to Diablo.

There are several interesting historical footnotes to this. The first was that the Diablo Canyon plant was approved, but in 1971 it was discovered to be near the Hosgri Fault, located about 4.5 kilometers offshore, after the plant had been permitted. This resulted in a $4.4 billion retrofit to make the plant better able to survive an earthquake on the Fault. Then in 2008 a second more minor fault, the Shoreline Fault, was found less than 1 kilometer offshore from the plant. There is now a conflict over whether the plant should be relicensed when it comes up for renewal.

The second footnote was that the Diablo Canyon controversy and many other internal issues ultimately resulted in the departure of Brower as the Club's Executive Director, as will be discussed shortly.

The third footnote is that it now appears that although I was on the winning side in the Diablo Canyon fight, no one knew about the earthquake faults that would later be found near the plant site or the effects of an earthquake and resulting tsunami experienced off the northeast Japanese coast in 2011, which resulted in catastrophic damage to nuclear plants that had been built on the nearby coast without adequate protection from such tsunamis. Fortunately, the Diablo Canyon plant was located 85 feet above sea level; the surge that damaged the Japanese Fukishima plant was about 23 feet, so Diablo is unlikely to be impacted by a comparable tsunami if there should be one near it. If I had known about the geologic faults, I would have opposed the Diablo Canyon site on these grounds. Unfortunately, neither I nor anyone else knew about the faults at the time. Whether a combined unfavorable report by all three of the Southern California authors would have changed the Club's position on the site, or whether the plant would be built if we had, is not known, but is highly doubtful given the then already well entrenched political viewpoints on the subject within the Club's leadership and the utility.

From my viewpoint, I was persuaded by three things: the overwhelming likelihood that the Diablo site would not be purchased for recreational use, the need to provide some additional sources of electricity for the rapidly growing California Coast, and the fact that nuclear power was being offered at very low cost at the time. I believed that the Club should not oppose new power plants in general and needed to work with those responsible for building them to minimize the inevitable local environmental damages that their construction would result in.

Based on more recent experience I believe that large current generation nuclear power plants are probably not wise in most cases because of the extreme toxicity of unintended nuclear releases using current generation plants and the apparent inability of humans to be careful enough and far seeing enough to avoid all such releases as shown by a succession of serious nuclear accidents in recent decades. Fossil fuel plants release pollutants too but they are carefully controlled in the US and are not toxic for thousands of years as in the case of current generation nuclear plants. But I believe that decisions concerning whether to use nuclear should be left to the marketplace assuming adequate regulation to avoid most environmental effects even though the market is far from perfect due to the long established need to regulate monopolies which most power companies are. Unfortunately, the US has not created a level playing field for nuclear plants thanks to the special subsidies provided by the Price Anderson Act. This is not a good way to allow the market to decide such issues and expect that the risks will be adequately accounted for. Unfortunately, one governmental intervention begets others. Without the Act I doubt that any US current-generation nuclear plants would have been constructed. In hindsight that would have been a better outcome in my view.

Successsful Efforts to Help Defeat a Major Heavy Rail Transit Proposal in Los Angeles

Liberals and environmentalists usually support mass transit in general and heavy rail transit in particular. It is highly doubtful that upper class liberals would be likely to use it, but it may fit in with their frequent environmental and anti-automobile ethos. The problem is that few others use it either, particularly in comparatively low population density areas such as the Los Angeles metropolitan area. The reason is that it is much faster and more convenient to use private automobiles, particularly in low population density areas. Autos provide door-to-door transportation when people want it rather than when it is offered by a transit agency. There is little environmental basis for preferring mass transit in most of the US in my view. Automobile emissions are now well controlled in the US, and often largely empty public transit vehicles directly or indirectly result in large emissions too.

In 1968 the Southern California Rapid Transit District proposed an 89-mile, five-corridor heavy rail transit network in the Los Angeles metropolitan area which they claimed would cost $2.5 billion and was to be financed by a 0.5% increase in the sales tax. I and a RAND colleague, Martin Wohl, prepared an extensive economic analysis which concluded that the project was actually economically infeasible, contrary to the studies financed by the supporters.[23] The *Los Angeles Times*, which was opposed to the project, heard of our paper and publicized it shortly before an election that would decide the outcome. I appeared on several local news television programs as well as at a rally staged by opponents. For these or other reasons the project was soundly defeated at the polls. It was and is my view that heavy rail transit is unsuited to the low density development which characterizes the Los Angeles Metropolitan Area simply because there are not enough riders to even remotely pay for such expensive facilities. This low density was primarily the result of the fact that most of the City was built after the advent of the automobile, which made it economically advantageous to build outward from the central business district rather than up within the district, as in New York and many foreign cities built before the advent of the automobile. As a result, no project would or will have the very high population densities required to even remotely pay for or make much use of a heavy rail transit system. Given the subsequent approval of California Proposition 13, which greatly decreased real estate taxes in the State, and the severe financial difficulties this created for state and local government finances, it is my view that approval of the major heavy rail system proposed in 1968 would have saddled the Los Angeles area with an added source of financial strain on top of the great strain it would soon experience with Proposition 13 without the heavy rail project.

---

[23] Carlin and Wohl, 1968.

Many years later, after I had left Los Angeles for EPA in Washington, DC, a short stretch of heavy rail transit was built but has never attracted much ridership, as I had expected. Even later an increase in the local sales tax was dedicated to transit development using light rail and existing rights of way where available. This is now being slowly developed as funds become available. At least until 2013, however, riders were largely on the honor system to pay their fares, and there is considerable question whether they paid very much. The system is now introducing fare cards which must be used, but encountering a number of problems. The honor system may have contributed to the burden borne by the dedicated sales tax. Light rail is not as rapid as heavy rail and will not pay for itself either in Los Angeles, but may be more cost-effective and less burdensome on taxpayers than heavy rail would have been. At least I hope so given the decision that has been made to pursue it. I wonder whether taxpayers will continue to pay for transit most of them will rarely use, particularly now that they will have no choice but to pay their fares.

Involvement with Internal Sierra Club Politics

During its first 60 years the Sierra Club was entirely a volunteer organization. It was not until 1952 that the Club hired its first paid employee, David Brower, as Executive Director. Prior to that and even into the 1970s, the power was with the Club's volunteer Board of Directors and a number of the Club's regional chapters, all run by volunteers. Brower had been active in the Club for many years and was well known in the San Francisco Chapter and in Club climbing circles. I found the Club to be actively engaged in two things I cared about, outings and conservation, so I ran for and was elected in 1968 to the Executive Committee of the Angeles Chapter (consisting of Los Angeles and Orange Counties in California) of the Sierra Club, the second largest chapter in the Club at the time. I served as Treasurer until 1970, when I was elected Chapter Chairman, where I served until my departure for EPA in 1971. At that time the Chapter was primarily an outings organization with a small group (including me) interested in what was called conservation.

In 1969 the internal Club fight over Brower, Diablo Canyon, and club management came to a head. Brower had supported the election of a slate of five candidates for the Board in 1968 who were both well known for their environmental efforts and sympathetic to him. Club elections are open to all Sierra Club members, who vote by secret ballot. In 1969 he did the same thing. If they were elected he would effectively control the 15 member Board. But in 1969 the more "conservative" majority on the Board also developed a slate of five candidates favoring their viewpoints. Thus the outcome of the fight between Brower and the more "conservative" majority depended on the 1969 membership vote for the Board.

I had followed this internal Club debate very carefully and attended most of the Board meetings in 1968. Several things influenced my views on the internal debate. Perhaps the most fundamental issue for me was that it became ever more evident that Brower's ideological basis for environmental advocacy was considerably more radical than mine and more radical than I thought was best for the environmental movement itself. My view was and is that environmental advocacy is best carried out by advocating a strong pro-environmental viewpoint but not so extreme that it will be unacceptable to society as a whole and thus have no influence on the outcome of an issue.

Advocating major changes in lifestyle, for example, that would be unacceptable to the majority of the population has always seemed to me to be a hopeless and counterproductive enterprise even if there were a reason to do this. (As noted elsewhere in this book, this is precisely what most US environmental organizations have done in recent years.) My judgment was that PG&E had decided to build a nuclear plant along the Coast and had enough support in the State and Federal Governments to do so, and the most that the Club could hope to do was to influence where it would be built unless they could show that there were substantial adverse safety aspects for the selected site. If Chernobyl or

61

Fukishima or possibly even Three Mile Island had already happened, then this would not have been the case, but they had not.

My judgment on Brower's degree of ideological radicalness, based on my attendance at most of the 1968 board meetings and my personal interactions with him, was later confirmed for me by his advocacy of much more radical views when he was later elected to the Club Board in the 1990s. He is quoted, for example, as saying "It seems that every time mankind is given a lot of energy, we go out and wreck something with it." He and others he was associated with consistently advocated more radical views that I did not share. I strongly felt that moderation rather than radicalism was the way to exercise influence as well as to advance the environmental cause.

Future events would show that my view was not widely shared by environmental organizations in the 1990s and beyond. But it was an easy sell in the Angeles Chapter at the time. I knew one or two Chapter members who advocated what I call radical environmentalism, but did not have any difficulty in avoiding their policy prescriptions since the Chapter leadership was not sympathetic at that time.

I fail to understand how ideological radicalism buys environmental success although it may be useful for rallying the already committed. The Grand Canyon dam campaign had been successful in no small part because it had broad public support. The main effect of radicalism is to turn off people who might support you if you showed that you understood their viewpoint and that of others in the public as a whole. I also believe that environmental decisions should be firmly grounded in good science and economics, not ideology.

A second issue of concern to me with regard to Brower was that I did not believe that a hired employee of the Club should attempt to take over the Board of Directors, who were his employers. At that point the Club would be run by its staff like so many other environmental organizations and contrary to the long traditions of the Sierra Club. I recognized Brower's many talents as a leader, a spokesman for environmental causes, and his energy and enthusiasm, which I had witnessed first-hand, and hoped that a compromise could be reached that would preserve his usefulness to the Club. But I feared that since Brower had chosen to directly challenge the "conservative" majority on the Board through the ballot box, a choice had to be made between Brower and the Board, however distasteful that might be. In this unpleasant situation I believed that the Board needed to control the Club rather than a hired employee who repeatedly disregarded the wishes of the Board. Unfortunately, the rift in the Club had gone too far for any reconciliation to be possible, particularly if Brower was unwilling to compromise, which he was. Sometimes difficult management decisions are messy but necessary.

Finally, I guessed that if Brower should leave, Michael McCloskey, then the Conservation Director, would become the Executive Director. I had gotten to know McCloskey over several years and regarded him as extremely competent and responsive to the Board and the volunteers. He did not have Brower's charisma, presence, or public following, but I thought he would do just fine, which he did.

So primarily for these reasons I decided to take a strong stand on the 1969 election and was able to get my viewpoint known by most of the active members of the Angeles Chapter. Whether the Angeles Chapter vote determined the outcome, I do not know, but I suspect that it may have. Brower had been active in the (San Francisco) Bay Chapter for many years before becoming Executive Director, but had very little contact with the Angeles Chapter in my experience, so had very little base of support there.

I believed that the role of environmental organizations should not be to oppose economic or energy development in general but rather to encourage careful consideration of the environmental aspects of what governments and institutions were doing to meet reasonable human needs. Although I agreed that the Board may have made a mistake (although I was not present and do not know all the circumstances) by approving a resolution that had the effect of approving the construction of Glen Canyon Dam on the Colorado

River, that did not imply that all compromises were either bad or should be avoided. Diablo was an excellent example, I felt, of a compromise that could and should be made given what was known at the time. In hindsight, everyone involved needed to pay much more attention to the location of earthquake faults and the inherent dangers of nuclear energy, but that was not an issue at the time due to a lack of knowledge, and was not pursued by either the Club or anyone else at this stage of the location decision.

I played a role in the 1969 Club election as a result of my involvement with the Angeles Chapter, which at that time was the Club's second largest chapter. Several of us in the Chapter's leadership felt strongly enough about the next Board election to actively oppose Brower's slate of nominees and favor the slate of candidates endorsed by the more "conservative" group on the Board who had favored approving Diablo Canyon and felt that Brower was pursuing a reckless approach to Club finances, particularly by aggressively pushing the publication of expensive but beautiful large format picture books on areas of major environmental interest (Brower had been a book editor prior to becoming the Club's Executive Director), and was not being sufficiently supportive of the Board's views towards "conservation" issues.

For these or other reasons, Angeles Chapter members appear to have voted strongly against the Brower slate, at least in part because of my opposition. Brower's departure from the Sierra Club led to the founding of Friends of the Earth (FOE) by Brower and a number of his supporters in the Sierra Club. The FOE Board also later removed Brower as the Executive Director for somewhat similar reasons, and he went on to found a second new environmental organization. So it is not clear how much influence this election ultimately had on the course of the environmental movement. But I had done what I could to keep the Club from straying too far towards environmental radicalism at that time and to preserve the important tradition of volunteer control of the Club.

Environmental Movement of the 1960s

The environmental movement as I knew it as an activist and Sierra Club chapter leader in the 1960s and early 1970s was primarily oriented towards preservation of wilderness and other natural or wild lands. It undertook a variety of "conservation" projects, but many of them were designed towards either creating new wilderness areas or national parks or preserving existing ones from intrusions of incompatible uses. This was the era when a number of large national parks were created, including Redwoods and North Cascades. Preventing dams from being built in the Grand Canyon fit right into the mainstream of the conservation organizations of the time since it involved preservation and ultimately expansion of an existing park. I never encountered anyone in the Sierra Club who opposed the Club's efforts on this campaign.

Lest I seem overly negative towards the Sierra Club elsewhere in this book, I have much to thank the Club for. Among other things, I met my future spouse through it, and it provided my introduction to the environmental movement, which turned out to influence my life's work.

4
First Thirty Years at the EPA

12 Years in the EPA Office of Research and Development

MY CAREER TOOK a major turn in 1971 after Richard Nixon formed EPA in December 1970, in response to Earth Day and rapidly increasing interest in environmental issues. EPA was formed from components from all over the Government that related to environmental quality. These components were from a variety of agencies and geographic locations, and the initial herculean task was to organize these components into a workable, single organization. There were a limited number of new functions and jobs created, but the bulk of the Agency was inherited from these components. Most of the Agency was organized along media lines—air, water, solid waste, etc, (generally called program offices)—just as the original components had been, but some were based on their function. One of these functional units was to become the Office of Research and Development (ORD). Another was the policy office. Through design or an effort to simplify the transition, the reorganization made no changes in the basic structure of environmental protection in the Government. The old units reported indirectly to the President, and the new EPA reported directly. This was not of great importance during the first 38 years of EPA's existence but was to play a crucial role after Obama's election in 2008.

Although new hires were not widespread, they were concentrated in the new offices created to oversee the Agency as a whole. This included the policy office, where the staff was predominantly fairly conservative and exercised considerable influence because successive administrators relied on them to provide a second opinion on most proposals by the program offices.

Implementation Research Division

One of the department heads at RAND, Stanley Greenfield, was nominated to be Assistant Administrator of what became the ORD at the new EPA. He had been aware of some of my environmental activities and brought to my attention a new job he was creating at EPA. I applied for the job and received a firm offer from EPA as Director of their new Implementation Research Division in what was to become ORD in Washington, DC. The Division was created to be a multimedia, interdisciplinary group with both physical scientists as well as economists to address similarly interdisciplinary, multi-media environmental problems in contrast to the media specific research that was previously carried out by some of the components that were combined to form EPA. This approach offered the opportunity to take a multimedia approach that was presumably one of the reasons for creating EPA in the first place rather than leaving the individual media components that were brought together in EPA in their original agencies.

I selected a number of areas for emphasis in the new Division, including the use of economic incentives for environmental protection control, the economic benefits of pollution control, energy use, and the use of the National Environmental Policy Act for environmental protection. Most of these areas represented largely new areas of research not previously pursued by the inherited components that came together to form EPA. As an economist I was interested in the use of economic incentives which offer economic advantages over the more common regulatory approaches usually used by pollution control agencies. But going back to my experience with the economic analysis of the Grand Canyon dams, I was particularly interested in making it possible for EPA to be able to explain what the economic benefits of pollution controls would be. It was very evident from our analyses that there were many benefits that people were willing to pay for but which could not be measured using market-based prices, the traditional approach used in economics. Although the Clean Air Act does not allow the use of economic factors in setting pollution standards, most of the other acts that EPA implements do allow the use of such factors. And even where economic benefits and costs cannot be used in setting a regulation, they can be used to determine whether a proposed regulation is or is not economically efficient. My hope was that such information would encourage EPA to take into account what the economic effects of their regulations might be. Energy use was an area of increasing interest in EPA and other Federal agencies in the early years, and I was active in the early efforts to coordinate this effort within the Agency. My Division wrote the first EPA report on the options available for energy conservation (the report found that there were a number of such opportunities), which I took an active interest in. One of EPA's major tools for environmental improvement was the National Environmental Policy Act (NEPA), and EPA had inherited some of the responsibility for administering it, but did not inherit any research components working on the subject since there were not any.

My primary interest, however, was in economic analysis. I was particularly interested in trying to make it possible for EPA to determine the economic benefits of what it proposed to do. I foresaw a time when cost-benefit analysis might actually be done on proposed regulations and believed that such analyses would both assist EPA to produce economically efficient regulations and to explain to the outside world why it believed that particular regulations were worthwhile. Unfortunately, the original organization of EPA did not clearly define which components would have responsibility for economic research. This problem persisted until well after the year 2000. The policy office was also greatly interested in economics and it required some effort on my part to keep them from acquiring the responsibility for the research component of that effort as well as the responsibility for current economic analyses, which they (quite reasonably) acquired in the early days of EPA. Their interest, however, was much more in the economic costs of regulations than in the economic benefits, which I regarded as the more interesting and largely unaddressed (except by the Implementation Research Division) research issue prior to 1981. Given my perspective concerning the importance of pursuing only regulations that met both economic and scientific criteria I believed that it was important for EPA to have the best methods available for doing what I believed to be the needed economic analyses. The predecessors of EPA had largely not explored the economic benefits of pollution control, so this was somewhat controversial at the time, as would soon become more evident.

Shift to Pollutant Assessment

Unfortunately, the Office of Research and Development's interest in the economic, other social, and interdisciplinary sciences lasted only as long as Greenfield remained the head of it. The next Assistant Administrator was a physical scientist with little interest in and perhaps even a dislike of them. I ended up in a division that handled most of what little economics research was done but whose primary responsibility was for providing ORD-wide oversight of the preparation of criteria documents to assess the scientific effects and other relevant aspects of a wide variety of pollutants under review or being considered for review. Since the emphasis was on the assessment function, my background in the physical sciences proved very useful, and I spent most of my time on it, including managing a large contract under which the National Academy of Sciences (NAS) prepared such reports for EPA. These NAS reports were not used as is but rather served as background information to determine whether regulatory action should be considered and as background information for preparing EPA criteria documents if EPA actions should be initiated. It is important to note that the NAS reports were not used as the EPA criteria documents, as the Obama Administration effectively did with the IPCC and related US reports when it came into office in 2009 and decided to issue a similar-in-content report for CO_2. But prior to that the actual criteria documents were prepared either directly by EPA employees, or in some less critical cases by contractors with continuing oversight by EPA employees and detailed review by various committees of the EPA Science Advisory Board (SAB) (which was also not done in 2009).

The Clean Air Act specifies that criteria air pollutants are to be reviewed every five years, but this schedule has rarely been met. In reality the assessment process has often taken eight to 10 years, particularly for major or controversial pollutants. If one of the Science Advisory Board (SAB) committees objected to a proposed criteria document, revision of it sometimes took several years. During the period from 1975 until 1983 I spent the majority of my time on entirely scientific issues concerned with assessment report preparation, during which time no one argued that I was unqualified to do such work. This involved selecting pollutants for review, supervising assessment preparation, and determining whether the assessments being prepared accurately depicted current knowledge concerning the pollutant being studied. The US Office of Personnel Management even determined at the time that I was fully qualified in the physical sciences.

Early External Grants for Economic Research

After considerable time the new ORD management finally decided that the way to handle economic research was to establish an external "center of excellence" through a grant solicitation and to devote it primarily to research on the economic benefits of increased pollution control. I thought that their area of emphasis was correct but that much of the language they insisted on using was inappropriate and unhelpful. Fortunately, I was to be the project officer and had great latitude in selecting, supervising, and paying for the research actually done. One advantage of this approach from their viewpoint may have been that under this arrangement ORD management would not have to allocate any of the personnel slots they had taken from economic and interdisciplinary research to such purposes in the future, thus allowing their use for their favored physical science research.

I primarily used competitive grant solicitations, which were very unusual in EPA at the time, but several decades later became the standard for almost all EPA research grants. After a national competition, the University of Wyoming was selected on the basis of a number of criteria. One of the most important underlying reasons was that they had more economists interested in and with a record of research in the field than any of the other applicants. Most applicants and most universities had only one or two such environmental and resource economists whereas they had half a dozen or so. The reason was that most universities only needed enough specialists in this field to teach a limited set of course offerings. The University of Wyoming, on the other hand, was particularly interested in resource economics since the state's economy primarily depended on natural resource development, and had hired a number of economists specializing in this field to correspond with the State's interest. Although the resource level was modest, this and earlier efforts under the Implementation Research Division represented some of the first major Federal investments in a field that ultimately became known as environmental economics. When I attended MIT, there was no course in environmental economics there or in most of the American or foreign academic worlds; now most graduate economics programs offer at least one course and often majors in this subject.

One interesting unintended result of this was the founding of the American Association of Environmental and Resource Economists, which became the principal professional association for what became known as environmental economists. This term was not previously widely used, but was to become a recognized specialty in economics in subsequent decades. One of the motivations for the founding was that the principal journal in the field was edited at Wyoming and the editor wanted to build a mechanism for transferring the editorship to other knowledgeable environmental economists. This possibility was discussed on one of my trips to the University as the EPA project officer for the grant. I agreed on the advisability of forming the Association, contributed to doing the work needed to form it outside EPA time, and was one of the small group that founded the new Association.

As a result of the research done under this early grant, the EPA policy office decided in the late 1970s that it was feasible to carry out useful economic benefits analyses and would expand its efforts to include a new group to determine the benefits analyses of environmental regulation in addition to their prior efforts which were confined almost entirely to the economic costs of regulation. This decision created a little competition for me but also a possible new home for my interests if things continued to go badly in ORD.

Changing to Full Time Economic Research Management

My EPA world changed dramatically, however, in early 1981, when the new President, Ronald Reagan, issued an Executive Order requiring cost-benefit analyses of all major Federal regulations soon after taking office. This is just what I had been trying to promote since I arrived at EPA in 1971, and just what I thought would and should happen, hopefully sooner rather than later. I used this Executive Order to argue that ORD needed to undertake substantial research as to how to do this since despite the earlier research I had supervised in ORD, the techniques for valuing environmental regulations were still controversial and much in need of improvement.

I argued that a budget of $4 million was needed for this purpose. Happily this initiative was rapidly approved both by EPA and the Office of Management and Budget (OMB) but was bitterly opposed by the then ORD management which was happy to have the added money but argued that it should be spent on more physical science research, which already received virtually all its resources. After a number of run-ins with ORD management, during one of which I was forbidden to speak to EPA budget office personnel about my successful budget initiative, a halt to all new grants for economic research, and even an attempt to "lose" one of my grant packages (I had kept a copy), I decided to try to move the program including the funds out of ORD into the policy office. The policy office was interested and ORD agreed to transfer the responsibility, people, and resources to the policy office in order to avoid an internal jurisdictional war with the policy office at the Agency level. Such voluntary resource transfers are highly unusual in bureaucracies (which generally try to preserve and expand their responsibilities and budgets), but new interim ORD management took a very broad view of what might be best for the economic research program. Once this program was established in the policy office in 1983, I spent most of my time on the management of economic research but retained my interest in the physical sciences. This economic research was done primarily through external grants and contracts rather than by in-house personnel as in the days of the Implementation Research Division.

Although much of the responsibility for economics had always resided in the policy office, the responsibility for economic research had always been in ORD, much to the policy office's dislike. This was due in part to my continuing interest and efforts to promote an economic research program in ORD despite the opposition of ORD management after Greenfield left.

Next 20 Years in the Policy Office

The policy office at EPA always had a somewhat conservative but in my view responsible outlook towards environmental regulation. In the early days of EPA it had somewhat overwhelming bureaucratic power since it was combined with the Agency's administrative functions, so controlled both policy and the budget. In general, it tended to adopt an objective, economic approach to all proposals, and during its early years wielded substantial influence over EPA regulations and was known as having more influence than most of the rest of EPA on most issues. This resulted from the insistence by the early EPA Administrators that most if not all proposed regulations be reviewed by and commented on by the policy office; their opposition often meant their death or at least substantial changes in them. Unfortunately for the Office, this incurred the wrath of the outside environmental organizations who repeatedly saw program office-proposed regulations they supported shot down by the policy office, often for good reasons in my view. So when President Clinton came into office, the environmental organizations managed to get one of their candidates appointed as Assistant Administrator for the policy office, as per an alleged pre-election deal with Clinton that they could do so. This Assistant Administrator, from the Sierra Club, carefully and methodically removed most of the policy office's power over regulations and its capability to review them, although he retained the responsibility to comment on regulations but on an equal basis with other offices. Many of the media-oriented subject matter experts, however, were transferred to other duties, so even this commentary and review responsibility was not as thoroughly carried out. And during the George W. Bush Administration the policy office rarely tried to overturn regulations proposed by other offices. One of the few instances when they did it was at the direction of the then Administrator who wanted the policy office to create a disagreement so that he could less conspicuously decide the outcome as he wanted to.

At the same time, as Agency employees hired in the early 1970s left or retired, they were gradually replaced by more environmentally radical new hires who were attracted to working at EPA by the opportunity to influence environmental regulations. So the Agency staff gradually became more radical at the same time that the policy office became less influential during the 1990s.

Fate of Economic Research

My move to the policy office had its pluses and minuses, of course. One of the minuses was that some of the senior policy office staff regarded the economic research funds as a slush fund to meet their particular policy interests of the moment rather than a serious effort to improve the methods used for determining the economic benefits of regulation. Another was that anytime there was a tradeoff between funds for short-term policy purposes and long-term research the policy purposes usually won. It did, however, end criticism by the policy office of how the research funds were spent, a continuing issue since the early days of EPA, since they now controlled the program. And it ended the constant interference by ORD management in economic research for which they had little understanding or sympathy. The policy office had an excellent and well trained economic staff as a result of their earlier decision to create a benefits group and understood the importance of benefits research. Finally, ORD went in an increasingly bureaucratic approach to economic research during the mid to late 1980s which made it very difficult to direct research to useful areas or produce useful output, and largely left the choice of topics of what little economic research was funded to outside researchers in the field— basically a token effort. Later I had difficulty in locating any research output that was done or even in finding out what was funded. I had anticipated this bureaucratic direction before leaving ORD and wanted none of it.

In the policy office I was able to organize a substantial research effort in environmental economics, which ultimately resulted in hundreds of reports primarily on how to carry out better economic analyses of environmental improvements. Although the formal program was ultimately killed by the Office of Management and Budget in the late 1980s, funding from a number of program offices as well as the in-house efforts by the policy office economics staff has provided a continuing series of related outputs. With the advent of the internet I was able to organize this research and make it readily available to the research community as a whole.

Adventures in the Policy Office

Since I had always wanted to combine the worlds of research and policy, the policy office held some attractions for me. The problem with the policy office was that immediate policy issues made it hard to safeguard research funds from use for other purposes since short term policy initiatives have a hard time competing with research that may not pay off until years later. This problem was much easier in the Research Office since they did not get involved in most policy issues so that I only had to argue that my proposed research would have a higher payoff than that proposed by others.

In the policy office I did have more involvement with policy, especially since most policy issues ended up there at some point. I was furthermore in the economics unit of the policy office, which commented on the economic analyses of regulations prepared by each program office. Thus the unit as a whole had some knowledge of what each program office was doing. When I got involved in policy issues, I sometimes sided with the environmental organizations and sometimes not, depending on the science and economics in each individual case.

As a result of being in the policy office, I was aware of an effort by the National Oceanic and Atmospheric Administration (NOAA) to use a controversial survey-based methodology for the valuation of non-market resources called contingent valuation in their regulations as to how to determine natural resource damages to be paid by those responsible for such damages. On this issue I worked with a number of environmental organizations to help defeat NOAA's proposal to give estimates based on contingent valuation a rebuttable presumption for determining such damages. In the case of the Exxon-Valdez oil spill, for example, the damages are the economic value of the adverse effects of the spill on environmental resources such as fish and wildlife. Since I had been working for a number of years developing the methodology for contingent valuation, I was well aware of the problems involved and did not believe that it was sufficiently developed to be given a rebuttable presumption in the regulations proposed by NOAA. My stand angered some of my non-EPA colleagues in environmental economics but became the EPA position on the proposed NOAA regulations despite strong opposition from another office within EPA. I believed that premature adoption of the new and very controversial methodology, which I had done much to promote in its early years, would cause more problems than it would solve. Soon after EPA's comments,[24] which I had written, were issued, NOAA revised and promulgated very different regulations that did not rely on the new methodology. Given my more recent strong stands against GWD/GEF, it is worth noting that on this occasion I sided with the environmental groups and worked with them to substantially revise a policy proposal that they strongly opposed.

Efforts to Determine Full Social Cost Pricing for Energy Sources

With the advent of the Clinton Administration there was increasing interest in the relation between energy use and the environment (after all, a former Gore aide, Carol Browner, was the EPA Administrator). I was successful in using a small portion of the funds made available for this purpose to determine the full social cost pricing of the major US energy sources. The purpose was to determine whether energy prices fully reflected their adverse environmental effects. If not, an economic case can be made that the government should adjust taxes and/or subsidies so that they would reflect them. If one source of energy is pricing its products so that they do not fully reflect these environmental effects (called externalities in economic jargon) then that source will be used more extensively than would be socially justified. The guidelines of the study made an important exception by not examining the climate change effects for the practical reason that no one could define what

---

[24] US Environmental Protection Agency, 1994.

these effects were. The results were still of importance because they suggested a "no regrets" solution that would hold without regard to whether there were or were not adverse effects of fossil fuel use on climate. It was a "no regrets" solution because there would be no regrets if there were no effects on climate. This turned out to be the case since it is now evident that if there is an externality resulting from energy use on climate change, it is a positive rather than a negative one (see Chapters 9 and 12), so any attempt to have included this effect would only made the study more uncertain and less useful.

The major conclusion of this study[25] was that the prices of the energy sources that seem most out of line with their non-climate environmental damages are those of coal and wood. Natural gas is a comparatively clean energy source that is currently taxed more than is warranted given the costs that its use imposes on society. Gasoline pays its own way in the sense that the current gasoline tax equals the environmental damages imposed. The gap between the environmental costs resulting from gasoline and the taxes already imposed was much less than for energy sources such as diesel fuel and heating oils. Moreover, all those adverse effects were dwarfed by the higher but highly uncertain environmental costs associated with coal. Now that we know that there are no significant adverse effects resulting from non-catastrophic increases in global temperatures (as discussed in Chapter 11), the results from this earlier study appear to be generally as valid today as they were then (although they could be usefully updated and expanded to additional sources and changes in taxation, of course).

It should be noted that the Obama Administration has attempted to derive a social cost of carbon to justify some of its climate-related regulations that includes highly speculative estimates of the effects of fossil fuel use on climate. For all the reasons discussed in Chapters 9 and 11 I regard these efforts as both wrong and useless.

---

[25] Viscusi *et al.*, 1994

Transition to Headquarters Control of ORD Grants and its Implications for Economic Research

Because of the way EPA was formed from various components of other agencies and the history of these agencies in locating their research efforts outside Washington, EPA inherited a wide assortment of laboratories in different states. Greenfield had attempted to bridge these differences with a complicated multi-modal management structure which essentially gave the field laboratories substantial influence over research priorities as well as implementation. Since there had never previously been a unified bureaucratic structure over all the research laboratories most of the resources in both people and external research dollars were handled at the laboratories and had remained that way through successive ORD managements.

But under the Clinton Administration a new Assistant Administrator was appointed with an academic background who strongly believed that EPA research quality would be greatly improved if externally funded research was administered by a centralized research center rather than the field laboratories. And he had some interest in economic research, which I pointed out to him was receiving at best token attention since my departure from ORD. He decided to transfer most external research dollars from the laboratories to a new headquarters office largely modeled after the National Science Foundation (NSF). The laboratories were extremely upset by this change but were unable to reverse it. One problem was that ORD no longer had any economic expertise at Headquarters; as a result, I became the de facto expertise even though I was located in the policy office. Another was that primary responsibility for the annual selection of grantees was alternately given to NSF and EPA for economics research on a rotating basis under a joint program arrangement. I was constantly trying to increase the specificity of the requests for proposals, and NSF was constantly trying to broaden them so as to give the academic research applicants more latitude. And I had no influence on the recipient selection, which was done using the standard peer review system long used by NSF and now by EPA. There was little question in my mind, however, that even though they were prepared by the same people, NSF Requests for Proposals (RFPs) attracted better qualified applicants than EPA RFPs, perhaps because the applicants trusted the NSF review process more than the EPA process. It was also my view that the NSF reviewers were better qualified than the EPA-selected ones were, so there was a tradeoff between RFP specificity and applicant quality. On balance I thought applicant quality was more important, although I would have much preferred both. In the end, however, ORD management decided that EPA control was more important and ended the NSF arrangement, hired its own economists, and no longer needed my assistance. This, however, left a jurisdictional conflict between ORD and the policy office over which one would sponsor economics research and how to coordinate the two efforts since I had brought the economics research responsibility with me from ORD many years before but had not removed it from ORD.

Private Trips to Polar Areas

Starting in 1995 and continuing to 2006 I made a number of privately funded trips to both the Arctic and the Antarctic out of interest in learning about them and because of their photographic possibilities. Polar wilderness areas are some of the most interesting anywhere. They combine magnificent scenery, unique wildlife, aurora, icebergs, and glaciers.

The connection between polar travel and the climate issue is that it strongly reinforces the observation that humans are simply not a major influence on climate. The whole climate issue arose from a conjecture that perhaps humans have become so dominant an influence on the natural world that we might be adversely impacting even the climate. One obvious question is whether that influence is positive or negative. The alarmists always seem to assume that it is negative but have not offered any solid evidence for this belief. For reasons discussed in Chapter 9, I believe it is very likely to be positive. But either way travel in wilderness areas in general and polar wilderness areas in particular cannot help but suggest that the influence of man on climate is very small. Although humans have demonstrable effects in warming the climate near large urban areas, these areas are actually rather small compared to the vast areas that remain as open ocean or as wilderness, particularly in polar areas. One can travel for days, as I have, in these areas and never even see another human that is not a member of the trip.

My trips included three trips to the Antarctic and five to the High Arctic. All but one was ship-based; the other one was a hiking trip in Eastern Greenland. Three were on a Russian ice breaker, and four were on ice-strengthened ships. I visited several research stations at both ends of the Earth and repeatedly asked whether they had observed any significant changes in temperatures. All reported that they had not. Obviously this was not a valid scientific conclusion, but it raised my suspicions about whether any significant warming was taking place. On all of these trips the major problem was not heat but rather cold and ice.

I was unusually lucky in that I was able to travel on a large, powerful Russian ice breaker for the more ice-hazardous trips. This ship was chartered by an adventure travel company after the demise of the Soviet Union and has recently returned to its previous work in the Russian Arctic, so is no longer available to transport tourists. Despite the continuing false predictions of an ice-free Arctic by GEF supporters, the only way to be relatively certain of reaching many destinations in the more treacherous of these areas of the world is by traveling on such a ship, Unfortunately, icebreakers have no keels, which makes them much more subject to high winds and waves in the open ocean. But my desire to visit and understand these areas made it all worthwhile, and I would not have wanted to miss any of these trips, some of which are no longer offered.

Most of the trips were carefully arranged to take advantage of what were expected to be the lowest ice levels of the year in the areas visited (which occur in the late summer). And the icebreaker trips were on an icebreaker in the second most powerful class of such ships and with helicopters on board so that the chances of being turned back or trapped in the ice would be minimized.

In 1997 I traveled halfway around Antarctica including a stop at the main US Antarctic base and the unearthly Dry Valleys, which have little or no snow or ice despite their location on mainland Antarctica. In 2000 I visited South Georgia Island and the Falkland Islands in the extreme South Atlantic. The northern coast of South Georgia is one of the most prolific wildlife areas in the world with many millions of penguins and other birds and seals who feed in the biologically productive seas north of the Island. It is much harder to reach but also far more prolific than the better known Galapagos Islands. The ice-strengthened ship I was on, however, sank several years later when it encountered sea ice near the northern end of the Antarctic Peninsula. No sea ice was encountered on the Falklands/South Georgia trip, so I believed it was fairly safe despite the age of the ship.

In 2006 I undertook an extensive private trip to Antarctica and Patagonia, including my second Antarctic ice breaker trip. It was and is my view that it is generally inadvisable to travel to areas with potentially heavy sea ice without an ice breaker with helicopter support. Despite the ongoing fears of global warming, the main problem on the Antarctic portion of the trip was too much sea ice rather than too little since sea ice around Antarctica has been increasing for a number of years (although largely unreported in the mass media and almost never mentioned by the CIC).

The main objective was to visit an emperor penguin colony on the Weddell Sea side of the Antarctic Peninsula, which is probably the easiest one for humans to reach. The ship had no difficulty sailing to the island where the colony was located thanks to a warm wind from the north. But when the wind shifted to the south after we arrived, the ice breaker soon became surrounded by heavy sea ice blown towards it by the wind, and the trip leader decided to leave the anchorage one day ahead of schedule for fear of being locked in the ice, just as Shackleton's ill-fated Endurance expedition had been, nearby, in 1915.

Even with the head start, our powerful icebreaker indeed had great difficulty leaving the anchorage and was in danger of being imprisoned in the ice with the nearest more powerful Russian icebreaker in the Russian Arctic. On the first day after leaving the anchorage the ship sailed only about 4 miles but advanced only about 1 mile in a straight line. There was a low cloud ceiling so the icebreaker's only remaining operating helicopter could not be used to see where there might be open water. On the second day, however, it was finally possible to see an area of open water from the helicopter after noon, and the ship finally managed to reach it late that day. In November 2009 a similar incident occurred at the same location except that the ship was delayed by eight days.[26] So there was certainly no evidence for global warming that I could see in either of these icebreaker trips to the Antarctic. In fact, Antarctic sea ice has been rapidly increasing in recent years, despite all the IPCC projections concerning global warming. The reality I experienced on these trips was increasing cold and ice, not melting and rising temperatures.

My views as to the dangers of Antarctic sea travel were further reinforced in late 2013 when a Russian ice-strengthened ship, the *Akademik Shokalskiy*, became stuck in heavy ice and requested assistance on the opposite side of Antarctica from the Weddell Sea in Commonwealth Bay while carrying a group of CAGW-inclined scientists, 26 tourists, and five representatives of CIC-oriented media to visit areas visited by Australian explorers in the early 1900s using wooden ships.

None of the three icebreakers initially sent to assist the ship could reach it even though it was much farther off the Antarctic Coast compared to the Australian expedition in the early 1900s. The apparent purpose of the 2013 voyage was to publicize the CAGW cause by finding the results of global warming, but perhaps the trip planners had failed to notice the substantial increase in sea

---

[26] Luck *et al.*, 2010.

ice near Antarctica in recent years. Even in 1997, when I sailed on a powerful icebreaker nearby, only icebreakers or ships escorted by icebreakers were able to reach the main US Antarctic station at McMurdo that season. Sailing in an ice-strengthened ship in waters with heavy sea ice can lead to disaster, as in the 2013 case, particularly when careful attention is not paid to weather forecasts and ice movement and with no capability to see ice movements from a helicopter. The irony of a ship looking for proof of global warming but getting stuck in ice is captured well by this cartoon, which draws the analogy with the CAGW hypothesis itself.

Josh's Cartoon on 2013Australian "Academic Junket" to Antarctica[27]

In the end, the *Akademik Shokalskiy* and one of the icebreakers sent to rescue it were able to escape the ice as a result of a change in the wind direction in early January 2014.

---

[27] Josh, 2013; Website: http://www.cartoonsbyjosh.com

My experiences on polar cruises reinforced my suspicion that the real enemy of man is not too much warmth but too much cold and ice. There is really very little even large icebreakers can do in very thick, multi-year ice. Sending ice-strengthened ships into more than moderate sea ice is dangerous. Polar sea travel requires not only the best equipment but also careful planning and execution and close attention to weather forecasts.

The 2013 Commonwealth Bay incident in Antarctica has some parallels to other recent incidents involving CAGW supporters in the Arctic. A case can be made that some CAGW supporters were so influenced by the CIC propaganda that they actually believed that the polar ice is melting away and that they can demonstrate that for the world's media by undertaking risky exploits which depend on there actually being polar warming. In many cases these efforts have ended in disaster. Paul Driessen offers some examples besides the Commonwealth Bay incident:[28]

> In 2007, Ann Bancroft and Liv Arnesen set off across the Arctic in the dead of winter, "to raise awareness about global warming," by showcasing the wide expanses of open water they were certain they would encounter. Instead, temperatures inside their tent plummeted to -58 F (-50 C), while outside the nighttime air plunged to -103 F (-75 C). Facing frostbite, amputated fingers and toes or even death, the two were airlifted out a bare 18 miles into their 530-mile expedition.
>
> The next winter it was British swimmer and ecologist Lewis Gordon Pugh, who planned to breast-stroke across open Arctic seas. Same story. Then fellow Brit Pen Hadow tried, and failed. In 2010 Aussie Tom Smitheringale set off to demonstrate "the effect that global warming is having on the polar ice caps." He was rescued and flown out, after coming "very close to the grave," he confessed.

---

[28] Driessen, 2014.

In the world of science, determining whether hypotheses are valid utilizes the scientific method, which requires comparisons between hypotheses and observations of nature. In the real world of polar travel, the validity of hypotheses as to what are prudent risks to take is determined much more brutally by Mother Nature. If your hypothesis proves wrong, you are likely to die, except in the modern world where fossil-fueled, CO_2-emitting rescue efforts may try to come to your rescue if notified of the problem in time. CAGW adherents seem to be particularly prone to actually believing their own propaganda and being proved wrong by Mother Nature, only to be rescued sometimes by fossil-fueled machines whose operations they are generally trying desperately to get governments to reduce to prevent CO_2 emissions.

After a glorious trip through scenically spectacular parts of Patagonia, I ended my 2006 trip in the capital of a South American country. There I visited an environmental economist I knew in the government. The economist's chief job at the time was to attempt to sell hydroelectric development strongly objected to by the country's environmentalists. I inquired why the economist's efforts did not seem to include using the CAGW hypothesis to argue the merits of hydro over other sources of electrical energy and received the unexpected response that there was little or no interest in the climate issue in this country so that it would not be likely to be a successful argument.

Earlier in the trip I had met a fellow tourist from Malaysia who worked for Toyota there. I inquired how the Prius was selling there and was told that it was not sold at the time because there was no demand for such energy-conserving, but significantly higher priced, vehicles. These anecdotes obviously do not prove how the rest of the world views GWD, but reinforced my growing suspicions that CO_2 emissions reductions would be a very hard sell outside the Organization for Economic Cooperation and Development (OECD) countries. This has proved to be the case since less developed countries have made it clear that they will only consider it in return for very large payments from the developed countries, and quite possibly not even in this unlikely case.

The polar areas are not very hospitable to humans or even to most animals that we know in temperate latitudes. With adequate preparations and particularly the use of powerful icebreakers they can usually be safely visited in the late summer months, which is largely what I did. I saw no harm and much to be said for warmer temperatures, however. The only native residents in the North American Arctic are the Inuit, who have learned to live off mainly the sea mammals they caught and are thought to have colonized the North American Arctic after the last ice age. The Inuit survived the Little Ice Age as a result of their sea mammal harvesting skills, unlike the Vikings who disappeared from Southern Greenland during that period, where they used European animal husbandry techniques. I enjoyed meeting the Inuits and seeing their settlements, but certainly would not want to live among them, particularly during the long, dark winters. And none of them complained to me about warming temperatures.

5
Final Ten Years at the EPA, or How I Came to Be a Climate Skeptic

My Early Interest in GWD

IN 2003 I was asked by the Library of Congress to present a public lecture on the economics of global warming. I declined on this particular topic on the basis that the issue was too political at EPA (obviously, I did not appreciate how much more political it would become after 2008), but offered and gave a presentation on the US experience with economic incentives for protecting the environment. This implied nothing about whether economic incentives such as cap and trade should be used to implement GWD, but did argue that economic incentives were more efficient than environmental regulations in many cases.

At least as important a reason for declining their initial suggested topic, however, was that I had not studied the issue closely enough to reach a thoughtful conclusion as to whether GWD made scientific or economic sense. I realized at that point, however, that this was going to be the most important issue of this generation in environmental economics and science, and that I simply had to become familiar enough with it in order to have a carefully thought out opinion on the issue. This led to a voyage of personal discovery on the climate issue, which was to consume most of my remaining seven years at EPA, result in my retirement, and consume at least the first five years of my retirement. Most of the remainder of this book will describe what I learned on this voyage.

It was actually a good thing that I did not undertake concentrated climate research prior to the time I did since there was a reorganization in the early 1990s inside EPA in which those working on climate in the policy office were transferred to the EPA Air Office. I can only imagine what would have happened if I had ended up there given subsequent events. I could only agree with the logic, of course, that climate should be dealt with by the Air Office, since it is primarily an air issue, but it had always been handled by the Policy Office for historical and personal interest reasons.

Despite my interesting travel experiences suggesting that some of the basic assumptions of GWD were implausible, I was if anything biased in favor of the CAGW hypothesis when I started trying to understand it in the early 2000s given my former role in the Sierra Club and a lifetime spent in environmental work. I had friends active in the environmental movement who had no doubts concerning the scientific validity of CAGW. Very few environmental economists seemed to have doubts either. The easiest thing to have done would have been to join the chorus, as a number of environmental economists (although usually with qualifications concerning how the standard GWD "solution" could be accomplished more efficiently, of course) and many climate scientists long since had. But I had a background in the physical sciences too. So I first wanted to examine the scientific issues to satisfy myself that my friends had gotten it right on this, probably the most important natural resource and environmental economics issue of our generation.

I encountered no opposition in my office at EPA to my increasing interest in this area. I took the precaution of getting to know the political appointee who oversaw the policy office during the George W. Bush Administration, and was my boss's boss. I soon found that he was a very strong climate skeptic but was not adverse if I held pro-GWD viewpoints nor opposed to my doing research on the subject. I found it refreshing that a George W. Bush appointee was open to staff interest in viewpoints that neither he nor the Administration held. I soon managed to get my standards of performance revised so that they explicitly required research on climate change. So at that point I was free not only to pursue research in the area but free to formulate either positive or negative views on GWD.

When I first started to study the global warming/climate change in the mid-2000s, my first concerns were about the proposed "solution," undoubtedly its very weakest part. It is not generally appreciated that weak as the scientific case is for CAGW, the case for the GWD proposed "solution" (government mandated reductions in global emissions of CO_2) is even weaker. The "solution" is really nothing short of laughable. The GWD adherents argue that somehow the world can and will manage to agree to reduce emissions by about 80 percent below 1990 levels (more than 90 percent from current world levels) even though new uses for energy are coming into widespread use every year and substitution of energy use for human and animal labor is what has made possible the substantial increases in the living standards over the last few centuries in the developed parts of the world, and even though there are no other reliable, economically viable energy sources.

My first discovery/conclusion as I started to research the issue was that the proposed GWD "solution" (decreasing GHG emissions) was unlikely to be either effective or efficient in reducing global warming even assuming that all the IPCC assumptions were correct. I talked with the head of a prominent environmental organization and a nationally known environmental reporter and learned that such conclusions were of no interest to them. I was astounded that this conclusion could be unrecognized and uninteresting. The head of the EPA policy office was interested in the subject, but had long held this view. I could not imagine that the environmental organizations, the Administration, the scientific journals, and the mass media were not really interested that the proposed climate "solution" would not achieve the advertised "benefits" or do so in the most efficient manner even assuming all the IPCC scientific assumptions. But that was the case.

After a little further research I determined that it was very likely that there actually was an effective and efficient (but highly controversial) solution to the alleged global warming "problem," called geoengineering. Geoengineering involves using deliberate human-induced alterations of the Earth or its surroundings to achieve particular human purposes. After some more discussions I soon learned that there was even less interest in geoengineering than in the uselessness of attempted GHG emissions reductions. The reason for this was that the environmentalists were totally opposed since it meant altering "nature" and did not involve their "solution," the mass media could not use it to show imminent danger since it could be implemented at the last minute rather than requiring action now, and the Bush Administration had no intention of implementing any "solution" to a problem they did not really think was a problem.

In spite of this, I decided to write up my findings in a professional paper since I had already invested enough time to learn all this. This paper[29] compared a geoengineering approach to the GWD "solution" under the assumption that all the IPCC assumptions were correct. It pointed out that if there were a need for reducing global temperatures, that the GWD "solution" was not reliable, effective, or the lowest cost way to achieve such an objective. I noticed that the market for such conclusions among professional journals was very small. I found this surprising as well, but soon began to understand that most climate journals would accept only pro-GWD articles and only one would accept anti-

---

[29] Carlin, 2007.

GWD articles. I was learning, but had much more to learn about the highly unusual climate change issue.

Four Very Dubious Largely Non-scientific Warmist Assumptions

When I first looked at some of the non-scientific assumptions that the CIC was making concerning their "solution" with the stated assumption that the science was valid, I soon concluded that four crucial non-scientific GWD assumptions concerning their "solution" were dubious at best, more likely wrong, and little more than wishful thinking. Most of the climate debate has concerned scientific issues, but the non-scientific assumptions made by GWD advocates are probably even weaker. These four assumptions were the following:

> O *It is realistic to drastically reduce emissions of GHGs using government-imposed regulations/taxes/incentives*
> O *An effective new international treaty can be reached to reduce GHG emissions*
> O *Funding can be found to "buy" support/"reimburse" less developed countries to reduce their CO_2 emissions*
> O *Major emitting countries would actually carry out whatever CO_2 reductions they might agree to*

I concluded that these four highly dubious assumptions, which are discussed in some of my early publications on climate,[30] were a fantasy and would doom any attempt at actual implementation of GWD. Taken together, the odds that all four of these crucial GWD assumptions would prove to be correct appear to be close to zero. These four indefensible assumptions have received much less attention than those concerning CAGW science, but need to receive it since they are even more implausible. In other words,

---

[30] Carlin, 2007a, and 2008.

even if the science were correct, the CIC "solution" dooms any effort to mitigate CAGW using it. Subsequent experience strongly supports my conclusions in this respect since these problems remain unsolved up to the current day.

I furthermore concluded that *if there were actually a CAGW problem*, global temperatures might be more effectively and inexpensively avoided by using geoengineering, but only after very careful study and attention to control issues.[31] This last conclusion is strongly opposed by the CIC (possibly because it would avoid the need they see to reduce CO_2 emissions—and thus reduce economic growth—which may be their real goal) as well as by many conservative groups, who wish to reduce the role of government, not increase it. But as argued in Chapter 9 it is currently a complete non-issue since there is no evidence for the CIC's fears concerning CAGW, and thus no need for any actions to reduce global warming, geoengineering or otherwise.

Introduction to Abandoned Windmill Junkyards

In 2005 I took a short vacation trip to the Island of Hawaii. Among many glorious destinations there I drove to scenic South Point, the most southerly part of the Island. But when I arrived, I found 37 rusted, unused, and unsightly wind turbines in the abandoned Kamaoa Wind Farm. The remains of the turbines were a major blight on beautiful South Point. This caused me to learn how they came to be there. They were built in the 1980s during a period when federal and state subsidies made it financially advantageous to build wind farms and not worry too much as to whether they ever generated much electricity. When the price of oil (Hawaii's main source of electricity) dropped in 1986 and the subsidies started to diminish, it no longer made economic sense even to repair the turbines when they broke down despite the strong winds that usually blow at South Point. The last of the 37 were finally removed

---

[31] Carlin, 2007.

in 2012 at a cost much higher than the value of the scrap metal after concerns were expressed by local residents.

Many more windmills have been abandoned elsewhere, particularly in California,[32] as illustrated in Figure 10-1. Some sources put the total number in the US at 14,000, but there is no easy way to verify this in part because the number fluctuates over time with new abandonments and new removals. Then as now windmills are largely dependent on governmental subsidies, and I expect that the new versions built in recent years may not survive any longer than their government subsidies despite their improved technology.

Introduction to CAGW Science Beginning in 2007-2008

I soon realized that while my deliberate prior assumption that the IPCC's scientific conclusions were correct did not diminish the importance of my negative findings concerning the assumptions concerning the GWD "solution," I did not want to continue making this assumption in future work and I simply had to learn enough about the IPCC scientific assessments to judge their validity. So I set to work to do so with the help and cooperation of a colleague, John Davidson, a physicist who had long specialized in energy issues. His ongoing assignment was to follow the scientific literature that might be of use to the EPA central economics group (formally known as the National Center for Environmental Economics) in the policy office, where we both worked, and he took particular interest in global warming/climate change. I (and other Center staff members) received brief summaries of and comments on new research from him that he considered relevant and significant on an almost daily basis and was continually updated on the papers reported mainly in the major interdisciplinary science journals. As I soon discovered, however, the primary limitation of his otherwise excellent efforts was that he rarely reviewed any studies not reported in these

---

[32] Walden, 2010.

journals. He had long used this approach to following science, but obviously it meant that articles that did not appear in these publications were not included in his reviews. And it soon became evident that the major interdisciplinary science journals were systematically ignoring the skeptic side of the debate on the CAGW hypothesis. I began to suspect that this was deliberate rather than accidental.

One way around this problem was to go directly to the leading researchers themselves rather than depend entirely on the major journal published literature. We decided that one way to learn more about both sides of the science of global warming/climate change was to sponsor a seminar series at EPA open to anyone interested with speakers representing both sides of the issue. Happily, our managers (this was during the George W. Bush Administration) agreed to this idea, and the first seminar was held in 2008. We took some pains to make as many of these seminars available on the EPA Website as possible as videos so that those that could not attend could also hear the presentations. What we really wanted was to have genuine give and take between the opposing sides so that we and anyone else could analyze the basis for their opposing viewpoints. Despite considerable effort on our part, however, we were unable to find CAGW-oriented scientists who were willing to debate skeptics, no matter how distinguished. We found a number of skeptics who were willing to debate alarmists. So we had to settle for opposing viewpoints being presented at separate seminars in such a way that each side had an equal number. We found it amazing that alarmists were unwilling to engage in such debates, but after a while it became obvious that we were not the only ones encountering this problem. Rather, there was a very strong tendency for alarmists to refuse to debate skeptics.

On the skeptic side one of our invitees was Professor Richard Lindzen, a distinguished climate scientist from MIT, who I learned from an alumni publication was coming to Washington to make a presentation before an alumni group. I soon learned from contacts in the EPA Office of Air Programs that he was persona non grata, but could not learn any non-pejorative reasons why, so we invited him anyway. I found his presentation persuasive, but some of the Air Office attendees were unconvinced. We insisted that he present only his views on CAGW science, not politics, and he followed our request. Similar science-only presentations were made by alarmist scientists, but I found their arguments much less persuasive.

Earlier, during 2007, we had flirted briefly with CAGW science, which was easy to do given the vast literature on the subject featured in the major interdisciplinary journals and science reporting in the mainstream media. But I soon started to have serious doubts. The first thing I noticed was that there was an obvious discrepancy between global temperatures at low altitudes gathered by satellites since 1979 and those gathered from ground-based temperature instruments. Although some differences could be expected since surface temperatures are not exactly the same as those at low altitudes, it was evident that even the basic trends differed. It seemed strange to me that the National Aeronautics and Space Administration (NASA), the US space Agency, depended on less than ideally situated ground measurements rather than NOAA satellites to determine temperatures, so I started to learn about the differences between the two and why the CIC scientific establishment might prefer ground measurements over satellite measurements.

Next I noticed that the satellite temperatures showed a definite pattern that coincided exactly with the fairly regular succession of ocean oscillations called El Ninos and La Ninas that make up the El Nino Southern Oscillation (ENSO) in the Pacific Ocean and amounted to changes of about $0.3°C$ in each direction—positive for El Ninos which take heat out of the ocean and put it into the atmosphere and negative for La Ninas, which do the opposite (see Figure C-14). They seemed highly unlikely to be related to changes in CO_2 levels since there was not a steady increase in temperatures similar to the steady increase with regular annual fluctuations in atmospheric levels of CO_2.

There had, however, been an unusually strong and out-ofsequence El Nino in 1998 which appeared to have increased global temperatures about one-third of a degree Centigrade for most of the following decade. So instead of a gradual increase in temperatures corresponding with the gradual increase in CO_2 as one would expect if atmospheric CO_2 determined temperatures, the reality was a minor step increase as a result of the 1998 super El Nino followed by what appeared to be a resumption of the normal succession of La Ninas and El Ninos at a slightly higher average temperature. This had not stopped the CIC from smoothing out their land-based temperatures using various averaging and graphing techniques apparently to obscure this and instead showing what appeared to be a steady, much larger increase instead, however. These graphing techniques clearly obscured the obvious and I believed important effects of ENSOs, which the IPCC tried and still tries to ignore.

Even worse for the IPCC assessments I noticed that in the fall of 2007 that the satellite temperature data were falling rapidly back towards their levels in the period before the strong El Nino of 1998. This raised strong doubts that the predictions of GWD adherents of rapidly increasing temperatures were correct. How could it be that temperatures were falling back towards pre-1998 levels with CO_2 levels steadily increasing if the IPCC conclusions were correct? The only way that the IPCC projections had any chance of being correct was if the 1998 super El Nino step increase was followed by additional such increases. But this was clearly not happening. Global temperatures appeared to be falling back into their pre-1998 pattern of ENSOs, although at a slightly higher average temperature. The IPCC projections showed a steady increase in temperatures, but the data showed just the opposite—a probable return to the succession of ENSO oscillations at a slightly higher level (see Figure C-12). It turned out that this was an important observation because 2008 brought a major La Nina thus suggesting that 3-5 year global temperatures are primarily influenced by ENSO, not CO_2 levels.

I believed the data, rather than the CIC interpretation of it, of course, and began to wonder whether there might be a deliberate attempt to mislead the public by very careful manipulation of the ground temperature data to give the impression of a seemingly steady increase when there was actually only a minor step increase followed by a resumption of a neutral ENSO. After all, the same national and international meteorological organizations that originally jointly formed (with the UN Environmental Programme) the IPCC also were responsible for compiling the worldwide ground temperature data. In other words, the data compilers were the interpreters too, which can lead to temptations to alter data to support your case in this case and many others.

Next I looked at temperature data from ice cores taken at the top of the Greenland ice sheet. This showed that temperatures had been irregularly decreasing there for over 3,000 years (see Figure 9-1). Since I knew from my travel in Greenland that Europeans had done better during the warmer parts of that period, I doubted that even moderate increases in global temperatures were anything to be concerned about. The real danger was what it always has been for humans—falling temperatures.

Global temperatures have been stuck in a tight range with the highest temperatures during the interglacial periods and the lowest during repeated ice ages, including the last one, when no humans are believed to have lived in the North American Arctic. Humans had managed to live in the Arctic only since the beginning of the current interglacial period by the most extreme adaptations involving hunting sea mammals and using them to meet most of their needs from food to clothing. And those that did not follow these adaptations (such as the Vikings in Greenland) did not survive. No human civilizations, I also knew, had ever taken root in the Antarctic before modern times. How could minor warming in these areas be anything except favorable? Humans have had a very hard time surviving the cold, particularly during the repeated ice ages experienced about every 100,000 years, and even the Little Ice Age just a few centuries ago.

I next looked to see if there was a strong relationship

(correlation in statistical terms) between global temperatures and atmospheric CO_2 levels. Looking for such correlations is normally the first thing scientists (and economists) do when trying to understand the relationship between major variables, and I was astonished to find that the IPCC did not appear to have done this. I soon came upon a study[33] on this question at the US level, which showed that the strongest relationships with temperature were not with CO_2 but rather various measures of ocean oscillations and solar variations (see Table C-1). The strongest correlation is between the ocean oscillations such as ENSO and temperatures, just as I had suspected from looking at the satellite data; the next strongest is with solar variations, and the weakest is with CO_2 in the period from 1979 on. In fact, CO_2 has no explanatory power over the period 1998-2007 according to this analysis. Although correlation does not prove causation, the failure to correlate can prove non-causation, which appears to be the case here. There are now much more powerful correlation studies, one of which is discussed in Chapter 9 and in Appendix C.2.

I was intrigued by the strong correlation with solar variations suggesting that there was also a strong relationship between global temperatures and the sunspot cycle. This suggested that the basis for the CAGW involving the effects of increasing carbon dioxide levels might be totally misplaced and that Earth's climate might be influenced primarily by solar variations and ocean oscillations. Once these questions had been raised I recalled my training at Caltech in general and my interactions with Richard Feynman in particular concerning the supreme importance of the scientific method and observable data in determining what is and is not valid science. So I set about applying the scientific method to determining the validity of the CAGW hypothesis. By late 2008 at the close of the Bush Administration it was quite evident that the GWD was simply that, a doctrine, in a desperate search for scientific credibility since it could not satisfy the scientific method (to be discussed in Chapter

---

[33] D'Aleo, 2008.

9). At that point everything fell into place in terms of understanding what was going on. Many of the observations that had troubled me because they did not fit the CAGW were correct and the CAGW hypothesis had no scientific validity. There was no real question about it. Although the Bush Administration was unwilling to publicly question CAGW, they did oppose the proposed CIC "solution."

In retrospect, 2008 was the pivotal year in my quest to determine where the truth lay in claims concerning the CAGW hypothesis. The most important initial influence was the rapid decline (relatively) in global satellite temperatures from 2007 into 2008 rather than the strong increases shown by the IPCC models. But every other aspect of CAGW science I explored then and subsequently led to the same conclusion. The data had convinced me. There was no going back unless I discovered contrary data, which I never did, and still have not. As explained in much more detail in Chapter 9 and Appendix C, the CAGW hypothesis that humans are causing a catastrophic increase in global temperatures in recent years is not supported by the available evidence. Hence it must be discarded or modified, just like Feynman had told us. It is irrelevant how many or how "important" the supporters of the hypothesis are (which in this case includes President Obama); all that matters is whether the best available data support the hypothesis. And after over 20 years of research and many tens of billions of dollars spent on it, it was long since time to abandon the hypothesis and find a hypothesis that actually has some validity.

Supreme Court Rules that Greenhouse Gases Can Be Regulated under the Clean Air Act

In April, 2007 the Supreme Court made a crucial 5 to 4 decision concerning controlling climate change. It ruled in *Massachusetts vs. Environmental Protection Agency* that EPA could use the Clean Air Act to regulate GHG emissions from motor vehicles if it determined that they caused or contributed to air pollution which may reasonably be anticipated to endanger public health or welfare.

They were not required to do so, but could do so. This ruling set the stage for attempts to regulate CO_2 emissions by EPA, without which the EPA would have had no basis for what it is now proposing to do. In my view the Court made a very serious error since the Act was clearly not intended by Congress to be used for regulating CO_2 and CO_2 is not a "pollutant."[34] The Clean Air Act barely mentions climate and the Act was clearly not written with the idea of regulating climate under the Act since a number of the emissions limits are inappropriate for regulating CO_2, but the Court effectively gave it to them anyway. The ruling greatly increased the stakes concerning such regulatory actions since the Court essentially rewrote the CAA to allow use of existing legislation to regulate CO_2 emissions without further action by Congress. In a polarized Congress that cannot effectively pass most controversial legislation this effectively gave EPA the power to regulate CO_2 with occasional added inputs from the courts. I did not realize it at the time, but this Court decision was to play a major role in my future as well.

The Bush Administration had argued that EPA could not regulate GHGs using the Act, but lost the case by a 5 to 4 decision. But realizing what the ruling could be used for, the relevant arm of the EPA Air Office decided rather independently to begin regulatory action using this decision and prepared a draft Technical Support Document (TSD) in 2007 for a proposed Endangerment Finding that would allow EPA to impose regulations intended to implement GWD. My EPA colleague, John Davidson, and I contributed to this draft and were listed as contributors. Strangely, the Air Office tried to send the draft directly to the Office of Management and Budget for review rather than going through the well-established formal procedures for submission by the policy office after review and approval by the rest of the Agency. OMB reportedly realized this and decided not to open the submission,

---

[34] For a more detailed discussion, see Lewis, 2014c, which includes an analysis of the Court's decision, which I agree with, as part of a larger analysis.

which meant that it died at that point for the balance of the Bush Administration. This strange behavior by the Air Office suggested to me that the Office had some high level radical environmentalists even during the George W. Bush Administration who were more interested in "saving the world" as they saw it than observing well established rules and regulations. Although the Bush Administration claimed that CAGW had some validity (I think to try to mollify the radical environmentalists), they were actually strongly opposed to it and publicly stated that they did not agree that EPA should regulate CO_2 emissions. In fact, there is some evidence that they had lost their first EPA Administrator, Christie Todd Whitman, over a previous disagreement on a related issue.

In the end, although EPA did include the Draft TSD in an Advance Notice of Proposed Rulemaking issued on July 30, 2008 to solicit public feedback on how EPA might implement GHG regulation, this curious episode concerning the Air Office's 2007 proposed Endangerment Finding was of importance primarily because it resulted in the preparation of a Draft TSD a year and a half prior to the arrival of the Obama Administration. The only rational explanation for the unauthorized submission was that influential staff in the Air Office thought they could embarrass the Bush Administration by making them turn down the proposed Endangerment Finding. I was told that OMB handled the problem by simply never opening the unauthorized submission, which avoided any entry into the regulatory paperwork system, the need for any official action, and resulting possible adverse publicity. But it also suggested the intensity of the pro-Endangerment Finding feelings within the Air Office and perhaps their frustration with the Bush Administration in blocking what they may have seen as the environmental "imperative" of regulating CO_2 emissions.

Massachusetts versus EPA applied to mobile source controls. EPA has interpreted the Clean Air Act as requiring stationary source controls once any "air pollutant" is regulated under any part of the Act; major stationary sources are subject to regulation under the Title I PSD preconstruction permit program and the Title V permit program.

Interactions with My Colleague

During this period, my immediate supervisor expressed some limited interest in what my colleague, John Davidson, and I were coming up with. One of our tentative conclusions was that the IPCC had dismissed solar influences which seemed to us to explain a number of observable phenomena such as temperature fluctuations correlated with solar cycles. I knew from discussions with my supervisor's supervisor, a political appointee, that he was a strong climate skeptic, and my supervisor almost certainly knew this as well. So this may explain why we were allowed to continue pursuing these ideas during the George W. Bush Administration. It was not long before I was a strong skeptic and John was becoming somewhat skeptical. This was quite a change for John because he had been an author of an early US Government review of the science, which had favored the CAGW hypothesis. I had made no public statements on the validity of CAGW, so it was much easier for me. But the more we read, the more skeptical we became.

So thanks to John's efforts, we reviewed a significant share of the research reported in the major multi-disciplinary journals. Soon we discovered that the climate world was really two worlds, what we called the red (skeptic) and blue (GEF) sides. We picked this characterization because it corresponded to the political characterization between red and blue states, which held for the climate issue among many others. All too soon it became apparent that there were few purple websites, magazines, newspapers, or even scientific journals—almost all were blue but a few were red. The mainstream media were particularly blue and almost never mentioned that there was even any controversy. This struck me as what one might be more likely to find on a political or ideological issue rather than a scientific one. But that was reality. Politics trumped science, which I strongly felt was wrong. I have no objections to expressions of political views, but I do object to using invalid scientific arguments in support of these views and calling it science.

My Political Concerns in the Fall of 2008

As the Presidential Campaign came to an end in late 2008 it became increasingly evident that Obama was likely to win and that he was taking a radical environmental outlook on energy and climate policy. It became clear that he would try to undermine the coal mining industry and the use of coal to produce electric power as many in the radical environmental movement had long wanted. The then EPA Administrator, Stephen Johnson, knew that he would be leaving office at the end of the Bush Administration and made a special effort to meet with and exchange views and information with the permanent senior EPA staff, many of whom he had known for many years since he was the first EPA Administrator chosen from the EPA permanent staff. So I inquired as to his views concerning the developing anti-coal agenda of a possible Obama Administration. His response was that the US coal industry had existed for many generations, had developed important connections in Washington over the years, and was a political fact of life that any Administration would have to come to terms with.

Shortly after the change in Administration I left on a near perfect, private, six-week around-the-world trip visiting Northern India and two of our children in Taiwan and San Francisco. The trip went very well and I gave little more thought to my concerns or Stephen Johnson's optimism until my return in early March, 2010. So when I arrived back in Washington in very late February I could not have been in a better mood. But I was in for a rude shock.

Obama Administration's Declarations on Transparency and Science Integrity

On January 21, 2009, just after his Inauguration, President Obama sent a memorandum to the heads of Executive Departments and Agencies stating:

My Administration is committed to creating an unprecedented level of openness in Government. We will work together with to ensure the public trust and establish a system of transparency, public participation, and collaboration. Openness will strengthen our democracy and promote efficiency and effectiveness in Government.[35]

The *Wall Street Journal* reported that this memorandum led to the following:

The nominee to head the Environmental Protection Agency (EPA), Lisa Jackson, joined in, exclaiming, 'As administrator, I will ensure EPA's efforts to address the environmental crises of today are rooted in three fundamental values: science-based policies and program, adherence to the rule of law, and overwhelming transparency.' In case anyone missed the point, Mr. Obama took another shot at his predecessors in April, vowing that 'the days of science taking a backseat to ideology are over.'[36]

Despite the early declarations of openness there has been continuing legal sparring during the last few years between conservative groups and EPA over the release of internal emails and other communications requested by the groups concerning climate issues under the Freedom of Information Act.[37] In general EPA has claimed that it was not required to release the requested material or so heavily redacted what was released as to make it useless for understanding what happened and why or maintained "secret" Email accounts for high level political appointees known only to insiders.

On March 9, 2009 the new President issued a memorandum

---

[35] Obama, 2009.
[36] Strassel, 2009.
[37] Martinson, 2013.

Final Ten Years at the EPA, or How I Became a Climate Skeptic

on science integrity. The memo begins as follows:[38]

> *Science and the scientific process must inform and guide decisions of my Administration on a wide range of issues, including improvement of public health, protection of the environment, increased efficiency in the use of energy and other resources, mitigation of the threat of climate change, and protection of national security.*
>
> *The public must be able to trust the science and scientific process informing public policy decisions. Political officials should not suppress or alter scientific or technological findings and conclusions. If scientific and technological information is developed and used by the Federal Government, it should ordinarily be made available to the public. To the extent permitted by law, there should be transparency in the preparation, identification, and use of scientific and technological information in policymaking. The selection of scientists and technology professionals for positions in the executive branch should be based on their scientific and technological knowledge, credentials, experience, and integrity.*

The memorandum also includes his instruction that:

> *Each Agency should adopt such additional procedures, including any appropriate whistleblower protections, as are necessary to ensure the integrity of scientific and technological information and processes on which the agency relies in its decision making or otherwise uses or prepares.*

According to the EPA Inspector General,[39]

---

[38] Obama, 2009a.
[39] US Environmental Protection Agency, 2013.

In response to the President's memorandum, the former EPA Administrator issued a memorandum in May 2009 to all the EPA's employees notifying them of the President's memorandum and that it provided important guideposts for how the EPA should conduct and use science. The former Administrator's memorandum stated that the President's memorandum provides the agency with a unique opportunity to further demonstrate a deep commitment to scientific integrity in the pursuit of the agency's vital mission of protecting human health and the environment. The former Administrator emphasized the agency should look for opportunities to strengthen existing policies and procedures that ensure scientific integrity within the agency.

The Director of the Office of Science and Technology Policy, part of the Executive Branch, issued a memorandum in December 2010 to provide guidance to agencies to implement the Administration's policies on scientific integrity. The Director instructed agencies to develop policies that, among other things, do the following:

1. *Ensure a culture of scientific integrity.*
2. *Strengthen the actual and perceived credibility of government research.*
3. *Facilitate the free flow of scientific and technological information, consistent with privacy and classification standards.*
4. *Establish principles for conveying scientific and technological information to the public.*

In response to the President's memorandum, the former EPA Administrator issued a memorandum in May 2009 to all the EPA's employees notifying them of the President's memorandum and that it provided important guideposts for how the EPA should conduct and use science. The former Administrator's memorandum stated that the President's memorandum provides the Agency with a unique opportunity to further demonstrate a deep commitment to scientific integrity in the pursuit of the agency's vital mission of protecting human health and the environment. The former Administrator emphasized the Agency should look for opportunities to strengthen existing policies and procedures that ensure scientific integrity within the agency.

In accordance with these memoranda, the EPA established its Scientific Integrity Policy in February 2012. The policy provides a framework intended to ensure scientific integrity throughout the EPA. The EPA's Scientific Integrity Policy also established a Scientific Integrity Committee, which is chaired by the scientific integrity official and consists of deputy scientific integrity officials from the agency's program and regional offices. Under the Scientific Integrity Policy, the Scientific Integrity Committee is charged with implementing, reviewing and revising, as needed, the policy governing specific areas of scientific integrity. Specifically, the Scientific Integrity Committee is responsible for:

- *Overseeing the development and implementation of training related to scientific integrity for all the EPA's employees.*
- *Generating and making publicly available an annual report to the EPA science advisor on the status of scientific integrity within the agency. This report should highlight scientific integrity successes, identify areas for improvement and develop a plan for addressing critical weaknesses, if any, in the agency's program and regional offices.*

According to the Scientific Integrity Policy, in advance of completing the annual report, the Scientific Integrity Committee is required to conduct an Agencywide annual meeting on scientific integrity that will include the involvement of the EPA's senior leadership, reports from offices and programs, and an opportunity for input from the EPA scientific community. The Scientific Integrity Committee is also expected to review the policy every 2 years for its effectiveness and adherence with applicable rules and regulations.

The Inspector General found the following in a review of EPA's response to the requirement to implement its scientific integrity policy in a report dated August 28, 2013:[40]

> *We met with the agency's interim scientific integrity official on March 14, 2013, to determine the status of developing and implementing training for the EPA's employees on the Scientific Integrity Policy. We found that the EPA has not developed or implemented agencywide training on the Scientific Integrity Policy. Although the policy has been in place since February 2012, the former interim scientific integrity official reported that the Scientific Integrity Committee had not completed development and implementation of an agencywide training program. He noted that part of the delay in developing the training was due to the fact that they invited union participation. He further stated that it has taken quite a while for the union to decide whether and how they wanted to participate in the training development.*
>
> *During our meeting with the former interim scientific integrity official, he could not provide any projected milestone dates or timeframes for when the committee will complete this training requirement. On May 1, 2013, according to the audit follow-up coordinator for ORD and the agency's Management Audit Tracking System, the estimated completion date for the agencywide training on the February 2012 Scientific Integrity Policy has been revised to December 31, 2013. However, neither the audit follow-up coordinator nor the Management Audit Tracking System entry indicated whether the agency's Scientific Integrity Committee was involved in establishing the*

---

[40] *Ibid.*

completion date for the agencywide training.

During our March 14, 2013, meeting, we also discovered that the EPA has not generated and made publicly available an annual report on the status of scientific integrity within the agency because the committee has not yet created it, as required by the Scientific Integrity Policy. The former interim scientific integrity official could not provide any timeframe for when the committee will complete the first annual report. The former interim scientific integrity official stated that the committee would have to develop and implement training on the Scientific Integrity Policy for the EPA's employees before they can complete the annual reporting requirement.

Conclusion

As a result of the Scientific Integrity Committee's lack of progress in implementing the training and annual reporting requirements, the committee cannot fully determine the EPA employees' compliance with the agency's Scientific Integrity Policy. In addition, required determinations of the effectiveness of the policy and the status of scientific integrity in the EPA are lacking and will continue to be delayed until the policy requirements are implemented. By implementing these key requirements in its Scientific Integrity Policy, the EPA would be acting in accordance with the President's 2009 memorandum for ensuring the integrity of the scientific process and further demonstrating the EPA's commitment to scientific integrity in the pursuit of the agency's vital mission of protecting human health and the environment.

Such is the fate of public relations ploys at the hands of the bureaucracy.

Obama EPA Decides to Use the Supreme Court Decision and Starts Pro-forma Review for Endangerment Finding Based on IPCC Reports

When I returned to Washington and EPA I soon learned that the situation had changed totally and that my worst fears from the previous fall were coming true, but even more rapidly than I had expected. The new Obama Administration had launched a crash effort to issue an Endangerment Finding under the Clean Air Act that would enable it to declare carbon dioxide and other greenhouse gases dangerous to public health and welfare, a necessary precursor to action to limit emissions of these gases. Previous such efforts on much more conventional pollutants have involved many years of effort by Agency staff, outside experts, and Science Advisory Board committees, as I well knew given my past association with the process, but the Obama Administration decided that a crash effort was required supposedly in order to serve as a threat of something worse that would happen should the Senate be so "obstinate" and defeat the Cap and Trade bill in Congress, which was the Administration's preferred policy option.

I held the strong view that Congress would reject the Cap and Trade bill primarily based on repeated assertions by Senator James Inhofe that the Administration could not get enough Republican votes in the Senate to overcome a likely filibuster. I regarded him as a very straight shooter with intimate knowledge of his colleagues' views who would only make such statements if he really believed what he said. I therefore believed that even though the Administration was pitching an Endangerment Finding as a fall back to threaten the opposition with, it would very soon become the major threat and perhaps the major goal of the Obama Administration's environmental policy. It was therefore crucial to do everything possible to avert an Endangerment Finding.

In the short run, such a finding would accomplish nothing except drive up the cost and decrease the reliability of electric power by outlawing new coal fired plants and probably forcing most or even all existing coal and possibly even natural gas plants to close for alleged GWD reasons. In the longer run it could be used to increase the cost of energy usage in other areas vital to the US standard of living and economic competitiveness. As an economist I regarded this as a very high cost to pay for no or more likely negative benefits. Even if the CAGW hypothesis were correct, the reduction in CO_2 emissions would be minuscule compared to the large increases occurring and likely to continue occurring in Asia, and there was every reason to believe that increased CO_2 emissions would be beneficial anyway. As a scientist I was appalled by the lack of any valid scientific basis for CAGW. Further, EPA was not making an independent review of the science but simply repeating the views of the IPCC and US Government reports based on the IPCC reports.[41]

I had spent my career trying to promote economic development, environmental protection, good science and economics, and rational analysis of multidisciplinary problems which I regarded as mutually supportive in the larger sense, and this egregious proposal would greatly harm all of these efforts. The Administration's proposal to issue an Endangerment Finding was highly likely to end up more than wiping out all the incremental gains I hoped I had helped to achieve during more than 37 years as a senior EPA employee and 6 years as a Sierra Club activist and leader. Finally, an Endangerment Finding was much easier to slow or stop before it was issued than afterwards given the presumption by most courts in favor of regulatory agency scientific and policy

---

[41] EPA stated on April 24, 2009, in EPA (2009c) that:

EPA has developed a technical support document (TSD) which synthesizes major findings from the best available scientific assessments that have gone through rigorous and transparent peer review. The TSD therefore relies most heavily on the major assessment reports of both the Intergovernmental Panel on Climate Change (IPCC) and the U.S. Climate Change Science Program (CCSP). EPA took this approach rather than conducting a new assessment of the scientific literature.

findings. The Endangerment Finding would have to distort the science involved in order to make its case.

I believed that scientific issues should be decided in a transparent process involving independent review by Agency scientific staff and the Science Advisory Board on the basis of objective analysis, not by politicians in the White House or the United Nations. I was only too well aware that my opposition to the Draft TSD draft was unlikely to change the Administration's misguided GWD policies but felt that I probably would not be able to forgive myself if I did not do what little I could to prevent an Endangerment Finding from being issued or at least modify it if it was. I was one of the few EPA employees unwilling to play along with the Administration's bureaucratic game, which was the route taken by almost every other EPA employee at the time. In my view, the Endangerment Finding would be the first and the most important step towards implementing GWD in the United States. Once it was issued, there would be a continuing series of EPA regulations that in the end would cost trillions of dollars paid by taxpayers and ratepayers for no or more likely negative benefits.

A substantial portion of the remainder of this book is the story of how these regulations unfolded since 2009, and why they are economically, scientifically, and legally suspect.

Apparently in order to expedite an Endangerment Finding, the Air Office decided to use the abortive 2007 draft TSD as the starting point for another effort. This time the Air Office followed normal EPA procedures, however, and started the normal internal review of its slightly revised version of the 2007 draft, but with a wildly accelerated time frame compared to any other major rule makings I knew of in the history of EPA or even of the 2007 draft TSD. An inter-office Task Force was formed including reviewers from each EPA Office. I attended every one of them through mid-March, although in many cases they were simply telephone conference calls.

Although the normal procedures for inter-office review were followed in early 2009, the ludicrously short review time on what was probably the most important rule making that EPA had ever undertaken in terms of its eventual cost and impact on the US economy, suggested that the review was intended as pro forma rather than any serious attempt to produce a scientifically accurate TSD. Specifically, all comments were due in one week after distribution of the Draft Technical Support Document for the Endangerment Finding on March 9, 2009. This week included extensive review time for managers and coordinators, who were to review all the reviews before submission to the Air Office, so that left less than five days to respond. Since the TSD draft was 57 pages of highly technical material, it was very doubtful that anyone not intimately familiar with the subject could even read let alone understand the draft in that period of time or compose thoughtful comments. The only advantage I had was that I had been researching the science and economics of climate change for several years, was very familiar with the IPCC arguments and the problems with them, and was familiar with the 2007 draft TSD on which this new draft was closely based. I had even been listed as a contributor to it!

It was all too evident even at this early stage that EPA was simply going through the bureaucratic motions and that the review exercise was simply that, an exercise to satisfy procedural requirements for a decision that had already been made at a very high level, almost certainly at the White House. I wondered how this might relate to the Administration's proclamation on scientific integrity, which was issued on exactly the same day as I received the revised TSD draft.

I initially assumed that John Davidson would support me in writing negative comments on the draft TSD as we had jointly authored a number of technical pieces, and started to draft my comments under that assumption but later in the week it became all too evident that there was no time for the coordination that would be needed and that I was the only staff member willing or able to run the risks involved in submitting negative (although intended to be helpful) comments on the TSD as a whole, so if such comments were to be made, I had to make them under my own name. But there was no time to carefully consider the options or their consequences. There was only time to write my comments, and very little of that. So I undertook the job, working almost non-stop during the allowed four to five working days after receiving the revised draft TSD for review.

In an effort to influence as many as possible of the other reviewers and avoid last minute surprises, I sent out various drafts to other interested reviewers and staff members I knew inside EPA at various times during the week as well as other related Email correspondence despite the obvious risk that I would be ordered to stop work on my comments. Given the timeframe, I had to make various compromises in my efforts compared to my normal standards and hoped that improvements could be made at the end, which did not prove possible.

The three main points in my comments on the draft TSD were that the CAGW hypothesis is invalid from a scientific viewpoint because it fails a number of critical comparisons with available observable data, that the TSD draft was seriously dated and the updates made to the 2007 version of the draft were inadequate, and that EPA should conduct an independent analysis of the science of global warming rather than adopting the conclusions of the IPCC and US Government reports based on their reports. These comments[42] can be found in full in Appendix A of this book.

With regard to my last main comment, an independent assessment is certainly needed and probably required by Section 202(a) of the Clean Air Act under which the Endangerment Finding was to be issued. Outsourcing such determinations to third party assessments is not or at least should not be permissible under the Act in my view. This issue received some attention after a 2011 report by the EPA Inspector General concerning the Endangerment Finding.[43] The IG faulted EPA for not following the strict peer

---

[42] Carlin, 2009.
[43] US Environmental Protection Agency, 2011

review guidelines issued by the Office of Management and Budget in developing the TSD since in the IG's view it was a highly influential scientific assessment (which it surely was) falling under the guidelines for such reviews "because it weighed the strength of the available science by its choices of information, data, studies, and conclusions included and excluded from the TSD." The EPA responded that these standards were unnecessary because the TSD was not a "highly influential scientific assessment," but rather "a document that summarized in a straightforward manner the key findings of the NRC (Natural Research Council), USGCRP (United States Global Change Research Program), and IPCC (Intergovernmental Panel on Climate Change)." Either the TSD was a "highly influential scientific assessment," which would require the use of the strictest peer review standards according to OMB's *Peer Review Bulletin*, or a summary of findings by non-EPA groups that would not require the strictest review standards but would then hardly seem to qualify as the Administrator's own judgment under Section 202(a) of the Act. EPA tried to have it both ways. If, as I recommended, EPA had conducted its own independent assessment this problem could have been avoided if they had also observed the highest review standards, which they clearly did not. Obviously speed was more important to the Obama Administration than observing either Section 202(a) or OMB's Peer Review guidelines. I found this irresponsible and reprehensible.

Late in the week of March 9, 2009, John Davidson set up a meeting with the authors of the TSD in the Air Office to outline our concerns about it. On the morning of the meeting, however, the Air Office participants cancelled it, saying that they were too busy. This was not a good development as we thought that we would be much more likely to influence them informally than through formal comments, but I could and did use the time to work on my comments. My comments on the TSD were submitted to my supervisor and the Center TSD coordinator, as required by the inter-office Task Force, on March 13.

Not only was the review of the draft TSD perfunctory at best, it contrasted sharply with other major EPA decisions that I had personally been involved with, which were subjected to extensive reviews often including the relevant Science Advisory Board committees. These last are time consuming and demanding of staff and reviewer time, but clearly justified in my view for such an important finding likely leading to trillions of dollars in costs to the economy. I admit, however, that if there were no skeptics on the Committee consulted, the result would have been the same. Very generally speaking, SAB committees tend to include environmentally-committed members and few that are openly hostile to what the Agency is trying to do.

EPA's Rejection of My Comments

I had known and worked with my supervisor, Albert McGartland, since the early 1980s. He had a reputation as being a quite conservative Republican cautious towards environmental regulation. My sense was that at this point he was mainly concerned about his longevity under the radical environmentalist Obama Administration, which had made it clear where it stood on global warming and pollution control in general even before taking office. My guess was that he may have been primarily anxious to keep his job, however, despite his presumed personal distaste for their policies. He may therefore have been particularly anxious to demonstrate a positive attitude toward their climate/energy policy and to avoid any suggestion that the central economics office might have any doubts about them.

So as expected, my supervisor's response to my request that he forward my comments to the Office of Air Programs was negative. Appendix D, a Congressional report prepared by the Minority Staff of two House committees on "The Politics of EPA's Endangerment Finding," provides a much more detailed and from my knowledge an accurate account of the events involved. On March 17 McGartland wrote stating that he would not forward my comments to the Office of Air Programs because:

The time for such discussion of fundamental issues has passed for this round. The administrator and the administration has [sic] decided to move forward on endangerment, and your comments do not help the legal or policy case for this decision... I can only see one impact of your comments given where we are in the process, and that would be a very negative impact on our office.

In other words, my comments were rejected at the Center level because they ran counter to EPA's draft position, which was supposedly what I was to comment on, because they might have a negative impact on his office within EPA (and thus perhaps on him). The draft TSD review was thus, as I had suspected, nothing more than a procedural formality since the decision had already been made, almost certainly in the White House. In a brief earlier verbal exchange with him he claimed that the IPCC report was written by climate experts and should be assumed to be correct. My comments in Chapter 9 and Donna Laframboise's[44] explain how wrong he was. I was a little surprised, however, by the apparent strength of his stated views since he had supervised global warming policy development for a number of years during the Clinton Administration and surely must have realized the weaknesses of the IPCC reports given his presumed political orientation and analytical background as an economist. My conclusion was that he was trying to head off my anti-CAGW views which he feared might endanger his credibility and thus his job under the new Obama Administration. He may have hoped that by supporting the Administration's GWD views he could persuade them of his allegiance despite his reputation as a political conservative.

A few minutes later he directed that I not spend any more EPA time on climate change. Five days earlier he had forbidden me to speak to or communicate with anyone outside of his Center on endangerment issues. Thus, the issue being commented on had

---

[44] Laframboise, 2011.

already been decided, and I was forbidden to tell anyone outside of his office about any concerns I might have about it or do any further work on the climate change issue as a whole. This was obviously in sharp contrast to the treatment of James Hansen in NASA during the second Bush Administration. In this case Hansen claimed that he had been muzzled when in fact there is no evidence that this happened since he continued to present his CAGW and GWD views publicly and continued to hold his supervisory job.

The practical effect of all this was that I was ordered by my supervisor not to do further work on climate and not to talk to anyone outside of the Center about my views on global warming/climate. I tried conscientiously to follow these orders even though I noticed that others had not received similar instructions inside or outside of EPA on this topic. As a result I attended no more meetings of the Task Force (since that would involve talking to people outside the Center about global warming) and largely confined any discussions of climate to my colleague, John Davidson.

The one possible exception, at least in my supervisor's view, concerned a seminar that John and I had organized featuring a pro-CAGW spokesperson from a major environmental organization primarily to help show "balance" in the seminar series. This was a Center-organized event, so in my view I was authorized to attend. I did so, and dared to question the speaker on factual matters concerning his glib use of carefully massaged ground-based global temperature data to make part of his standard CAGW viewpoint. One of my questions was why the speaker had used only ground-based temperature data rather than satellite data, which is all too typical of presentations by climate alarmists. It appeared, however, that his knowledge concerning the satellite data consisted of the opinion that it was less reliable than the ground-based data, but that he could not explain why this might be so. In fact, just the opposite is the case since ground-based data assumes that global temperatures can be built-up from a diverse and changing set of ground temperature measurement stations primarily located in larger population centers despite increasing urban heat island effects as urban areas expand.

Most of the attendees were Center staff although it was open to others. Apparently someone, I suspect a Center staff member who attended the seminar, mentioned my attendance and questions to my supervisor, who called me in that afternoon and in front of a staff witness brought in apparently to back his version of events in case of later questions, read me the riot act about having violated his gag order concerning not discussing climate outside of the Center. He made it clear that the penalties for alleged further violations of his gag order would be severe, presumably removal for insubordination. My defense was that the event was a Center-sponsored event and thus exempt from his previous gag order. But his interpretation was that his gag order included Center-sponsored scientific presentations by outside CAGW supporters.

So much for free scientific debate and discussion in an Administration with explicitly specified pretensions of openness, transparency, and scientific integrity. One must not even raise inappropriate factual scientific questions concerning CAGW science at a scientific seminar that I had helped to organize and sponsored by the Center I worked in. Even the reliability and basis for surface temperature data used to support a proposed regulatory policy costing trillions of dollars must not be questioned. The very important underlying issue here is how the GWD adherents have manipulated the surface temperature data (which they unfortunately control through the national meteorological agencies that make up the World Meteorological Organization, which in turn established the IPCC with the UN Environment Programme) for the last quarter of the 20th Century to make it appear to be the rapidly increasing "hockey stick" shown in all their graphs. Surely the extent of global warming is an important scientific issue concerning GWD. The reality is discussed in Appendix C. My guess was that my supervisor feared that my questions might get back to the Air Office and suggest that he was not supportive of their GWD proposals even though the questions were from me, not him. I suspected that he was playing along with the CAGW "science" and did not want anyone in his organization to be identified with even scientific questions concerning it. I fear that his reading of the real views of the Obama Administration towards transparency and scientific integrity were correct, despite all their rhetoric concerning transparency and scientific integrity.

Caltech Reunion

In May, 2009 I decided to revisit Caltech for a well-attended reunion of my Caltech class. The first thing I noticed on arriving at the Institute was that the San Gabriel Mountains were clearly visible. The ozone pollution, I was told, had been greatly reduced, and that appeared to be the case. The second thing was the huge expansion of the Caltech campus. Perhaps partly as a result of the large Federal expenditures on science in the intervening years, the campus had expanded by a factor of perhaps four since I had attended with numerous new research and academic buildings.

My main surprise, however, was that I did not meet any classmates at the Reunion who questioned the science supporting GWD even though most of them had spent their careers in the sciences or engineering. This led to a number of interesting discussions, of course. I was simply astounded that those who had gone through the same education as I had and had also had contact with Richard Feynman seemed not to be skeptical, as I thought all scientists should routinely be on all scientific issues. In fairness, of course, they had not had the opportunity to spend the previous few years as I had examining and writing about the evidence and had presumably been exposed to the pervasive GWD propaganda. One of my best friends in the class had even written an article that assumed that the CAGW view of climate was correct and was a little concerned that possibly his assumption was incorrect on the basis of my comments at the reunion. Late in the weekend I did meet several graduates from other classes who agreed entirely with my views. And in 2010 I met Harrison Schmitt, the astronaut, who graduated two years before I did from Caltech and strongly agreed

Final Ten Years at the EPA, or How I Became a Climate Skeptic
with my views.

CEI's Release of My Comments and EPA Emails

On June 23, 2009, just before the June 26 vote by the US House of Representatives on the Cap and Trade (tax) bill, the Competitive Enterprise Institute (CEI), a Washington, DC-based conservative non-profit organization, released a copy of an intermediate version of my TSD comments to the press along with copies of a number of the even more damning emails from my supervisor. The version of the Comments released suggested that one or more of my fellow EPA employees to whom I had sent earlier drafts (before the gag order) might have been the person who gave them to CEI since I had not given them to anyone outside EPA. I received word of this in a phone call from a reporter from Dow Jones, who wanted to know whether the emails were authentic. I asked him to send me copies so I could compare them with my Email records. He did so, and I determined that they were entirely accurate. A few days after this I received a phone call from my supervisor asking if I could forward him copies of the emails as he could not access them at the moment. And, he added, seemingly almost as an afterthought and with no explanation, that I was now free to talk to the press about all this as long as I stated that I was expressing my own views and not those of EPA. A few days later, his supervisor also called me with a similar message. Apparently once CEI had released the information EPA management must have decided that gag orders were no longer an effective public relations tool. As a result, I then felt free to respond to numerous requests for interviews both on and off camera, including on Fox and Friends and the Glenn Beck show on Fox News. My observations concerning these interactions with the press will be discussed in Chapter 7.

An interesting question is how much the release of my comments on the draft TSD affected the ultimate outcome of the GWD. This will probably take some time to determine, but it may be relevant that few if any Republican members of Congress have publicly endorsed major GWD-related policies since its release, in contrast to John McCain's opposite viewpoint during the 2008 election campaign.

Vote on Waxman-Markey Cap and Trade Bill in the House of Representatives

On March 26, 2009, the House of Representatives scheduled a vote on a Democratic bill to reduce US CO_2 emissions by 83% by 2050 entitled the American Clean Energy and Security Act[45] better known as the Waxman-Markey Cap-and-Trade bill . Some of the principal features included the following:[46]

> 1. Sec. 116 would establish first-ever new source performance standards (NSPS) for power-plant CO_2 emissions under §111(b) of the Clean Air Act. The provision would require new coal power plants permitted after Jan. 1, 2009 and before Jan. 1, 2020 to achieve a 50% reduction in CO_2 emissions. The only technology capable of meeting the requirement is carbon capture and storage (CCS), which can make a new coal plant more than four times more costly than a new natural gas combined cycle (NGCC) power plant (see Table 2 of this EIA report[47]).
>
> 2. Title III...would establish a cap-and-trade program covering all major emitters. Stationary sources would have to achieve CO_2-equivalent greenhouse gas emission reductions of 3% below 2005 levels by 2012, 17% below by 2020, 42% below by 2030, and 83% below by 2050.

---

[45] US House of Representatives, 2009.
[46] Lewis, 2014a.
[47] US Energy Information Administration, 2013.

3. Title I would require utilities to supply increasing percentages of electricity from a combination of efficiency upgrades and renewable sources (6% in 2012, 9.5% in 2014, 13% in 2016, 16.5% in 2018, and 20% in 2020-2039).

4. Sec. 144…would require states to establish and enforce electricity demand-reduction goals, described as "the maximum reductions that are realistically achievable with an aggressive effort to deploy Smart Grid and peak demand reduction technologies and methods."

The bill passed by a very narrow margin largely along party lines with 219 votes in favor including 8 Republicans and 212 opposed including 44 Democrats. Numerous efforts were made by the Obama Administration to get a similar bill passed by the Senate, but failed, and the legislation died at the end of the 111th Congress.

If it had been enacted it would have required CO_2 emissions per person to be reduced to approximately the level in 1875 by 2050. This was the level shortly after the Civil War,[48] and would have been totally devastating to the US economy in my view.

EPA's Disinformation Campaign

After the CEI leak EPA responded to CEI's allegations that EPA had suppressed my Comments by attempting to discredit me by claiming that I was "not a scientist" and was "not part of the working group dealing with the issue" (see Appendix D). *The New York Times* (see Appendix B) reported that I had "never been assigned to work on climate change" and "was not assigned to submit comments on the document." Various other blue media and blog sites also followed EPA's lead. Rather than responding to the specific issues I raised in my comments, EPA and the CIC may have decided that it

---

[48] Hayward and Green, 2009.

was better to question my qualifications and my assignment to working on climate in hopes that the scientific and procedural issues I raised were too complicated for the average reader to be likely to try to reach an independent judgment on.

The principal point of the campaign appeared to be that I was unqualified to raise scientific issues concerning climate since I was allegedly not a scientist and that my comments were not prepared in response to the internal staff review of the TSD. The EPA Press Office specifically took this viewpoint despite contrary evidence subsequently reported in the House Minority report (see App. D).

This disinformation campaign raises a whole series of questions. If I was scientifically unqualified, why had I been specifically required in writing to work on climate science issues since at least the mid-2000s as part of my official duties? If I was apparently judged qualified to do the research then, why was I suddenly unqualified? Why had I been listed as a contributor to the 2007 draft TSD if I was unqualified? Why had I been allowed to supervise production of scientific assessments similar in purpose to the TSD for a period of seven years in the 1970s and early 1980s? If I was unqualified now, surely I was even more unqualified then. And why had the US Office of Personnel Management determined that I was fully qualified in the physical sciences? And were my years at Caltech studying physics apparently totally wasted?

Or is it possible that there was collusion between the consistently CIC-friendly *New York Times* and EPA to attempt to discredit me so as to limit the public relations problems raised by the CEI leak? EPA claims that I was unqualified just happened to be picked up in a "news" article in the *Times* to be found in Appendix B, and soon to be echoed around the vast CIC blogosphere echo chamber.

With regard to the EPA claim that I was not asked to prepare comments on the TSD, EPA responded on September 3, 2009 to a letter from Representative Joe Barton that "Yes...Dr. Carlin was one of several members of the NCEE workgroup that reviewed the draft TSD for EPA's proposed Endangerment Finding for greenhouse gases."[49] As noted in Appendix D, the House report concluded that I "actively participated in NCEE's review of the TSD and climate issues in general." They further reported that my "contributions on climate change were not in question prior to this controversy. Before attempting to discredit him, EPA even listed him as coauthor/contributor of the 2007 and 2009 TSD reports."[50]

Unfortunately, I am not alone in being on the receiving end of CIC smear attacks against skeptics they disagree with. The best known of these attacks was against Dr. Willie Soon in February, 2015. This too was initiated with an article in the *New York Times*[51] and may have been prompted by a scientific article[52] Soon had co-authored that the CIC disliked. Evidence has emerged that an Obama political organization had advance knowledge of this attack and is strongly supporting other such attacks against a number of climate heretics by congressional Democrats[53] including demands for correspondence and financial support information. As will be discussed in Chapter 9, I believe that movements should be judged in part by the character of their public communications and behavior. Political and journalistic attacks on scientific skeptics is reprehensible and inexcusable.

Creation of Individual Climate Website

In mid-2009 a long-time acquaintance suggested the potential usefulness of the creation of my own Website on climate (http://www.CarlinEconomics.com) so as to promote my science and economics-oriented approach. As a result of my work on the EPA website and my then familiarity with the "red" and "blue" climate websites, I immediately recognized the merit of his idea, and accepted his kind offer to create one for my use.

---

[49] See Appendix 20 of US House of Representatives, 2009.
[50] Appendix D.
[51] Gillis and Schwartz, 2015.
[52] Monckton *et al.*, 2015.
[53] Goddard, 2015a.

Climategate

In late 2009, just prior to the Copenhagen Summit designed to lead to a worldwide agreement to control CO_2 emissions, someone leaked the contents of more than 1,000 emails and about 3,000 other documents found on a server used by the Climatic Research Unit (CRU) of the University of East Anglia, the leading British center promoting CAGW, during the period 1996 to 2009. In late 2011, just prior to a UN climate conference in Durban, South Africa, the same individual appears to have leaked another 5,000 emails. Finally, in May, 2013, the remaining emails were made accessible. The contents of the emails raise many questions as to why the data seem not to support the CRU's repeated public claims of unprecedented global warming, various tricks they had used to "encourage" the data to reveal the desired conclusions, their efforts to prevent publication of the views of skeptics in reputable journals, and deletions of data and other efforts so as to not make it available under Freedom of Information laws so that they would not fall into the hands of skeptics. I regarded these emails and other material as evidence of scientific fraud. I had long since suspected all this, but it was revealing to see it in their own words. How anyone could continue to believe the CAGW propaganda after this episode puzzled me.

Issuance of the Endangerment Finding and the Technical Support Documet

As part of the run-up to the December, 2009 Copenhagen Conference, which the Obama Administration apparently hoped would lead to a new worldwide agreement on governmental control of CO_2 emissions, EPA issued its Endangerment Finding and TSD on December 7.[54] I could not find any significant changes that had been made as a result of my efforts but was surprised that I

---

[54] US Environmental Protection Agency, 2009b.

was still listed as one of the "EPA authors and contributors."[55] But I sincerely doubted that the issuance would affect the outcome of the Copenhagen conference, and it clearly did not.

Collapse of the 2009 Copenhagen Conference

Perhaps the most important event was not what was happening in the US Government but rather what was happening in Copenhagen in late 2009. Although he UN conference in Copenhagen was attended by many world leaders, including President Obama, the conference was unable to solve the basic problem that had bedeviled the UN climate efforts since the Kyoto Protocol was drafted in 1997: How to persuade the less developed countries to undertake CO_2 emissions reductions which were very much against their interest since it would restrict their use of energy, which is vital to improving their living standards. At Kyoto the problem had been papered over by exempting the less developed countries from the requirement to reduce emissions. But since the less developed countries were rapidly becoming the major emitters and certainly the countries with the most rapidly growing emissions by 2009, this problem had to be addressed if CO_2 emissions were ever to be reduced in any meaningful way. Continuing to exempt emissions from less developed countries would come nowhere near the emissions reductions allegedly "needed." Payments from the developed countries to the less developed were a theoretical possibility, and very large but vague promises were made by some developed countries at the Conference, but it was obvious that they were unlikely to be fulfilled, and would almost certainly not change the views of the recipients even if they were.

The opposition of the less developed countries is and always has been the major unsolved problem of the CIC. Actions by the US Government to reduce US CO_2 emissions are highly unlikely to ever solve this problem, despite the continuing statements to the

---

[55] US Environmental Protection Agency, 2009a.

contrary by the Obama Administration.

Retirement from EPA

All of this had transpired so rapidly that I had not even thought about what I might do as a result. I did not leak my comments or the emails to the Competitive Enterprise Institute, so had not been forced to consider what might happen if I did. Prior to the leak I had reached an uneasy truce with my supervisor after explaining that I intended no harm and simply wanted to assist EPA to produce a more accurate TSD. But after the leak and particularly after my acceptance of opportunities for news media exposure and after the EPA disinformation campaign it was evident that whatever truce there might have been had died. Further, my supervisor made it clear that he intended to use whatever supervisory levers he had to encourage me to leave. He could not fire me without cause, but he could make it very miserable for me to stay, primarily by loading ever more extensive make-work projects on me so that I could not possibly do them all, presumably with the result that he could then claim that I had not met all the requirements he had imposed.

One obvious alternative I was considering throughout this period was to retire from EPA. I had been eligible for several years, but had stayed on because I felt I was doing interesting and worthwhile things. Although the gag order had been lifted, I was still barred from any further work on climate change as an EPA employee, but could do so as a retiree. Perhaps most important, I did not want to work for an Administration which had such a radically different view of what was best for the environment, the economy, scientific integrity, and transparency. I did not believe in what EPA was trying to do on energy and the environment under Obama, and life is too short to work for people who are trying to make life worse rather than better for the country and world. So it was a fairly easy decision to retire in very early 2010. EPA could then no longer impose their censorship and make-work requirements, and I could work on research that I found the most interesting and useful. This book is one of the results.

130

6
Lessons Learned Since Retiring from the EPA

Introduction to the Skeptic World

SOON AFTER LEAVING EPA in very early 2010 I was invited to join the Cooler Heads Coalition, the main coordinating group for climate change skeptic groups in the US. I received no compensation for this, of course, and had almost no knowledge of their role. What I soon discovered was that although much more efficiently run than the numerous Sierra Club conservation meetings I had attended in the 1960s and early 1970s, it is basically the same idea—keeping everyone up to date so that they understand what is going on and is likely to happen soon so that they hopefully can function effectively in promoting their cause and will not get in each other's way. To some extent these coordination meetings are less necessary than in the case of the Sierra Club 45 years ago because numerous skeptic blogs keep everyone interested up to date on the latest events and can be supplemented with mass emails as needed and weekly and daily updates, which also exist. The meetings are certainly much shorter and to the point than the Sierra Club meetings, but bear no relationship to the supposedly sinister conspiracies by evil polluter paymasters repeatedly pictured by the CIC. The focus is on upcoming Congressional action or hearings concerning climate change and energy issues of possible interest to the attendees.

Despite the depiction of skeptics by CIC propaganda as a monolithic coalition of carefully orchestrated groups responding to the whims of big oil or big coal or some other "evil" polluters, what I found was a group of like-minded groups and individuals who are strongly opposed to current climate and energy policies espoused by radical energy environmentalists because the participants believe that they will have a disastrous effect on the welfare of the American people and the economy and are not supported by the science. It is true that the attendees generally share a conservative political philosophy (possibly because they generally support a decreased role for government while the radical energy environmentalists uniformly espouse a greatly expanded role). Each group represented in the Cooler Heads Coalition pursues its own vision of how to achieve its ends, with no visible attempt to impose a structure or tight coordination of their efforts.

The purpose of the meetings is to share information, not to pass out orders or funding from "evil" funders, assuming that such even exist. The meetings are distinctly low budget. No food or even water is usually offered despite the fact that they are held at lunchtime. Time is allocated based on the chairman's view of current priorities such as upcoming Congressional votes on issues directly related to climate change politics. Although he sometimes expresses his views as to what he thinks these priorities should be, the attendees are completely free to express their views and even to follow their own agenda, and often do. The Chairman has a very modest budget, which is used in part to hold presentations by visiting climate skeptics to Congressional audiences.

If, as claimed by the CIC, the skeptics are so well funded, why are their principal coordinating meetings so low budget? Where are the retreats to Cancun or Bali, all expenses paid by wealthy polluting corporations, like the CIC repeatedly supports, often with public funds? The answer, of course, is that those fighting the CIC are relatively starved for funds despite the all-pervasive CIC-myth to the contrary. Surely after over five years of involvement with the skeptic world, during which I have met and talked to most of the principal players, surely someone would have slipped up and told me about the latest lavish get together is some exotic place or the well-paid jobs going unfilled for lack of suitable candidates. Heartland's James Taylor argues that the net annual expenditures by conservative think tanks on the skeptic cause is no greater than $46 million per year.[56] I suspect it is much less.

The more interesting question is why there is not more funding, not why there is so much. Like so many of the CIC public-relations-inspired claims, a surplus of "evil" funders is simply propaganda, probably perpetuated to explain their lack of progress towards their GEF nirvana, the strong opposition from skeptic citizen-scientists, and the desire to persuade alarmists of the alleged righteousness of their cause. Yet every alarmist I know believes this myth. By far the biggest money is actually on the CIC side, much of it from taxpayers, ratepayer payments to CIC-affiliated corporations, sympathetic governments, and environmentally-oriented foundations.

Although a number of the permanent staff employees of skeptic-oriented organizations such as the Competitive Enterprise Institute and the Heartland Institute are indeed paid (but also work on other issues), my observation is that they are not getting rich and in most cases are motivated in their energy policy efforts by their strong commitment to supporting good science and economics and good (but smaller) government. They and the many citizen-scientist volunteers are united by a strong feeling that they are doing essential work in the public interest against overwhelming odds. After all, fighting bureaucracies like EPA, the White House, and the IPCC with budgets in the billions except for the IPCC, which is in the millions, most of the mainstream media, major environmental

---

[56] Taylor, 2014.

organizations, and environmentally oriented foundations on a shoe string is not exactly a picnic. Funding is scarce and difficult to get based on actual discussions I have been involved in among some of the organizations sponsoring skeptic efforts. No expensive public relations firms or their products are in evidence. It is all done in-house. Does everyone agree with everyone else? Of course not. The remarkable thing is that there is as little disagreement as there is and that the skeptics are as effective as they are given their overwhelming financial and staffing disadvantage.

I have yet to meet anyone active in the skeptic effort who I would judge is doing it primarily for the income (where there is any). So where does the bulk of the funding come from? In many cases the "funding" consists of time volunteered by retired or tenured scientists and energy specialists who are very concerned about the damage GWD/GEF will do to the US and the world. As volunteers their activities are based primarily on their personal perceptions as to where their efforts will be most effective, so tight coordination is really not that useful, nor in evidence.

Another myth deliberately created by the CIC is that skeptics are "flat-earthers," "deniers," and ideologues who are anti-science. These views have been expressed by President Obama and Secretary of State Kerry, among others. This certainly differs from my experience, where active skeptics are largely trained in the physical sciences or engineering. In other words, they usually have an education that requires an understanding of the scientific method. A recent survey of participants on online forums on skeptic blogs found that they included the following:[57]

- Around half of respondents had worked in engineering and a quarter in science.
- Around 80% had college degrees of which about 40% were "post graduate" qualified.

---

[57] The survey was undertaken by Mike Haseler, and reported in Nova, 2014.

- Respondents were asked which areas they had formal "post-school qualification". A third said "physics/chemistry. One third said math. Just under 40% said engineering. 40% said they had post school training in computer programming.

The survey also found the following with regard to their views on climate:[58]

96% of respondents said that atmospheric CO_2 levels are increasing with 79% attributing the increase to man-made sources. 81% agreed that global temperatures had increased over the 20th century and 81% also agreed that CO_2 is a warming gas. But only 2% believed that increases in CO_2 would cause catastrophic global warming.

The survey originator reached the following conclusions from his data:[59]

Above all, these highly qualified people—experts in their own spheres—look at the published data and trust their own analysis, so their views match the available data. They agree that the climate warmed over the 20th century (this has been measured), that CO_2 levels are increasing (this too has been measured) and that CO_2 is a warming gas (it helps trap heat in the atmosphere and the effects can be measured). Beyond this, the survey found that 98% of respondents believe that the climate varies naturally and that increasing CO_2 levels won't cause catastrophic warming.

---

[58] Haseler, 2014.
[59] Ibid.

I am happy to say that I am not aware of any skeptic websites which pour out the daily *ad hominem* attacks and vitriol so much in evidence on too many of the CIC sites. The one minor exception I know of is a website called DrRoySpencer.com, whose owner finally tired in 2014 of being called a "denier" and once and only once suggested that alarmists should be referred to as global warming Nazis.[60] Many skeptic websites actually discuss the science involved from a sophisticated and objective viewpoint. Many are owned by individuals and usually ask for donations to supplement their own contributions, but often offer remarkably high scientific content.

Invitations to Major Climate Conferences

In 2010, 2011, 2012, and 2014 I was invited to present my views on EPA's role in climate change/global warming before the principal public forum for skeptics in the United States, the International Climate Change Conferences sponsored by the Heartland Institute. At the 2014 Conference I was awarded a Climate Science Whistleblower Award[61] for my efforts to point out the scientific errors in the draft Technical Support Document for the EPA GHG Endangerment Finding. As of mid-2014, there have been nine of these conferences held in Chicago, New York City, Washington, DC, Las Vegas, Sydney, Australia, and Munich, Germany. These have been reasonably well but by no means extravagantly funded, but have never strayed from their information provision role into political tactics or campaign ideas. These conferences are run by Heartland's President, Joseph Bast, his lead climate change staffer, James Taylor, and supporting staff, and financed by anonymous donors. Bast has repeatedly stated that none of the funding comes from environmentally "evil" sources, and I believe him. Joe takes a personal interest in climate change science

---

[60] Spencer, 2014.
[61] Heartland Institute, 2014.

despite having also worked with the Sierra Club as a volunteer many years ago, and sees to it that most viewpoints within the skeptic community are represented at the conferences by well-informed scientists. The speakers and attendees seem to be quite conservative politically just like the skeptic community as a whole.

Heartland has made a serious effort to invite CAGW supporters to speak, but has almost always been turned down. One CAGW supporter did speak at one of the conferences (ICCC6) which I attended but felt that he had to "dumb down" his presentation to the point that many attendees, who are generally very well informed on climate science, were unlikely to find it of much interest. He made a good faith effort, but convinced no one. Generally speaking, his argument was that because CO_2 is a GHG therefore there has to be a serious warming problem as India and China develop. The counter-argument, of course, is that the effect appears to be so minor compared to other effects (such as solar variability) and the climate system so complicated that it will make little difference in global temperatures.

My learning experience with regard to climate change/global warming did not end with my departure from EPA. Rather, I have had the time and energy not only to follow the energy-environment discussion but even to think about what it all means. John Davidson and I had concluded early on that much more needed to be known about the role played by the sun and other astronomical bodies in climate change. We had attempted to contact anyone in government who was working on this, but found very few. What interest there was appeared to be from discouraged and disheartened individuals concerned about their personal future. We similarly looked at the academic world, which had a little more interest but not very much. What is needed, I believe, is an explanation that can explain the Little Ice Age, the Medieval Warm Period, and the minor increases in temperatures in the late 20th Century. I then began to explore the onset of the Little Ice Age and realized that this may well have been a dress rehearsal for the real thing—a new ice age—which could recur within the next few millennia and possibly as early as a few centuries from now, as in previous interglacial periods.

In early 2011 I received an invitation to make a presentation on geoengineering at a conference organized by the Russian Federation Russian Academy of Sciences and the Federal Service for Hydrometeorology and Environmental Monitoring and in Moscow in late 2011. I decided to try to use Scafetta's research (to be discussed in Chapter 9 and Appendix C) to reach some conclusions as to when and under what circumstances geoengineering might be needed, particularly in the event of significant global cooling. I found that the Russians were receptive to some geoengineering approaches (while playing a careful game of seeming support for the IPCC leadership and viewpoints).

The Moscow conference was of considerable interest to me for a number of reasons. The first was my earlier work on Cold War strategic issues at RAND. I actually had an opportunity to meet one of the people involved in that conflict and even to tour the Kremlin, the historic home of the Russian Government. I also had an opportunity to experience a non-UNFCCC conference partially sponsored by the IPCC, something which few other skeptics have ever been able to do from the inside. I understand that this was just one of numerous alarmist-oriented conferences sponsored around the world. This one cost the Russians quite a bit, but may be similar to others held both by the UN and national governments. Moscow in November is hardly a prime destination, but I lucked out on the weather since there was only rain and no snow. Rajendra Pachauri, the IPCC Chairman and other IPCC and UN climate officials addressed the Conference and were generally treated as royalty by the Conference organizers. I even met Pachauri's young Italian (female) personal assistant, who was travelling with him.[62] The Russians were clearly playing a careful game, giving recognition to

---

[62] Pachauri resigned as Chairperson of the IPCC on February 24, 2015 as a result of sexual harassment charges by employees of an Indian NGO he headed, including a personal assistant who accompanied him on airplanes.

the IPCC while also inviting at least one US skeptic with knowledge of geoengineering to participate.

It is rumored that Russia provided the crucial final ratification of the Kyoto Protocol despite some scientific skepticism concerning CAGW in part because they had been promised membership in the World Trade Organization in return. The Russian officials I spoke to at the Conference were disappointed that as of then they had not been able to join WTO, although that finally occurred in August, 2012.

I was encouraged that at least one senior Russian official (Yuri Izrael), whom I met, was able to include an extensive section on geoengineering in the Conference program, presumably because he believes it is an interesting approach and had enough bureaucratic influence to insist on it. Some of the pro-GWD participants were not very enthusiastic about either geoengineering or my participation. Presumably other countries have IPCC-co-sponsored conferences where GWD-inclined speakers routinely appear. I give the Russians great credit for inviting a skeptic, for their hospitality, and for having an open mind on GWD. Most GWD supporters strongly dislike geoengineering, perhaps because it could be used to reduce global warming without reducing GHG emissions. My impression was that Izrael, on the other hand, does not believe that humans play an important role in global warming[63] but does believe that geoengineering is potentially useful. I give the Russians in general and Izrael in particular credit for keeping an open mind. Under Izrael's direction, the Russians have actually made a start with a geoengineering field experiment.[64]

About the time of the Moscow conference, Nicola Scafetta sent me a preprint of an article[65] he was publishing detailing how he believes solar cycles influence Earth's climate, and how planetary motions influence the solar cycles. Since I had been working with

---

[63] This observation is supported by Izrael, 2005.
[64] Izrael, 2012.
[65] Scafetta, 2012.

his earlier research, I immediately recognized the potential significance of his findings (which will be discussed in Chapter 9). My presentation was subsequently published by the Russians.[66]

2011 Journal Publication and the Cook Paper

Shortly after leaving EPA I started work on an overview paper on climate science and economics which I submitted to a peer reviewed journal at the end of 2010. After three months of interactions with three anonymous reviewers it was published on April 1, 2011.[67]

From a policy perspective, the paper's conclusions included the following:

- *The economic benefits of reducing CO_2 emissions may be about two orders of magnitude less than those estimated by most economists because the climate sensitivity factor is much lower than assumed by the United Nations because feedback is negative rather than positive and the effects of CO_2 emissions reductions on atmospheric CO_2 appear to be short rather than long lasting.*

- *The costs of CO_2 emissions reductions are perhaps an order of magnitude higher than usually estimated because of technological and implementation problems recently identified.*

- *CO_2 emissions reductions are economically unattractive since the few benefits remaining after the corrections for the above effects are quite unlikely to economically justify the much higher costs unless much lower cost geoengineering is used.*

- *The risk of catastrophic anthropogenic global warming appears to be so low that it is not currently worth doing*

---

[66] Carlin, 2012.
[67] Carlin, 2011.

anything to try to control it, including geoengineering.

From a historical perspective, the paper built on my Comments on the Draft EPA TSD[68] by presenting an expanded version of a few portions of that material in journal article format, incorporating many new or updated references, and explaining the implications of the science for the economic benefits and costs of climate change control. It is also particularly noteworthy for appearing in a peer-reviewed journal rather than the "gray literature," such as a report to EPA, where many skeptic analyses end up—something that warmists never fail to point out. Although this article was not written for EPA, it has major implications for the scientific validity (or lack thereof) of the December 2009 EPA Endangerment Finding and the economics that EPA and many economists have used to justify current efforts to regulate the emission of greenhouse gases under the Clean Air Act, cap-and-trade schemes, and other approaches to controlling climate change.

From a scientific perspective, the paper starts with a detailed examination of the scientific validity of two of the central tenets of the AGW hypothesis. By applying the scientific method the paper shows why these two tenets are not scientifically valid since predictions made using these hypotheses fail to correspond with observational data.

From an economic perspective the paper then develops correction factors to be used to adjust previous economic estimates of the economic benefits of global warming control for these scientifically invalid aspects of the AGW hypothesis. It also briefly summarizes many of the previous analyses of the economic benefits and costs of climate control, analyzes why previous analyses reached the conclusions they did, and contrasts them with the policy conclusions reached in this paper. It also critically examines the economic costs of control.

From a methodological perspective, the article argues that

---

[68] Carlin, 2009.

economic analyses of interdisciplinary issues such as climate change would be much more useful if they critically examine what other disciplines have to say, insist on using the most relevant observational data and the scientific method, and examine lower cost alternatives that would accomplish the same objectives. These general principles are illustrated by applying them to the case of climate change mitigation, one of the most interdisciplinary of public policy issues. The analysis shows how use of these principles leads to quite different conclusions than those of most previous such economic analyses.

John Cook's Classification of My Paper

While I was writing this book a group of alarmists headed by John Cook published a paper[69] claiming that 97% of published paper abstracts in peer-reviewed journals from 1991-2011 matching the topics "global climate change" or "global warming" that expressed a position concerning CAGW endorsed the "consensus" position that humans are causing global warming." One of the papers they examined was mine,[70] which they had classified as "explicitly endorses CAGW but does not quantify or minimize."

When asked whether this is an accurate representation of my paper I responded as follows on May 27, 2013:

> *No, if Cook et al.'s paper classifies my paper, 'A Multidisciplinary, Science-Based Approach to the Economics of Climate Change' as 'explicitly endorses AGW but does not quantify or minimize,' nothing could be further from either my intent or the contents of my paper. I did not explicitly or even implicitly endorse AGW and did quantify my skepticism concerning AGW. Both the paper and the abstract make this clear. The abstract includes the following statement:*

---

[69] Cook *et al.*, 2013.
[70] Carlin, 2011.

Lessons Learned Since Retiring from the EPA

The economic benefits of reducing CO_2 emissions may be about two orders of magnitude less than those estimated by most economists because the climate sensitivity factor (CSF) is much lower than assumed by the United Nations because feedback is negative rather than positive and the effects of CO_2 emissions reductions on atmospheric CO_2 appear to be short rather than long lasting.

In brief, I argue that human activity may increase temperatures over what they would otherwise have been without human activity, but the effect is so minor that it is not worth serious consideration.

I would classify my paper in Cook *et al*'s category (7): Explicit rejection with quantification. My paper shows that two critical components of the AGW hypothesis are not supported by the available observational evidence and that a related hypothesis is highly doubtful. I hence conclude that the AGW hypothesis as a whole is not supported and state that hypotheses not supported by evidence should be rejected.

With regard to quantification, I state that the economic benefits of reducing CO_2 are about two orders of magnitude less than assumed by pro-AGW economists using the IPCC AR4 report because of problems with the IPCC science. Surely 1/100th of the IPCC AGW estimate is less than half of the very minor global warming that has occurred since humans became a significant source of CO_2.

When asked whether I had any further comment on the Cook *et al.* (2013) paper, I responded as follows:

> *If Cook et al.'s paper is so far off in its classification of my paper, the next question is whether their treatment of my paper is an outlier in the quality of their analysis or is representative. Since I understand that five other skeptic paper authors whose papers were classified by Cook et al. (Idso, Morner, Scaffeta, Soon, and Shaviv) have similar concerns to date, the classification problems in Cook's paper*

may be more general. Further, in all six cases the effect of the misclassifications is to exaggerate Cook et al's conclusions rather than being apparently random errors due to sloppy analysis. Since their conclusions are at best no better than their data, it appears likely that Cook et al's conclusions are exaggerated as well as being unsupported by the evidence that they offer. I have not done an analysis of each of the papers Cook et al. classified, but I believe that there is sufficient evidence concerning misclassification that Cook et al's paper should be withdrawn by the authors and the data reanalyzed, preferably by less-biased reviewers.

One possible explanation for this apparent pattern of misclassification into "more favorable" classifications in terms of supporting the AGW hypothesis is that Cook et al. may have reverse engineered their paper. That is, perhaps the authors started by deciding the "answer" they wanted (97 percent) based on previous alarmist studies on the subject. They certainly had strong motivation to come up with this "answer" given the huge propaganda investment by alarmists in this particular number. So in the end they may have concluded that they needed to reclassify enough skeptic papers into "more favorable" classifications in order to reach this possibly predetermined "answer" and hoped that these misclassifications would go unnoticed by the world's press and governmental officials trumpeting their scientifically irrelevant conclusions. Obviously, whether this was actually done is known only to the authors, but I offer it as a hypothesis that might explain the apparently widespread and one-directional misclassifications of skeptic papers. Mere sloppy analysis should have resulted in a random pattern of misclassifications.

The political significance of Cook *et al*'s paper is that it may be the basis for Obama's claim in a major speech on June 25, 2013[71] and for EPA Administrator Gina McCarthy's claims on September 18 that 97 per cent of scientist agree that human activities are changing the atmosphere in unprecedented ways. Apparently the statements by me and five other skeptic authors concerning Cook *et al*'s mischaracterization of our papers did not have much impact on their characterization of the percentage of scientists supporting AGW.[72] Their emphasis on an alleged percentage consensus strongly underlines either their total misunderstanding of science or their deliberate attempt to manipulate science for their policy ends. Politicians' use of Cook *et al*. despite its many deficiencies pointed out by many skeptics does not show great scientific understanding in my view or judgment on their part. Subsequent analyses by skeptics suggested that the paper was intended for propaganda purposes and used definitions that made the conclusions reached meaningless.[73] A comprehensive 2014 study concluded that there is no 97% consensus on global warming in the various CIC studies and that it is "not even close."[74] JoAnne Nova has compiled a scathing analysis of Cook *et al*.[75] Legates et al wrote a comprehensive rejoinder.[76] Richard Tol has published[77] a further refutation of Cook *et al*. Finally, PopularTechnology.net has compiled a list of 97 articles refuting the paper.[78]

In late 2013 I published another journal article,[79] which questioned the ill effects of the precedents being established by the CIC in terms of future politically motivated scientific and economic

---

[71] Obama, 2013.
[72] PopularTechnology.net, 2013.
[73] Montford, 2013.
[74] Friends of Science, 2014.
[75] Nova, 2014a.
[76] Legates *et al.*, 2013.
[77] Tol, 2014a.
[78] PopularTechnology.net, 2014.
[79] Carlin, 2013.

proposals.

The Decision to Write This Book

In 2012 it became evident that the climate wars were far from over, particularly after the US 2012 election, and that there was and is monumental misinformation, disinformation, and misunderstanding concerning the CIC and their GWD/GEF ideology. I was having continuing concerns about the adverse economic and environmental effects that would occur if this ideology should become institutionalized in regulations governing the production and use of energy. The proposed CIC solution was much worse than the alleged problem.

I accordingly decided that my contribution towards solving these problems might include a book—this book—explaining my side of the story, the science, and the history. As of the completion of the manuscript I have received no compensation for my efforts on the book—nor do I expect to receive any other than a minor portion of possible book sales.

Major Developments in the Rest of the World

In many ways Europe led the US in trying to implement GWD, primarily because of the unwillingness of the Clinton and the George W. Bush Administrations to allow major changes in US energy policy as well as the overwhelming opposition of the US Senate to the Kyoto Protocol in 1997. This all changed with the election of Obama in 2008, who attempted to follow the European model and even briefly held up Spain as his ideal, but could not get approval of cap and trade by the Senate and lost control of the House of Representatives in the 2010 Congressional election in significant part because of voter discontent with Democrats from red/coal states who had voted for cap and trade.[80]

---

[80] Lewis, 2014a.

More recently there have been interesting developments in Europe and Australia. In 2013, the Labor Government in Australia headed by Julia Guillard, which had pushed through a carbon tax in 2012, was replaced by one headed by Kevin Rudd. An election followed in September, 2013 that resulted in a loss by the Labor Party and the formation of a new Government headed by Tony Abbott, who promised to repeal the unpopular tax, and did so in July, 2014. Guillard has blamed the tax as a major factor in her and Rudd's defeats.[81]

The developments in Europe were also important. The Spanish Government dramatically curtailed its generous subsidies for renewables because of its severe financial problems, which may result in financial disaster and bankruptcy for more than 50,000 investors in renewable energy. Britain, apparently reacting to a groundswell of opposition to windmills blighting the countryside, announced that it would not build any more on land. The Czech Republic decided to end all subsidies for renewables by the end of 2013.

A major part of this story was Germany. Many businesses and homeowners installed solar panels, wind turbines, and other "renewable" sources in recent years to take advantage of mandatory purchases using generous feed-in tariffs, which resulted in much higher rates for everyone else (except "heavy users"), particularly renters, who usually also have lower income. On at least one hour on one sunny, windy day renewables have generated more than 67% of German demand.[82] But all this capacity proved useless in December, 2012, and January, 2013, and in the second week of December 2013. In the latter case, there was very little renewable energy generated because of fog and wind doldrums. Conventional power plants (including nuclear) had to carry nearly the full load.[83] In the year earlier case, grid operators had to import nuclear power

---

[81] Gillard, 2013.

[82] Wetzel, 2013.

[83] *Ibid.*

from the Czech Republic and France and power up an old oil-fired power plant in Austria.

Facing an election in 2013 with some of the highest electricity rates in Europe, which have doubled since 2000, the German Government reduced its feed-in tariffs for solar electricity. The German solar industry is facing financial collapse and huge losses. Instead, the Government is now frantically supporting the construction of new coal-fired plants including some soft coal/lignite plants—yes, the type of plants that US environmental organizations most hate, but worse because some of them burn even dirtier lignite. An unsolved problem is the electric grid instability resulting from heavy reliance on "renewables." Sudden changes in the electricity supplied to the German grid have resulted in significant damages to companies whose processes cannot tolerate even momentary interruptions.[84] And all this because the Government insisted on intervening in the energy market by ending nuclear power and promoting renewables. German utilities are threatening to close down natural gas-fired generating plants since they are losing money on them because of increasing solar/wind use on clear/windy days.

The US should now be able to more clearly see what awaits the US if the Obama Administration should be successful in its GWD/GEF/EWD aspirations, although this has long been clear. Germany has ended up with a renewable energy subsidy of about 8.8 US cents per kwh paid by non-high energy users or about US$350 per year for the average German family,[85] ever increasing electricity prices, the second highest electricity rates in Europe, which are three times higher than US rates, the necessity of keeping conventional plants on standby, a coal power plant construction boom, much greater electric grid instability, angry neighbors who are called on to transmit North German solar power to Southern Germany, supply their power to Germany on calm, overcast days,

---

[84] Schröder, 2012.
[85] Based on Deutche Welle, 2013, and Neubacher, 2012.

cope with the greatly increased system instability resulting from all these power surges, and German companies moving or thinking of moving energy-intensive industries overseas.[86]

And what has Germany gained? The potential knowledge that on the best hours of the best days up to 67% of its power is from renewable sources, even though solar/wind power accomplishes exactly the same amount of useful work when available, is indistinguishable by the user from any other source, and requires standby fossil fueled plants to be kept ready to substitute for it when clouds appear or the wind dies. And the substitution of renewables has very little effect on CO_2 emissions and no measureable effect on a non-problem, global warming. It should be noted that Germany's CO_2 emissions are increasing despite flat economic output and declining population.[87] In late 2013 the Merkel Government decided to reduce its subsides for renewables because of their high cost.

What were the thorough Germans thinking? Why does the Obama Administration want so badly to do the same for the US if it is allowed to do so? Clearly substitution of government fiat for a market economy in energy results in disaster even in the hands of the ultra-thorough Germans. Using a "free" resource ends up being much more expensive economically and environmentally much worse than using non-renewable resources in the German case and very likely in the US as well. If the Germans cannot get government to make green energy work, how could the US?

2013 IPCC Report

In September, 2013, the IPCC released their new AR5 report for Working Group 1 (physical sciences). Despite the fact that virtually every major prediction based on their climate models in previous

---

[86] A former US Ambassador to Germany and now US Senator summarizes how all this came about in Germany and the results in Coats, 2014.

[87] Carlyle, 2013. See also Bastasch, 2014.

reports has proven to be grossly inconsistent with observational reality, including their oft projected rise in temperatures with increased CO_2 levels since 1998, the report claims even higher confidence (now 95% certainty) in their conclusion that higher temperatures are primarily due to human activities. This conclusion depends crucially on the assumption that their models have accounted for all natural variability that could account for the very limited warming observed prior to 1998. But the new report speculates that this is not the case since the missing heat is alleged to have disappeared into the oceans, which could not occur if their models had taken all natural variability into account. So their tentative explanation for the missing heat ends up undermining their basic assumption.

I note that one of principal changes made in the report at the Stockholm meeting of governmental representatives in September, 2013, was apparently to bury any real discussion of the problems with the reality of the model results and the global warming hiatus. This tells me that the governments involved were not ready to seriously consider calling off the CAGW false alarm, presumably because they think they can still get their way, possibly because of Obama's plans to use the EPA to institute GEF in the US. They obviously plan to continue as before regardless of the overwhelming evidence on the other side. The IPCC has dug itself into a very deep hole and has chosen to continue digging. Since it is obviously incapable of using observational data to correct errors in its hypothesis, it is not capable of carrying out meaningful scientific assessments. I see no further use for the organization, and question whether it ever had any.

The International Results of Endorsing EWD Science and CAGW

The major reason that GWD/GEF has not been implemented on a worldwide basis is not the opposition of skeptics to the science involved (as widely alleged by the CIC in the US), but rather the opposition of the less developed world to mandatory reductions in

CO_2 emissions because it would make it more difficult for them to achieve further economic development.

The failure of the developed world to argue against the views of many less developed countries concerning the alleged effects of CO_2 emissions on extreme and unusual weather has resulted in an unnecessary additional impasse between the developed and less developed worlds over "compensation" for the damages of such weather events allegedly caused by developed world CO_2 emissions. Obama's decision to very publicly endorse EWD, despite the absence of any scientific basis for it, appears to have contributed to this push by less developed countries. Instead of taking measures to reduce the adverse effects of these weather events, they are demanding a new UN agency to calculate and distribute loss and damage benefits using large additional payments from the developed world, and this was agreed to (with US and other developed nations' approval) at the 2013 UN Warsaw COP Conference as a new UN legal framework, the Warsaw International Mechanism for Loss and Damage Associated with Climate Change Impacts.[88] And instead of making the scientifically supportable viewpoint that there is no scientific basis for EWD, the US and other developed countries are left arguing when and how much their payments should be. Such payments for weather damages could well cost many trillions of dollars if the developed world should ever agree to actually pay them. Imagine FEMA expanded to the less developed world but with full compensation for extreme weather losses/damages. This will not only create unnecessary ill will towards the developed world (when the payments are not made) and provide a false incentive to delay measures to reduce the effects of such events, but is also likely to make the Administration's efforts to achieve a new worldwide agreement to replace the expired Kyoto Protocol even more difficult than it already is. In fact, I believe that the agreement of the less developed world is very unlikely. This has strong implications

---

[88] United Nations Framework Convention for Climate Change, 2013.

for the Obama Administration's efforts since even with the full cooperation of the Republicans in Congress and the Supreme Court, the global CIC cause is likely to be hopeless. Even the IPCC has not endorsed the scientific basis of EWD, but the US nevertheless argues for this scientifically unsupportable and potentially immensely expensive doctrine. Its adoption by the Obama Administration may or may not improve their chances of US public support, but encourages the less developed countries to demand even higher payments from the developed world.

Similarly, the US has not argued against the scientifically unsupportable CAGW, has contributed towards a $35 billion "Fast Start" during 2010-12, promised a $3 billion contribution in 2014 to the UN's Green Climate Fund, and promised that it will contribute to a $100 billion per year transfer from the developed world to the less developed starting by 2020 under the Fund to "take action on climate change." The only possible source for such funding is the taxpayers of the developed world. In addition, the US agreed to a resolution at the 2013 UN meeting at Warsaw that would allow each country to make its own "contribution" towards CO_2 reductions under a proposed Kyoto successor protocol. I would be very surprised if the less developed countries and their ally China (now the world's largest emitter of man-made CO_2) will decide to make very large contributions despite the fact that the less developed countries are currently the major source of most of the increases in allegedly bad CO_2 emissions. India has already indicated that this is their policy,[89] and China has agreed to stabilize emissions only about 2030.[90]

Could it be that the less developed countries are trying to use

---

[89] Mohan, 2014. It is also reported that Indian Prime Minister Modi refused to agree to support a peak year for Indian CO_2 emissions during Obama's visit to India in January 2015 (see Chauhan, 2015), and strongly supports greatly increased use of India's large coal deposits for energy development.

[90] Nakamura and Mufson, 2014

CAGW and EWD science as the basis for world income redistribution at the expense of the developed countries? As of late 2014, Canada, Russia, and Australia have announced that they will not be involved in the proposed Kyoto successor, and Japan will greatly reduce its future CO_2 reduction targets. The best thing would be for the US to do likewise. The Obama Administration's argument for aggressive EPA regulations to reduce CO_2 emissions appears to be that this would show leadership vis-à-vis the rest of the world towards contributing to the Administration's GWD/GEF goals. A more likely outcome is that the US would incur large costs and the less developed world would do little more than make token reductions (under the Kyoto Protocol the less developed countries were not obligated to make any reductions at all), just as they would without EPA regulatory efforts. The cost for a developed country to be involved in a possible new Protocol (or whatever it may be called to escape US Senate review) could be very large—both for adaptation and mitigation assistance to less developed countries and for "loss and damage" from extreme weather events.

Current Situation in the US

In broad terms proposed EPA regulations to impose GWD-inspired controls on CO_2 emissions started out in 2009 as an alleged fallback in case the cap and trade bill did not pass Congress, but quickly became the main Administration effort when it became evident to them that cap and trade could not pass the Senate because of determined opposition by Republicans, as I had expected. So late in 2009 and early 2010 EPA launched a regulatory effort based on its Endangerment Finding issued late in 2009 to control CO_2 together with new and continuing efforts to reduce various pollutants that would have adverse impacts on fossil fuel energy use, including the following:[91]

---

[91] Many of these and their economic effects are discussed in US Senate, 2012.

- *Drastic reductions in the mileage standards for new automobiles*
- *Use of government auto firm bailouts to "encourage" production of electric vehicles and smaller vehicles*
- *Mercury Utility MACT*
- *Cross-State Air Pollution Rule*
- *Coal combustion residuals (coal ash or fly ash)*
- *More stringent effluent guidelines for power plants*
- *Regional haze reductions from power plants*
- *More stringent ambient air quality standards for ozone, which could lead to reductions in nitrogen dioxide emissions from coal plants.*
- *More stringent national ambient air quality standards for $PM_{2.5,}$ which would lead to further reductions of sulfur dioxide and nitrogen compounds and energy use.*

More recently, Obama has added:

- *Tougher efficiency standards for heavy trucks*
- *Regulations limiting CO_2 emissions from power plants and prohibiting new coal power plants*
- *Phase out of hydrofluorocarbons (HFCs) previously promoted under Montreal Protocol and Significant New Alternatives Policy (SNAP) program*
- *Regulation of methane emissions from and oil and gas production*

In many cases, however, EPA appeared to delay initiating or implementing those regulations that might prove particularly burdensome to the regulated industries or states prior to the 2012 election, in all probability because the Obama Administration did not want to overly endanger its reelection chances.[92] The President appears to have declared an end to this apparent delay in his speech on June 25, 2013[93] which announced his new Climate Action Plan, so that a tsunami of new regulations have been proposed or can be expected during 2014 and 2015. Obama had already threatened in his 2013 State of the Union speech to impose new regulations if Congress did not pass his desired cap and trade legislation. His June 25, 2013 speech announced the imminent implementation of this threat.

The President's June 25 speech also put him firmly on record as fully supporting GEF/EWD in all respects, including deriding the scientific opposition in quite derogatory terms. Contrary to Obama's characterization of himself as a uniter in his 2008 campaign, this put him on record (although no one had any doubts prior to the speech) as firmly on the alarmist side and with no intention of trying to understand or work with the opposition. He instead stated that he would use his personal power over the Executive Branch to push through climate policy changes through EPA regulations that he knows he cannot obtain from Congress and is no longer trying to do so. This open declaration by Obama that he will attempt to go around existing laws and the Constitution to accomplish GEF/EWD is a direct challenge to the rule of law and the Constitution.

It is not possible to be a "uniter" if you state at the outset that your opponents are wrong and must not be listened to because they hold views you claim have long been rejected by science. This almost certainly means that no reconciliation will be possible on GEF/EWD during Obama's remaining time in office and that the Obama Administration will continue and expand its efforts to reduce US CO_2 emissions by their claimed direct Executive regulatory authority even though there is little hope of an effective new international agreement—which the US Senate would not

---

[92] *Ibid.* Also see Eilperin, 2013.
[93] Obama, 2013.

ratify anyway. This has to be the worst of the many fantasies embodied in GEF/EWD.

The President claims that the alarmists are correct and anyone who does not share their views is a member of a "Flat Earth Society."[94] In other words, the President does not want to hear or understand the views of science skeptics—he has already determined scientific truth and that is that. This statement also means that Federal climate funds are likely to continue to flow only to alarmists and their renewable energy fantasies, who therefore will have no incentive to reconcile their views with the skeptics or the overwhelming scientific evidence based on the scientific method. In 2013-4 Obama's public statements on climate became increasingly belligerent towards anyone who held a contrary viewpoint. He repeatedly tried to belittle and make fun of them. Why he expected that this would do anything other than enrage the opposition is unclear. This was hardly the actions one might expect from a politician who promised to be a "uniter." Even more unusual is that his sharpest attacks were on scientific issues, not political ones. Scientists who have already endangered their careers by not adopting the "party line" are not likely to change their minds because of presidential attacks on their scientific judgments.

The only rational conclusion is that Obama wants to drive a wedge between the public and the skeptics. But the public includes roughly 43% of the population that agrees with the skeptics and 9% who are unsure.[95] Using fallacy and abusive language is not likely to gain him respect among the 52% that he would need to win over if the issue were to be decided democratically. But since he has already decided the issue and the needed action by executive decree/proposed EPA regulations, maybe he does not think he needs to win over the 52%—they will just have to fall into line.

All this is against a background in which there has been no statistically significant trend in global temperatures using any of the

---

[94] Obama, 2013.
[95] Sappenfield, 2014.

major global temperature datasets (including the major ground-based ones controlled by the CIC) from at least 1998 until at least late-2014 while the IPCC projections are being left further behind with each passing year (see Figure C-7). Apparently the President hopes that people will not figure this out or understand its significance. In other words, the climate emperor has no clothes, and Obama continues to say ever more forcefully that he does and that action is so urgent that there is no time to wait for Congress so as to avoid further polarizing the US political system on this issue. Instead, EPA must rewrite the Clean Air Act by Executive actions to allow EPA to impose many of the provisions of the failed Waxman-Markey bill directly without Congressional action.

As discussed in more detail in Chapter 11, the overview is that in a number of areas, particularly air pollution (but not all—particularly the less visible ones) EPA has already largely reached and passed the point of diminishing returns to further tightening on non-CO_2 related air regulations, barring new science that argues otherwise. On CO_2-related regulations there is no scientific or economic basis for doing anything in the first place (as discussed in Chapters 9 and 11 and Appendix C). But given the de-facto nearly absolute power of EPA under Obama, subject only to so far impotent Congressional and court overrides and theoretical Congressional constraints on appropriations not applied as of 2014 during the Obama years, some or all of these proposed regulations may be promulgated during the last two years of Obama's second term.

The Obama Administration's true colors with regard to open science were made even clearer on July 31, 2013, when the newly appointed Interior Secretary, Sally Jewell, told her Department that she hopes there are "no climate change deniers" in Interior.[96] This appears to be a clear effort to intimidate any employees who might even consider challenging the Administration's support for the IPCC's version of climate science.

---

[96] Lewis, 2013.

The primary current climate-related threat in the US is the Obama Administration's use of its control over the EPA to impose GWD/GEF/EWD by administrative fiat based on a 5-4 Supreme Court decision and without even a favorable vote in Congress. In my view this is far worse than any global warming/climate change/extreme weather that Obama alleges will result from increases in carbon dioxide emissions and for which there is little or no evidence and no solution as proposed. If EPA should be successful in its key GHG regulations in closing existing coal power plants there is little doubt that other vital parts of the economy will also be regulated in years to come as attempts are made to reduce CO_2 emissions in both electricity and non-electricity generating sectors of the economy as well. The net result would be comprehensive Federal control over the private economy using fossil fuel energy. So much for the virtues of private enterprise and environmental Federalism, long the basis for US environmental policy!

The Obama Administration claims that such CO_2 reductions by the US will somehow induce the less developed countries to do the same. To make this claim more concrete, I note the following. On September 18, 2013 the EPA Administrator, Gina McCarthy, admitted at a Congressional hearing that the purpose of existing and new EPA CO_2 regulations was to improve the chances that other countries would agree to reduce their emissions rather than providing any measureable benefits to the US public.[97] I sincerely doubt that such present and proposed actions by EPA will persuade less developed countries to do much of anything of this sort. Citizens of a number of these countries such as India are so desperate for electric power that they often resort to stealing it by

---

[97] Lewis, 2013a. The key statement by McCarthy is as follows: "What we're attempting to do is put together a comprehensive climate plan, across the administration, that positions the U.S. for leadership on this issue and that will prompt and leverage international discussions and action."

158

plugging in unauthorized cables to their street or their neighbors' power lines. The idea that these countries will agree to actually reduce energy use because the US may reduce its CO_2 emissions through EPA regulations because of UN reports they have never heard of is ridiculous. And if they actually did, I question whether this is a reasonable goal of US policy.

The Obama Administration strategy may be to include the less developed countries in a new Paris Protocol by encouraging them to make vague promises to level off their emissions sometime far in the future. This is one interpretation of the message provided by the US-China agreement of November 12, 2014.[98] Thus the Obama Administration position appears to be that they are willing to force potentially competitively crippling and continuing increases in US electricity prices and dangerous reductions in electric power reliability in exchange for nothing of any real value from other nations in terms of enforceable reductions in CO_2 emissions. The major exception is the European Union, which has made commitments, but which are carefully conditioned on similar action by other nations.[99] Many of the European countries are fast losing their green zeal as their electricity prices continue to increase and their international competitiveness declines.[100]

The only remaining issue is whether the rest of the US political and legal system will allow the Obama Administration to unilaterally impose their EPA regulations on CO_2 emissions or will insist that Congress must be given a role. Given the independent go-it-alone policy pursued by the Obama Administration in other areas such as healthcare insurance and immigration, the issue is fast becoming not CO_2 reductions or immigration or health insurance, but the rule of law. As in the case of CO_2 regulation, they are trying to use every available legal loophole they can find and many that they cannot to impose their will on the US public despite opposition

---

[98] White House, 2014.
[99] Peiser, 2014.
[100] Ibid.

from Congress.

The US Government under President Obama has taken an increasingly reckless (or should I say brazen) approach towards GEF/EWD in 2013 and 2014. The IPCC is mentioned less and less. The Administration seems to believe that official pronouncements can and should be used to determine climate science and the policies "needed" to solve alleged climate problems without any serious attempt to provide a scientific basis for either in the case of EWD. The public is apparently supposed to take these pronouncements on faith just because the President states them. On June 13, 2013, the President stated that extreme weather is made worse by increasing CO_2 and likened skeptics to members of a "Flat Earth Society."[101] As noted earlier, Secretary of the Interior Jewel stated on July 31 that that she hopes there are "no climate change deniers" in Interior. In 2014, Obama's science advisor stated that "weather practically everywhere is being caused by climate change," and that the drought in California is tied to climate change[102] even though one climate alarmist has pointed out that "a US Government report noted 'droughts have, for the most part, become shorter, less frequent and cover a smaller portion of the US over the last century.'"[103]

On February 16, 2014 Secretary of State John Kerry also claimed that climate skeptics were members of the "Flat Earth Society" and declared that "We should not allow a tiny minority of shoddy scientists" and "extreme ideologues to compete with scientific facts." In other words, although science requires critical inquiry and debate, Kerry believes that skeptics who disagree with him on scientific issues should be blackballed and subject to *ad hominem* attacks. After outlining a litany of recent extreme weather events, he claimed that "climate change can now be considered

---

[101] Obama, June 2013.

[102] Barron-Lopez, 2014.

[103] Pielke, Jr., 2012, citing a US Government report citing Andreadis and Lettenmaler, 2006.

another weapon of mass destruction, perhaps the world's most fearsome weapon of mass destruction."[104] I do wonder if Kerry, Gore, or Obama have ever taken a course in science or understand what the scientific method is. The more pessimistic possibility is that they know but think that most of the rest of the population do not and will not figure it out.

As discussed in Chapter 9, it is easy to show by applying the scientific method that Obama and these senior officials are simply wrong with regard to the science. This is ideology/religion and politics with only the thinnest pretense of science. The public is simply supposed to accept the personal opinion of the President, the highest ranking US Government science official, and the Secretary of State that the science is as they say it is, not what the scientific method determines. Those who disagree will not be allowed to work at least at the Department of the Interior and likely elsewhere.

In May 2014 the Obama Administration released a new *National Climate Assessment*[105] (NCA) which reached a new low for the scientific inaccuracy and irresponsibility of its climate catastrophism. Despite no global warming for at least 17 years, this report is much more extreme and unscientific than even the reports of the IPCC, which it sometimes contradicts. The general theme is that human-caused climate change is now occurring, that this is proved every time a new major storm causes significant damage, that it is caused by human-caused CO_2 emissions, and that immediate action is required to prevent further damages from extreme weather (in other words, the EWD). Fifteen climate skeptics have issued a rebuttal to the NCA.[106] Easterbrook compares a number of the claims and assertions in the NCA with observable data.[107] A US Senate minority report has criticized it.[108] The NCA

---

[104] Kerry, 2014.
[105] US Global Change Research Program, 2014.
[106] D'Aleo *et al.*, 2014.
[107] Easterbrook, 2014.

has descended to the level of taxpayer-financed blatant political propaganda with only the thinnest pretense of science. This is far from an isolated instance; rather, it has continued in other Federal climate pronouncements.[109]

Obama and EPA have taken to referring to CO_2 emissions as "carbon pollution." This strongly suggests that the Administration is being disingenuous by attempting to compare CO_2 emissions to emissions of "conventional" pollutants that have proven adverse effects to humans and ecosystems even though CO_2 is an odorless, colorless gas essential to life on Earth and has been at far higher levels during much of geologic history. Animals, including humans, exhale it; plants die if the levels are much lower than they currently are. "Carbon pollution," not accidentally in my view, congers up images of carbon soot, which CO_2 has no relation to. Could it be that the Administration realizes that GEF is not selling and the only way to sell government-mandated CO_2 reductions is by trying to confuse the public into believing that CO_2 is real pollution rather than just another environmental scare campaign? One of their other tactics, promoting EWD as the basis for reducing CO_2 emissions, suffers from similar problems since no serious scientific pretext has been made for it. The Administration appears to being not only disingenuous but also attempting to deliberately mislead the public as to what the science says. I call this scientific fraud.

This controversy is getting increasingly ugly. Science should never be dictated by politicians if we are to preserve open scientific inquiry and even democracy itself. The history of the Soviet Union and Nazi Germany tells us how this is may end; the problem here in the US first became evident to me in March, 2009, while working at EPA and has been getting progressively more evident since then. I may have been the first victim, but others may follow. It is reprehensible for non-government groups to personally attack climate skeptics, as they have been doing for some time. It is

---

[108] US Senate, 2014b.
[109] Goddard, 2015.

unforgiveable for high public officials to say what scientific beliefs government employees should have in order to be employed, to belittle those who hold different ones, and to issue reports and PR releases containing demonstrably false and misleading science in support of their desired policies. It is all too evident that the Obama Administration and much of the the the rest of the CIC would support repeal of the First Amendment with respect to climate viewpoints contrary to theirs. In early 2015 major media outlets, some Democratic members of Congress, and organizations representing the President publicly attacked the funding of those who do not share their views on climate science.[110] A number of prominent alarmists have advocated penalizing sceptics for their views.[111] Fortunately, they have so far been unable to silence those who do not share their view of climate science, but it is not for want of trying.

New EPA Power Plant Regulations Challenge the Rule of Law in 2014

In January, 2014, EPA re-proposed Standards of Performance for New Stationary Sources: Electric Utility Generating Units[112] which restrict CO_2 emissions from new fossil fuel power plants and effectively prohibit new coal plants and sets a new source performance standard for CO_2 comparable in stringency to that proposed in Sec. 116 of the Waxman-Markey bill (see Chapter 5). This regulation would effectively prevent the construction of new coal power plants on the basis that carbon capture and storage (CCS) is a proven technology. This is far from the case since no subsidy-free, commercial-sized CCS plant is now operational anywhere in the world (no project receiving Federal funds is eligible for demonstrating a new technology), and the most

---

[110] Rust, 2015; Goddard, 2015a.
[111] WND, 2015.
[112] US Environmental Protection Agency, 2014.

advanced demonstration project is heavily subsidized and way over budget. This appears to be one of several potential legal liabilities for this regulation.[113] It is not only a very expensive technology but indeed a very dangerous one because CO_2 is heavier than air and if it should accidentally escape from underground storage facilities and collect in an undisturbed place it can (and has under different circumstances in Africa) kill large numbers of people who will suffocate from lack of oxygen. To propose using such an unproven, very expensive, climatically ineffective, and potentially very deadly technology to solve a non-existent problem has to be one of the more outrageous regulatory proposals EPA has ever made. The real purpose of the regulations on new sources of GHGs is to allow EPA to promulgate standards for reducing emissions from existing power plants, which is the CIC's real objective, since under the Clean Air Act these cannot be promulgated until ones for new plants are in place.[114]

EPA claims that the regulation will have no costs and no benefits so it is all right economically. I beg to differ. If it has no benefits, it should not be promulgated, particularly if such promulgation will make it possible to regulate existing coal power plants. There is no such thing as a free EPA regulation. Why do we need more government interference in the energy production marketplace for no (their claim) or more likely negative benefits? Why do we want to endanger the lives of potentially large numbers of people who may literally suffocate to death if there should be an accidental release of CO_2 from these as yet unproven carbon storage facilities?

In June, 2014, EPA proposed Carbon Pollution Emission Guidelines for Existing Stationary Source: Electric Utility Generating Units.[115] Using an obscure and little used section of the Clean Air Act, Section 111(d), EPA proposed to require reductions

---

[113] See Lewis *et al.*, 2014.
[114] Wayland, 2013.
[115] US Environmental Protection Agency, 2014a.

of CO_2 by an average of 30% compared to what EPA considers each state "should have been" in 2005 by 2030. It

> require[s] each state to adopt a "best system of emission reduction" (BSER) constructed out of four "building blocks": (1) Increase heat-rate efficiency of coal power plants; (2) increase base load generation from NGCC [natural gas combined cycle] power plants while decreasing generation from coal power plants; (3) increase generation from wind, solar, nuclear and other low- or non-carbon energy sources; and (4) decrease electricity demand by industry and other end-users.[116]

The most important reductions in CO_2 are expected to come through (2), thus continuing the Administration's "War on Coal." To implement these strategies, states may use various largely legislative options including cap-and-trade, renewable electricity mandates, and demand-side management programs. This is an imaginative but legally highly doubtful approach[117] to the requirements of Section 111(d) since previously BSER has referred to specific technologies or practices to reduce pollution inside a plant boundary. This promises to be another area where litigation is certain to follow.

Most of the major features of Waxman-Markey appear in one or the other of the two new proposed regulations.[118] In other words, Obama is attempting to impose many of the features of the failed Waxman-Markey bill directly through imaginative regulations relying on an alleged "clerical error" by Congress in its revision to the Clean Air Act in 1990, thus bypassing Congress and effectively

---

[116] Lewis, 2014.
[117] Tribe and Massey Energy Corporation, 2014, and Lewis et al., 2014. The latter lists 10 ways in which the draft regulation is unlawful, of which only a few will be discussed here.
[118] Lewis et al., 2014.

rewriting some sections the Clean Air Act to make this possible. The Constitution does not give the President the power to overturn a decision of Congress (not to pass Waxman-Markey) by rewriting a previous law to do just that through regulations. If this should be approved by the Supreme Court, there would not appear to be any real need for Congress since the President may thus have acquired the ability to rewrite any law the President chooses. These attempts at rewriting are simply breathtaking in their audacity, high-handedness, and apparent illegality, and appears to lay the groundwork for just what the Constitution tried to avoid—an imperial Presidency.

The major cost of the second proposed regulation is to close many US coal-fired power plants and replace them with wind/solar or natural gas plants or even do away with them through "demand management." Even if the proposed regulation is struck down by the courts it has already created huge uncertainties in the electric power industry, which will drive up costs just because no one can predict what will actually happen. EPA is making several imaginative new interpretations of the Clean Air Act; their use of Section 111(d) is particularly problematic legally. It will take many years for these attempted rewrites to be tested through the courts. An adverse ruling for EPA should mean that an existing plant can continue to operate while the regulation is revised and re-promulgated, which can take even more years. Meanwhile the power companies will be left to ponder whether a useful repair should be made to an old coal plant. Power companies are quite likely to conclude that it is better to wait, probably for years, while all this shakes out. The economic costs of these proposed regulations will be enormous even if no plant is actually closed as a result. Both the economic costs and benefits will be discussed in general in Chapter 11 and in detail in Appendix C.4.

Although these regulations for existing plants are extremely flexible according to EPA, they include specific but varying targets for CO_2 emissions reductions which each state is required to meet; whether this is legal is also likely to be tested in the courts. The variations in the CO_2 reductions are to be determined mainly as follows:

> *Two variables in particular affect both a state's 2030 standard and the expense required to meet it: (1) How much of the state's current generation comes from coal, and (2) how much idle natural gas combined cycle (NGCC), renewable, and nuclear generation capacity exists to meet consumer demand as the state ramps down and phases out coal generation.[119]*

The states must propose implementation plans; those from red states may be less than supportive of EPA's proposals. Under the law, EPA can (if Section 111(d) should be determined to be applicable by the courts) disapprove state plans and impose its own, most likely mainly on red states which in many cases have not thus far passed various CO_2 emissions limitations such as cap and trade, Renewable Performance Standards (RPSs), and demand reduction programs.

So EPA is basically trying to COMPEL the red states to close coal plants or adopt legislatively the usual litany of environmentalist attempts to rig the markets for electricity generation so as to favor their ideologically preferred "renewables" (except hydro, of course, which the environmentalists also disapprove of) already found largely in some blue states. In other words, EPA is providing red states a menu of "outside the plant fence" legislative options which in EPA's view they must select from to reduce existing power plant CO_2 emissions to the extent that EPA has specified for each state or EPA will do it for them administratively with various penalties. Many, if not most red states, have very strong adverse views.[120] If states comply with EPA's demands, one likely alternative is to require substitution of wind and solar for much lower cost and

---

[119] Lewis, 2014d.
[120] US Chamber of Commerce, 2015.

more reliable coal-based electric generating plants with the result that ratepayers will pay much higher prices for electricity through their electric utility bills.

There are a number of reviews of the legality of EPA's proposed CO_2 reduction power plant rules. Probably the most devastating one as of late 2014 was by Laurence Tribe,[121] Professor of Constitutional Law at Harvard, and perhaps the leading liberal constitutional law scholar as well as a former mentor of Barack Obama.[122] Tribe believes, as I do, but his former student, Barack Obama, evidently does not, that "burning" the Constitution is not an appropriate course of action to pursue in this or any other circumstances. He finds that the proposed existing power plant rule violates the 1st, 2nd, 5th, and 10th Amendments, and in March 2015 congressional testimony is akin to "burning the Constitution." In Tribes' words:

> The central principle at stake is the rule of law—the basic premise that EPA must comply with fundamental statutory and constitutional requirements in carrying out its mission....It is a remarkable example of executive overreach and an administrative agency's assertion of power beyond its statutory authority....The Proposed Rule lacks any legal basis and should be withdrawn....The Proposed Rule invades state regulatory control in an unprecedented manner under the Clean Air Act and raises grave constitutional questions. It seeks to commandeer state agencies in violation of core structural principles of federalism and the Tenth Amendment....At bottom, the Proposed Rule hides political choices and frustrates accountability. It forces states to adopt policies that will raise energy costs and prove deeply unpopular,

---

[121] Tribe and Peabody Energy Corporation, 2014. For a more popularly-written version of Tribe's arguments see Tribe, 2014a.
[122] Arnold, 2014a.

*while cloaking those policies in the garb of state "choice"—
even though in fact the polices are compelled by EPA.*

One of his carefully documented concerns is that states would no
longer be sovereign with regard to enacting CO_2 emission reduction
laws; they would be mere puppets of the EPA's whims. He quotes
from a previous Supreme Court decision striking down such a
proposal as follows: "Where the Federal Government directs the
States to regulate, it may be state officials who will bear the brunt
of public disapproval, while the federal officials who devised the
regulatory program may remain insulated from the electoral
ramifications of their decision."[123]

Yet through June, 2014, the Supreme Court has backed the
radical environmentalists in almost all its climate-related decisions.
A June, 2014, decision does suggest, however, that the Court may
be starting to try to hold the Administration to the letter of the law
rather than accept EPA's attempts to revise it administratively.

One potentially very important detail is that EPA has based its
legal justification for its proposed regulation of existing power
plants on the claim that the House and Senate versions of Section
111(d) are in conflict with each other and that they therefore have
the authority to determine what to use.[124] Tribe strongly disagrees.
EPA's viewpoint is also at great variance with that of several other
legal scholars[125] and at least 14 states,[126] which believe that EPA's
interpretation is illegal. The legal issue is whether the literal reading
of this Section, namely, that it *expressly prohibits EPA* from
regulating "any air pollutant...emitted from a[n] [existing] source
category which is regulated under [the national emission regime in
Section 112 of the CAA]" is to be followed or whether EPA has

---

[123] Tribes' quotation is from *New York v. United States*, 505 U.S. at 169.
[124] Keim, 2014.
[125] See Haun, 2013, and Potts, 2014.
[126] See eenews.net, 2015. For a more understandable summary for the
non-lawyer, see Yeatman, 2014b.

the authority to ignore this prohibition so that it can do what it proposes, which is exactly the opposite.[127]

I believe that Tribe makes a very strong case that EPA's proposed regulation on existing power plants is unconstitutional. But I question whether skeptics should depend on the Supreme Court to come to the rescue since the Court has been generally supportive of EPA's climate regulations to date and created many of the current problems in their *Massachusetts vs. EPA* decision.

As mentioned, one of the other legal problems is how EPA can force reluctant states to enact legislation reducing CO_2 emissions outside plant boundaries. Within plant boundaries the CAA sets down how this is to be done, but outside plant boundaries reductions would have to be enacted by state legislatures. If the state legislatures refuse to do so, as many red states may well do, the issue will be whether EPA has the real power to make them enact legislation that they refuse to do? In my view they do not.[128]

One thing is certain, however, EPA has created a major legal controversy that will probably take some time to be settled unless Congress should settle the issues in the meantime. Obama has done nothing to try to unite the parties involved, as he promised in general to do in 2008, but instead has picked sides on this issue from the beginning, repeatedly publicly denigrated the opposition, and expended resources on a lengthy legal battle that may not be resolved while he is still in office. It may not be accidental that the proposed regulation appears to be crafted to hit the red and coal states harder than his base of support in the blue states.

---

[127] The prohibition is contained in Section 111(d)(1)(A)(i), and can be found at 42 U.S. Code §7411. In adding this provision, Congress reasoned that §112 standards required ultra-stringent controls for hazardous air pollution from existing stationary sources, and this rendered 111(d) standards superfluous, given the fact that controls for the former regulatory regime would capture pollutants subject to the latter.

[128] Glaser *et al.*, 2014. See also Tribe and Peabody Energy Company, 2014.

Lessons Learned Since Retiring from the EPA

It may be that the Obama Administration knows all these legal problems but is primarily trying to fool the less developed countries into supporting a follow-on to the Kyoto treaty by thinking that the US will actually implement government-imposed CO_2 reductions in the near future. If these countries actually believe this they are making a grave error. The EPA proposed CO_2 regulations are a legal house of cards or possibly even a bluff. But then the science on which it is supposedly based is too.

If the proposed regulation were fully implemented one climate model used by EPA (and accordingly very likely to greatly overestimate the temperature effects of added CO_2 emissions) estimates that global temperatures would decrease by $0.018°C$ by the year 2100.[129] Even if the CAGW hypothesis were correct, the daily, weekly, monthly and yearly changes in global temperatures makes this simply not measureable and clearly not worth even trying to do in terms of reducing CO_2 emissions (assuming that was worthwhile in the first place).

The absurdity of effectively and unilaterally trying to ban new and closing many existing US coal plants without any agreement by less developed countries to make comparable changes to reduce CO_2 emissions becomes even more evident when it is compared to the coal plants being constructed, planned, and used in the rest of the world, particularly in China, India, and even Germany. China added the equivalent of the entire US coal-fired generating capacity from 2005-9.[130] From 2010-13 it added half the coal generating capacity of the US again. Germany is adding to its inventory after a disastrous experiment with renewables and as a result of nuclear fears brought on by the Fukushima disaster. China burns 4 billion tons of coal a year; the US burns less than 1 billion; and the EU about 0.6 billion. What the US does about building new coal plants or continuing to operate existing ones will have no measureable effect on atmospheric CO_2 emissions or atmospheric levels; the

---

[129] Knappenberger and Michaels, 2014.
[130] Larsen, 2014.

171

main effect will be to increase the cost of electricity in the US (since coal is the lowest cost source in some areas), decrease the flexibility of the power industry to reliably meet US needs, and increase the chances of grid failure and load shedding. And as discussed in Chapter 9, lowering CO_2 emissions will have little or no effect on currently non-existent warming and climate change and no effect on extreme weather, as claimed by the Obama Administration. As discussed in Chapter 11 there is good evidence that modest increases in global temperatures would actually be economically beneficial. So that leaves high costs borne largely by the poor, decreased US competitiveness, a new drag on the weak US economy, and (as discussed in Chapter 11) negative economic benefits.

Just as I observed in India many years ago, Obama proposes that Federal bureaucrats with no real knowledge or understanding of how the economy or the electric power sector works will determine an arbitrary level and sources of electric power available in each state in the US. There will be increasing costs to these proposed regulations for each delay they cause in repairing and improving the US electric generating system, even if no coal plant is ever actually closed as a result.

In announcing the proposed regulations intended to reduce CO_2 emissions from existing plants on June 2, 2014, the Administration emphasized that reducing CO_2 emissions will reduce asthma, with corresponding graphics showing sick children. This appears odd because the purpose of the regulation is to decrease CO_2 from existing power plants. CO_2 has never been shown to have adverse effects on asthma, and reducing it will not reduce asthma. Although it has been remarkably difficult to determine how EPA and the President arrived at their asthma reduction claims, their justification appears to be that the proposed regulation, if ever determined to be legal, would allegedly have a number of health "co-benefits," supposedly including asthma reductions.[131] These "co-

---

[131] For a summary of the controversy see Corsi, 2014.

benefits" will be discussed in Chapter 11.

Marlo Lewis, however, argues that this is an unsupportable claim since air pollution from coal plants have been decreasing for years while asthma has been increasing, so decreasing these pollutants further as a result of the "co-benefits" of CO_2 reductions from the coal power sector appears very likely to have little connection with reducing asthma.[132] New research suggests rather that asthma is related to poverty.[133] The proposed EPA regulations on existing power plants would increase poverty because of the resulting higher electric rates, so presumably asthma. If so, asthma reduction is just like reductions in extreme weather—false promises as to what EPA can deliver. Asthma is believed to be related to allergies to compounds made by living organisms such as those created by dust mites in houses, not to changes in the conventional pollutant loading in the outside air such as those coming from coal power plants. Claiming that asthma will decrease if CO_2 emissions decrease further appears to be less than honest, but that is what the EPA and the Administration have resorted to to try to sell their CO_2 regulations. It may even be that reducing home ventilation in order to save energy, as strongly advocated by many GWD-inspired environmental organizations and the Federal Government, may be a causative factor in the rise of asthma in recent years.

A serious question is why, if EPA knows what causes asthma and what steps it can take to reduce it, have these steps not long since been taken? The Clean Air Act specifies that EPA is to protect public health with an adequate margin of safety. There was no need to wait on a highly problematic regulation on emissions of CO_2 from power plants to take action. All it would need is scientific proof that a given pollutant causes asthma. Is the proposed existing power plant regulation the most effective way to reduce asthma? Surely if asthma can be reduced by EPA at all there must be other

---

[132] Lewis, 2014.
[133] Keet et al., 2015.

ways that would more directly reduce asthma. Why have these not been used? The whole asthma PR campaign has the feel of a misleading public relations blitz, just like so-called "carbon pollution" and extreme weather reduction; in all three of these apparent PR campaigns the Administration is making promises that it cannot fulfill by any currently known means.

The Administration is attempting to sell its CO_2 regulations by putting forth increasingly misleading statements and claims which have no basis in science or reality.[134] Our public officials want to waste enormous sums paid for by electricity ratepayers on GEF/EWD and to get public support for this by misleading the public as to what it will buy other than political patronage. The increasing lengths they have gone to do so suggest increasing desperation on their part. With only two years more in office and no global warming for 17 or more years, they undoubtedly realize that this is their last change, so why worry about accuracy, honest public statements, and scientific integrity?

As discussed in some detail in Chapter 11, there is simply no economic justification for these regulations, which includes one that is extremely expensive; the propaganda used to justify them is simply that—propaganda. The language used by the Administration to justify their effort to shut down existing coal power plants and replace them with ones emitting a little less CO_2, all at the expense of the ratepayers, strongly suggests that they are more interested in selling their program than being honest with the American people. Others have claimed to observe similar behavior with regard to other Obama Administration programs and policies. I seem to recall something about "If you like your health plan, you can keep it."

Another economic cost not included in EPA's analysis is that

---

[134] These misleading claims may be related to an EPA memo dated March, 2009 obtained under FOIA recommending just such a sales campaign for public relations purposes. Since the EPA sales effort has little if any relation to scientific reality it does not reflect well on the integrity of EPA's campaign to sell its CO_2 control efforts. See Burnett, 2015.

of the increased risks to the safety of the US electric grid from removing coal power plants. This will be discussed in Chapter 11 along with the crucial part played by coal plants now targeted for closure in keeping electricity available in the Northeastern States during the worst of the winter of 2013-14.

The Obama Administration has recently extended its war on fossil fuels to the oil and gas industry by proposing regulations limiting methane emissions from the industry. This appears to have little scientific basis[135] or practical sense.[136]

Larger Implications of the Approach Taken by EPA to Reducing CO_2 Emissions

Although the future of EPA's regulation of CO_2 emissions from power plants has numerous potential legal problems, some of which have just been discussed, this is very small compared to the potential complications and unwanted outcomes that could result. The Clean Air Act requires EPA to invoke other sections of the Act if it determines that a pollutant endangers public health and welfare. EPA determined in late 2009 that GHGs do so, despite my extensive disagreements with the scientific basis for doing so and those of many other skeptics. EPA may not choose to invoke these other sections, but it can be compelled to do so, most likely through suits filed by radical environmental groups.[137] EPA has few if any defenses against such suits.

The ultimate threat is that EPA could be compelled to establish a National Ambient Air Quality Standard (NAAQS) for GHGs. Since nothing EPA does will reduce atmospheric CO_2 levels to any measureable extent, a NAAQS standard set below then current levels would represent a legal time bomb against any CO_2 releases since no level below current levels could ever be attained

---

[135] Singer, 2015

[136] *Wall Street Journal* Editorial Board, 2015.

[137] Yeatman, 2014d.

through EPA actions. The obvious result would be a mandate for the deindustrialization of the US since the US economy is dependent on the use of inexpensive and reliable energy. The economic effects of such an outcome would be literally catastrophic to the US economy and population. The only likely ways to avoid such an outcome would be EPA withdrawal of its Endangerment Finding, adverse court decisions on the legality of the Endangerment Finding, or Congressional action; Congressional action was not feasible prior to the 2014 mid-term elections, but may be possible in 2015 and 2016. Several radical energy environmental groups are already threatening to sue EPA to reduce CO_2 emissions from airplanes despite Obama Administration opposition, so this is not an idle threat.[138] In brief, the Supreme Court 2007 decision and the Endangerment Finding have opened up a legal can of worms with potentially catastrophic consequences. EPA is either unaware of this or, more likely, chooses to ignore it.

Since even the proposed new EPA regulations on power plant CO_2 emissions would have no measurable effects on atmospheric CO_2 levels even on the basis of the consistently incorrect IPCC models, the only way to measurably reduce CO_2 emissions using the CIC "solution" would have to involve worldwide CO_2 emissions reductions, particularly by the most rapidly growing large economies, India and China. This assumes, of course, that atmospheric CO_2 levels are significantly influenced by changes in CO_2 emissions rather than temperatures, a matter of significant doubt, as discussed in Chapter 9. But even making this assumption, significant reductions in world CO_2 emissions will only occur if other large emitting nations agree to it. The normal way to accomplish this would be through an international treaty. Even if signed by most nations, such a treaty must be ratified by the US Senate by a vote of at least 67 senators in order for the US to be a party to such a treaty. The prospects of this are close to zero. And without US involvement and large payments from the developed

---

[138] *Ibid.*

world, approval of an effective treaty by less developed nations, are very dim.

As of late 2014, the current political situation with regard to GWD/GEF/EWD in the US, which is discussed in much more detail in Chapter 8, is that of an overwhelming advantage to the Obama Administration. The President has announced that he will pursue GWD/GEF and now EWD using his alleged executive powers since Congress refuses to act as he wishes it to. Some of the more liberal Congressional Democrats support him in this. The Congressional Republicans are fairly united in opposing Obama on this issue, but as of late 2014 had not been able to muster the votes to kill his plans, partly because they have been the minority party in the Senate until 2015 and because they lose key votes in the Senate even in their own party.

The Obama Administration continues to support GWD/GEF/EWD in every way it can without Congressional approval. The President has now added strong verbal support as well in a major speech on June 25, 2013. His main lever is control of the development of regulations, particularly at EPA. This is a slow process and can and almost certainly will be opposed by numerous court cases filed by the opposition, but in the end the Administration may be able to ultimately get what it wants if not prevented from doing so by Congressional action, or the end of Obama's Presidency, or definitive court action. Even if there are unfavorable court decisions, the Administration always has the option to rewrite the proposed regulations and try again. So time is on the Administration's side until it leaves office. The Administration has not faced serious opposition in their drive to implement GEF through EPA as of late 2014, but may now confront considerably greater opposition from the new 114th Congress.

As of mid-2014 the Administration's strategy seems to be to first prevent the construction of any new coal-fired power plants and to close many and hopefully in their view all existing coal plants. Whether they would then go after natural gas fired plants is unclear, but if they do so they could force the US to rely on renewables, which is presumably one of their objectives. This is about as far away from reliance on the markets to guide energy investments, as I advocate, as it is possible to get, and would have costs in the trillions because of the cost of building new energy generation facilities, the much higher costs and lower reliability of individual renewables and lower grid reliability as a whole, with no benefits that I know of other than the alleged but unproven health "co-benefits" discussed in detail in the Chapter 11. I regard this as a high risk strategy that would result in a huge loss for most Americans, particularly the less affluent, thanks to the resulting rapid increases in electricity prices and reduced grid reliability. Presumably the Administration regards this as their only option to achieve GWD/GEF/EWD. This is just what I was trying to prevent in early 2009, although the outcome so far is even worse than I expected.

The primary current threat in the US is that the Obama EPA will force states to impose cap and trade, renewable portfolio standards, and other attempts to rig the electricity generation markets towards renewables. In other words, the Administration seeks to get states, particularly red states that have resisted the CIC proposals, to impose through regulations many of the features of the Waxman-Markey bill that it could not get approved in the Senate.

One of the weaknesses of this EPA/NRDC scheme is that it requires some cooperation by coal and red states. This may not be forthcoming and may result in moving the political battle to the individual state legislatures in these states.

As of late 2014, the legislative/legal situation is a long way from being settled, and may not be resolved for a number years. One of the key decisions will be whether the courts will prevent EPA from proceeding with their proposed CO_2 power plant regulations while the courts sort out the legality and constitutionality of EPA's very legally shaky proposed regulations. If they do so, this would push implementation into the next administration. In the immigration issue, a district court issued such a stay in early 2014, but Congress decided not to take legislative action in early 2015.

Dim Prospects for an Effective New International Climate Agreement

The CIC somehow believes that in a world that still relies on warfare and more recently terrorism to settle many of its differences, the countries of the world will come together on a unified basis to greatly reduce CO_2 emissions even though war is much more widely recognized as being harmful than CO_2 is, and no unified solution has been found to avoid war. In short, it is a utopian dream that is unlikely to ever come about and would probably not be effective in achieving its climate objectives even if it somehow did since the reductions must be made decades in advance of the alleged, but badly understood, adverse effects, and the basis for selecting the targets so far in advance is based on invalid science.

The breakdown of the Copenhagen conference in 2009 was just one of many indications of the total unreality of this GWD "solution." The breakdown of the Kellogg-Briand Treaty of 1928 outlawing the use of war as an instrument of national policy shortly before World War II illustrates the fate of a much more important such utopian dream a number of decades earlier. Yet it is the government-mandated CO_2 reduction "solution" which is the one constant in the shifting sands of GWD ideology. The science or the arguments may change, but the "solution" never seems to, possibly because it is the real objective of the movement rather than their changing stated objectives of preventing global warming, climate change, or extreme weather.

One of the more interesting questions is just what the Obama Administration plans to do if and when it manages to impose regulations that will bring CO_2 emissions from fossil fuel-based power plants down to the levels decreed in their power plant proposed regulations? The hints by Administration spokespersons are that they plan to use this to try to persuade other developed and particularly less developed countries to reduce CO_2 emissions. Further reductions in US CO_2 emissions will not reduce global temperatures in any measureable way, so presumably they plan to use these controls as bargaining chips to get less developed nations to reduce their emissions. The history of recent UN climate meetings, however, shows that the less developed countries will at most offer vague, unenforceable emission reductions in return for US promises of reductions through EPA regulations. The EU and the US hope that a new agreement will be reached at a Paris UN meeting in late 2015. As of late 2014 the European Union had decided to make further requirements for the use of renewable sources but only if there is a legally binding global agreement at the UN Paris conference.[139]

The Obama Administration has not stated what it will do if no effective, legally binding agreement is reached. The most likely outcome in that case is that they will proclaim that whatever unanimous but vague compromise may be reached in Paris is a dramatic breakthrough that will save the world from the supposed scourge of increasing CO_2. Australia, Canada, and Russia appear to have no interest in a follow-on to the Kyoto Protocol. Germany is trying to reduce its renewable subsidies and building coal plants. And less developed countries appear to have no interest in cutting emissions themselves in the near term, particularly without firm guarantees of the annual $100 billion in developed country support promised at Copenhagen. The Republican majority US Congress is unlikely to agree to any new Protocol either. Finally, the 2014 Lima conference agreed that all countries would receive an invitation to define a carbon reduction target of their own choosing, with no specific consequences if they do not live up to their chosen targets. Chinese and Indian delegates successfully demanded that every

---

[139] Peiser, 2014. One of their proposals is to achieve 60% emissions reductions by 2050. An analysis (Lewis, 2015a) of the implications of this for less developed countries reveals that this would lead to absurd results.

"shall" be changed to "may."[140] If there is a new international agreement, it is highly likely to be so watered down as to be meaningless. And for this the Obama Administration proposes to impose real costs on the US economy and population.

So one likely outcome of a possible successful use of EPA's alleged powers to mandate GWD/GEF/EWD in the US is that US emissions would be decreased below what the US is already doing (currently the "best" in the world); but the main effect would be to drive up US energy costs and decrease US international competitiveness, just as has happened in Europe, where electricity prices have soared and energy intensive industries are fleeing.[141] If both the US and the EU choose to voluntarily decrease their emissions further under a new UN agreement or independently, they will be the only ones incurring any significant costs for mitigation, reduced economic competitiveness, and possibly for payments to the less developed world. These expenditures will not produce any measureable decrease in global temperatures, and may have no effect at all. The less developed world would enjoy increased economic competitiveness and is likely to pay little or nothing. One of our major economic competitors, China, would incur few if any of these costs until about 2030, assuming that they live up to their vague agreement with Obama.

A further but related problem for the Obama Administration is that many Asian nations are tiring of the Administration's and other Western developed countries' insistent anti-fossil fuels approach to their desire for the energy they need for rapid economic development. The most visible aspect of this tiring is the creation of a new Asian Infrastructure Investment Bank backed by China. One of the apparent purposes of this Bank is to circumvent the US attempts to direct lending away from coal and other fossil fuel sources of energy which the radical environmentalists and the Administration are trying to forestall through lending

---

[140] *Wall Street Journal* Editorial Board, 2014.
[141] The European experience is recounted in some detail in Peiser, 2014.

181

restrictions.[142] These lending restrictions are imposed by both the World Bank and US Government-funded lending institutions are also an issue with regard to the new UN institutions for this purpose. Some academics have proposed exemptions to these restrictions,[143] but the obvious answer is the same as in all the unintended results of GEF—the restrictions need to be abandoned and rapidly expanded fossil fuel use by the less developed countries encouraged, not prevented. It is unconscionable for the US and other developed countries to try to limit CO_2 emissions by less developed countries trying to provide access to fossil fuel energy so necessary for a better life for their citizens as a result of an ideology that is scientifically invalid. Yet this is exactly what is happening.

The Obama Administration is reported to be proposing various schemes to circumvent Congress' constitutionally required approval of treaties, possibly by trying to "update" the 1992 UNFCCC Treaty and making CO_2 reduction targets voluntary.[144] As in the USEPA proposed power plant CO_2 regulations, the Administration is trying to break new constitutional ground with the effect of further enhancing Presidential power. The US courts may have the last say.

---

[142] Porter, 2015.
[143] Moss, *et al.*, 2014
[144] Davenport, 2014a

7
Lessons Learned about Journalism

ONE OF THE principal players in the CIC has clearly been the mass media since they have generally supported GWD/GEF openly and enthusiastically.

Lost Objectivity by the Mass Media

I had, of course, long been aware that the media had gradually been changing during the late 1900s, but the extent of this change was brought home to me rapidly by my experiences after CEI's release of my draft Comments on the Draft TSD[145] in late June, 2009.

I received numerous requests for interviews after the release, but they were predominantly from what might be considered right of center media. I did receive some phone calls from liberal or "left of center" media, but they were almost uniformly confrontational and obviously searching for some weaknesses in my efforts or other alleged faults. In retrospect it was easy to predict where individual reporters would come down. If the publication they worked for had expressed skepticism concerning the GWD they were likely to be supportive, and vice versa. Even beyond that, if the publication was

---

[145] Carlin, 2009.

known to be "liberal" or to support mainly Democratic candidates for public office, they either showed no interest or could be counted on to be critical of my efforts. So Fox News and the *Wall Street Journal* were uniformly sympathetic and the *New York Times* critical after an initial factual report. The same can be said for blog sites. Their coverage was uniformly related to their climate political coloration. Although I attempted to say the same thing to everyone that inquired, this was probably an error since it was soon evident which reporters would adopt which positions based on who they worked for. The same can be said of most other media, especially *Wikipedia*, which had and still has a very strong pro-GWD orientation. A better approach would have been to make sure all the important points were made to the favorable reporters and to take a strong defensive posture with the unfavorable ones simply on the basis of the type of publication they worked for. But I had not been involved with the mass media for almost 40 years and had much to learn about how it now operated.

I once asked a prominent skeptic if there were any websites that attempted to be impartial. He responded that he only knew of one, which listed the GWD favorable and unfavorable arguments side-by-side. Every other climate site I know of largely reflects one side or the other.

The most surprising case from my then naïve viewpoint was the *New York Times*. My parents had gone out of their way to buy it every day in a city where it was difficult to find or buy serious newspapers, and long supported their extensive coverage and relative impartiality. But in this case after running a very brief factual piece early in the coverage period, and asking for documents through the Freedom of Information Act (which they actually received, unlike the case of most conservative media in other contexts), the *Times* coverage featured a clever article late in the coverage period which attempted to support the paper's party line through innuendo and aspersions and appears to have reflected mainly the EPA "line" at the time. I sensed what they were trying to do and carefully outlined my case to the reporter involved, but he

184

was obviously not listening to my viewpoint. See Appendix B for a point-by-point rebuttal to the *NYT* article.

Principal EPA/Liberal Media Attacks and My Responses

After a while it became clear what the main points in the left wing attacks were. One could tell just by reading the first sentence of any article which side they were going to take even if it was not obvious from the publication/blog in which it had appeared. I was a somewhat unusual case for them in that I had never been employed by any of their right wing "bogeymen," their usual assertion with regard to many of their opponents. In fact, EPA has long had some of tightest conflict of interest rules in the Federal Government. This was further reinforced because I was a senior EPA employee and required to fill out the same complex financial disclosure form used for political appointees. I was specifically prohibited from accepting funds from sources that would possibly bias my decisions for at least the previous 38 years, even if such funds had been offered, which they were not. No even semi-serious attempt was made to argue specifically where they believe I might have erred in the scientific judgments in my Comments. The principal points included the following:

- *I had not recognized the ultimate authority of the IPCC on all things climate*
- *I was an economist and therefore unqualified to judge the science*
- *My comments did not meet journal standards*
- *I had used references that had not been published in peer-reviewed journals*

Unrecognized Authority of the IPCC in All Things Climate

Although this was not stated in these words, this was obviously my major sin. This is basically an appeal to authority argument. This carries no weight for me. I am interested in where my arguments may possibly have been incorrect and why, but not whether or not particular alleged authorities agree or not, particularly a politicized organization with a consistent record of bad predictions. This particular criticism of my Comments often took the form of arguing that I had "cherry-picked" my data. So if only I had understood the totality of the IPCC reports I would have realized how they all fitted together to make the CAGW case. On the contrary, I in part attempted to select those aspects of the climate debate that were most telling as to whether or not the CAGW hypothesis satisfies the scientific method.

I Was Allegedly Unqualified to Write My Comments

This and related *ad hominem* attacks were perhaps the leading specific argument by both EPA and the liberal/left media. In retrospect, it is easy to see why they took this viewpoint. Despite the obvious lack of scientific qualifications of Al Gore (a politician and former reporter) and the IPCC's former head, Rajendra Pachauri (a retired railroad engineer now accused of sexual harassment), the pro-CAGW groups soon adopted this attack as their standard fare. In fact, I could immediately tell where any given piece was going to come out even if I ignored the source by whether it identified me as an economist or scientist. I regarded this as particularly egregious on the part of EPA since they had access to my personnel file and to other EPA employees with considerable knowledge about my activities over the previous 38 years. Obviously the then current necessity to mitigate the embarrassment of the moment overcame any desire they might have had for accuracy.

Lessons Learned about Journalism

As detailed in Chapter 3 above, it happens that I had worked for EPA for 7-8 years primarily as a physical scientist, my position description was ambiguous on this issue, and I had been found to be fully qualified as a physical scientist by the Office of Personnel Management. In fact, I had worked in and briefly headed a Division whose primary responsibility was the development of "criteria documents," whose purpose was to summarize the scientific and technical knowledge available on particular compounds being considered for possible regulation. This involved overseeing the preparation of over a dozen such reports. These criteria documents are entirely analogous to the Draft TSD I was commenting on except for the pollutants considered. I had not worked on any criteria documents on carbon dioxide other than the 2007 OAP draft, it is true, but no one else at EPA had either prior to the preparation of this early predecessor of the Draft TSD.

The reality of the situation is that both EPA and the liberal/left media that came to their support had long since adopted a public relations approach of defending GWD/GEF while totally ignoring the scientific content of what was being discussed. No one who was not funded by the CIC was and is regarded as having sufficient expertise to comment on climate issues regardless of their background or scholarship on the subject. That way all other expertise can be dismissed as uninformed or irrelevant or even better, in the pay of "evil" polluters. Unfortunately, my case was far from an isolated instance. The CIC had and still has substantial resources devoted to pursuing this misleading public relations strategy. I had been aware of this strategy as a result of reading some of the "blue" websites, but was nevertheless unprepared for the viciousness and baselessness of their attacks.

Another characteristic of the exchange over my Comments was the resources that were poured into it by the CIC. Why were so many people spending so much time attacking anyone who differed from GWD orthodoxy? Why not instead carry out a reasoned debate on the scientific and economic issues raised rather than viciously attacking the motives of the author or his/her alleged

187

lack of expertise? Just what was going on here? Could it be that the CIC public relations juggernaut was being lavishly financed by someone or some groups beyond their main (unwitting) supporter, the American taxpayers? There were repeated allegations on "red" websites that the opposition blue sites were substantially funded by George Soros and other wealthy liberals, but at the time I did not have knowledge whether this was accurate (more recent evidence is discussed in Chapter 11). Or could it be that the CIC could not defend their science and found it was more useful to launch *ad hominem* attacks on skeptics than engage in a losing scientific discussion?

Comments Did Not Meet Journal Standards

In an apparent attempt to capitalize on the fact that the IPCC had had about two decades to write five versions of their reports whereas I was limited to only 4 or 5 days to write my comments, my opponents charged that I had not written a journal quality set of comments. It is obviously true that nothing of this length or complexity can meet journal standards in less than a week of effort. Nor were they prepared to meet such standards. My comments were intended to provide helpful comments to the OAP that would aid them in revising their TSD. The revised draft TSD as a whole prepared by OAP was later peer reviewed prior to its issuance in December, 2009, but only by "safe" GWD supporters inside government despite the OMB requirement for broad review by diverse reviewers.

Interestingly, the EPA Inspector General found that the TSD peer review did not meet the Federal requirements for reviews of major policy-related reports.[146] EPA responded that the background document for the Endangerment Finding did not meet the definition of a major policy-related report since it was merely a compilation of what others (particularly the IPCC) had already said.

---

[146] USEPA, 2011

This raised the question of whether EPA had met the legal requirements for independent review of proposed regulations. How could it be that one of the most controversial and crucial scientific findings ever made by EPA was not a major policy-related report but was nevertheless the basis on which future regulations that could cost trillions of dollars would be based? Unfortunately, these important issues have generally not been discussed in the mass media or the "blue" Websites, and can generally only be found on a few "red" Websites.

Use of References that Had Not Been Published in a Peer-Reviewed Journal

This was entirely true, although many of them had been. Once again the problem was that the CIC had received more than $39 billion from US taxpayers alone to make their case, while skeptics/realists had generally been denied access to publishing in most major journals and usually had no way to be paid for their largely volunteer efforts. And we now know that the IPCC reports did not even come close to their alleged use of only peer-reviewed literature since a significant proportion of their references did not meet that standard.[147] Even if it were the case, to argue that all or most referenced articles should have been based on peer-reviewed articles was simply another self-serving argument by GWD supporters to take advantage of their overwhelming funding advantage. In science, truth is not determined by publication in peer-reviewed journals but rather by comparisons between a hypothesis and relevant real world data. Further, it is not possible to prove that something is wrong using exclusively material that argues that it is right, so skeptics have a difficult job anyway.

---

[147] Watts, 2010.

The Increasing Importance of the Internet

Although also widely understood by many, I was also surprised at the extent to which both sides in the GWD war relied on the internet to get their message across. This was particularly the case for the skeptical blogosphere, probably because of the alarmists' almost complete control of the mainstream liberal media. The polarization is even greater than in the mass media. It does not take much time to learn the GWD leanings of each site. One prominent site, WattsUpWithThat.com, even listed the general views of each site towards AGW on its blogroll until recently. I characterized the sites as red for anti-GWD and blue for pro-GWD, corresponding to the mass media characterizations of red and blue states in terms of support for Republican and Democratic candidates. Another difference between the red and blue sites is that the red sites concentrate more on the science whereas the blue sites are more (but not exclusively by any means) devoted to cheerleading and all too often epithets and *ad hominem* attacks on skeptics.[148] The more scientifically-oriented red sites sometimes even take on the role of scientific journals as a result of comments offered by readers.[149] So though the posts at these blogs are usually not subject to prior peer review, as in most journals, the reader comments are often quite pointed and entirely public, and thereby serve the same purpose, only often better since it is done publicly and is subject to immediate rebuttal.

Special Problem of *Wikipedia*

I had been brought up using hard copy encyclopedias and had little experience with the Web versions until my interest in climate developed. I had always found the printed versions reasonably

---

[148] For an illustration of the approach taken by one of the more prominent blue sites in their own words, see the Climategate Email quoted in Ball, 2013.

[149] Sharman, 2013.

impartial and balanced in their treatment of the topics they discussed. I was therefore quite surprised to learn that *Wikipedia*, the predominant online encyclopedia, had a strong coloration at least when it came to the global warming issue. Other skeptics had long noted the same problem. So when one of the *Wikipedia* editors wrote me asking for a picture, I pointed out my concerns to him. He responded that he was well aware of this issue, but *Wikipedia* had its rules that had to be followed. The rules appeared to be that statements by government agencies and the *New York Times* were to be trusted over anything else that might appear. Since both sources were taking a similar line, if not collaborating in attempting to denigrate my competence to judge CAGW "science" and my efforts to do so, *Wikipedia* found it easy to pursue the particular coloration on this issue that they obviously wanted to anyway. Others have found that entries involving global warming/climate change are consistently and rapidly changed so as to uphold the party line by a *Wikipedia* editor.[150]

Whether a strongly biased online encyclopedia (at least in the climate area) is better than no online encyclopedia is an open question in my mind.

Perspective on the Media's Role in the GWD/GEF

The liberal mass media have and still are playing a very important role in CIC by publicly supporting GWD/GEF and even EWD. Without their continuing uncritical repetition of whatever the alarmists say, no one would have heard of the alleged climate threat or the CIC "solution." And politicians would not have considered supporting GWD/GEF. So why did the liberal mass media provide this critical support? Why did they not fulfill their important role of impartially examining the claims of all sides on public policy issues such as this one?

In the case of the British Broadcasting Corporation (BBC),

---

[150] D'Aleo, 2014.

charged with being impartial and supported by mandatory public contributions, it has recently been revealed that the former Labor government deliberately attempted to influence coverage through using foreign aid funds to provide a GWD seminar using representatives of environmental organizations and other CIC supporters to make the case to high level BBC executives that skeptic viewpoints should be deliberately excluded from climate coverage.[151] Subsequently, the BBC took a very consistent pro-CIC approach and among other things did not believe they were required to present skeptic views to balance alarmist views. The BBC subsequently attempted to cover this up over a period of six years, only to have it confirmed as a result of legal proceedings under the British equivalent of the US freedom of information act.

The situation in the US is not much better in terms of its results, although slightly more transparent. A recent study found the following:[152]

- *As of March 6, 2014, it had been over 3-1/2 years since either ABC or CBS had reported the views of a CAGW skeptic scientist, and over 9 months for NBC.*

- *In October, 2013 the Los Angeles Times letters editor decided the newspaper would no longer print letters taking a CAGW skeptic position.*

- *The science forum of news aggregator website reddit, with over 4 million subscribers (twice as many as the New York Times), has decided not to publish CAGW skeptics' comments.*

- *Similar decisions have been made by the Sydney Morning Herald, CNN's Reliable Sources program, Popular Science, and others.*

---

[151] Rose, 2014.
[152] Allen and Seymour, 2014.

So the media are actually helping the CIC to avoid any direct confrontations or even discussions between alarmists and skeptics, when they should be trying to encourage it so as to better inform the public as to what the differences are. Besides deliberate government and private news media attempts to skew media coverage, another obvious reason for such behavior is that potential scary stories sell their product. Few want to read about happy outcomes or continued good weather. But articles that publicize unwanted outcomes, or better yet, catastrophes, are of more public interest. Another answer in the US case appears to be that the mass media reporters and editors are predominantly composed of liberal Democrats employed by liberal Democratic-leaning mass media and the liberal Democratic establishment is largely supporting the alarmists, perhaps because of the views of the more elite segments of the liberal Democratic movement. So there is no market for independent critical reporting except in the case of the much smaller, more conservative, usually Republican-leaning press. This, in turn, reflected the general split among the population as a whole. Rather than providing an independent, critical viewpoint, the mass media has increasingly reflected what they think their readers would like to read. Thus liberal Democratic mass media are often read primarily by liberal Democrats and conservative Republican media are often read by conservative Republicans. Readers prefer media that take their general viewpoint rather than taking an objective, independent view, and the internet has made newspapers and blog sites catering to every shade of opinion easily available.

It is worth noting that the IPCC process was clearly set up along old media lines rather than new media lines. Each stage of their media communications process is controlled by trusted gatekeepers who can be relied on to follow the IPCC line. In the last few years this has been supplemented by internet sites such as RealClimate.com, which was the CIC answer to the proliferation of science-oriented skeptic websites,[153] but this site is not known for

---

[153] Based on Climategate Emails.

including anything beyond the GWD/GEF party line or including unfavorable comments on that line. This is in marked contrast to some of the skeptic "media" composed mainly of internet sites where a wide diversity of comments both pro- and anti-CAGW are often tolerated. It is becoming harder and harder for the mass media to control public opinion since anyone who has access to the internet can read all shades of opinion within seconds free of charge. Without these new internet sites, the liberal mass media would almost certainly have achieved their GWD/GEF goals by now.

The internet has very low entry costs and is open to anyone who wants to get public support for their views and can afford the modest cost of buying and maintaining a website. A few of the skeptics, such as Anthony Watts (http://WattsUpWithThat.com) in the United States and Joanne Nova (http://JoanneNova.com.au) in Australia, quickly learned how to use it well. The trick is to provide a large assortment of new entries of interest to readers and a relatively neutral sounding board for those who want to express their views on each particular topic discussed. In an earlier day the support of the mass media would have been enough to insure national support of a particular policy viewpoint. But thus far this has not happened in the case of GWD/GEF and almost certainly will not in the future.

8
Lessons Learned about Government

A SECOND MAJOR CIC player has been GWD/GEF-leaning governments at all levels, especially some of the US blue states, the Obama Administration, many Western European governments, the European Union, and the United Nations.

The politics of environmental improvement have changed greatly since I initially became involved with environmental issues in the 1960s. At that time environmental improvement had bipartisan support in the US. Then California Governor Reagan even held a large environmental conference in Los Angeles to proclaim his strong support. The Angeles Chapter of the Sierra Club, the Club's second largest at the time, had at least as many Republicans as Democrats as best I could tell. The major US environmental laws passed in the late 1900s were largely bi-partisan in their support.

As a result of the leftward drift of the environmental movement since then, the movement now largely identifies itself with the Democrats, usually very liberal Democrats. In the case of GWD legislation there is now a sharp difference in support, with Democrats generally favoring such legislation and Republicans

195

generally opposed.[154] This is undoubtedly a major change, which is consistent with the radical increase in partisanship between the two parties, but is particularly striking given the early history of bipartisan environmental legislation. I question whether the increasing identification of the environmental movement with the far left of the Democratic Party has or will lead to passage of GEF legislation. The only hope for implementing GEF was to hope that Liberal Democrats remained in control of the Executive Branch and at least the Senate, that the President continued to try to push through EPA regulations that they favor over the objections of Congressional Republicans, and that the courts continued to approve EPA's GEF regulations. Any short-term hopes of this, however, were dashed by the results of the 2014 Senate elections. The environmental movement spent quite sizeable resources in hopes of avoiding this outcome, but without evident effect. Their problems will become increasingly serious in the future as the courts begin to grapple with the many novel and probably illegal attempts to remold the Clean Air Act to do what it was not intended to do.

Senator Inhofe and other Republican senators and representatives appear to have a good understanding of the scientific and economic issues surrounding CAGW. Senator Inhofe has even written a book about his views.[155] President Obama's forceful presentation of his GWD and more recently EWD views seems unlikely to change the Republicans views—in fact, just the opposite.

As outlined in Chapter 6, Obama has now stated that he will

---

[154] A national exit poll conducted by Edison Research among voters in the November, 2014 US election found that 86% of voters who identified themselves as Democrats said climate change was a serious problem while 12% said it was not. Of those who identified themselves as Republicans, 31% said climate change was a serious problem and 67% said it was not. See Rappaport, 2014.

[155] Inhofe, 2012.

push his GWD/GEF/EWD agenda forward through actions by EPA and other government agencies that he controls and in spite of opposition from Republicans in Congress and by "red" and energy states. This is likely to lead to even greater political acrimony on the subject, the possibility of a strong backlash against EPA and other Federal environmental regulatory agencies, a split in the Democratic Party between its liberal powerbases on the East and West Coasts and its blue collar energy state wings, and the risk of a prolonged legal controversy because of EPA's attempt to rewrite the Clean Air Act by executive fiat so that it can implement GEF/EWD without approval by Congress.

The environmental movement and the Obama Administration appear to believe that they are saving the world, that this is urgent, and that it is worth gambling their future and the future of the US electric power system to achieve GEF/EWD in the near future. But what the environmental movement may well end up doing is to decrease its future influence by trying to achieve radical energy policies that have provoked strong opposition by Congressional Republicans and many voters, particularly in red and energy-producing states. As noted in Chapter 6, the movement and the Obama Administration already gambled once on the Cap and Trade/tax bill and as at least a partial result the Democratic Party lost control of the House of Representatives in 2010. In 2014 they went on to gamble the Democratic Party's control of the Senate since many Democratic senators from red and energy states were attacked on the basis of their actual or alleged support for GEF. A key White House aide even went so far as to state that Republican efforts to oppose major new EPA regulations "have a zero chance of working."[156]

As discussed in Chapter 6, President Obama appears to be more interested in pursuing his ideas for change than in uniting political Washington, as he promised in 2008, and observing the constraints imposed by the Constitution intended to create a

---

[156] Manning, 2014.

deliberately conservative system. His expansive view of his powers has now led him to propose or implement changes in the law not only in climate/energy policy, but also in health insurance[157] and immigration policy. In fact, his enthusiasm for unilateral action appears to have expanded as time has gone by and no one has effectively pushed back on his attempts to rewrite and selectively enforce the acts of Congress to favor his policies. The obvious question is why we need Congress if the Executive Branch can rewrite acts of Congress to do what Congress has refused to do or what existing law forbids? Is this what the framers of the Constitution had in mind? Clearly it is not. The US Government was designed by the Constitution to be a conservative institution under which change was purposely made difficult and slow.

So what started out as a scientific speculation as to whether the Earth might be warming because humans are emitting more carbon dioxide in order to achieve a far higher standard of living has now been transformed by the Obama Administration into part of its multi-pronged effort to challenge the constitutional powers of Congress in a number of fields. In late 2014 this confrontation became much more public when Obama initiated executive action on the issue of immigration,[158] but the outcome may influence the outcome on climate and health insurance as well since the legal and constitutional issues have some similarities. This may result in serious political, legal, and constitutional confrontations. The outcome of the 2014 mid-term election surely is not optimistic for

---

[157] Obama's expansive view of his authority to rewrite laws to achieve his desired ends has resulted in somewhat similar problems in the case of health insurance, where the Circuit Court of Appeals for the District of Columbia ruled on July 22, 2014 in *Halbig v. Burwell* that Obama had illegally spent funds and imposed taxes not authorized by the Affordable Care Act. See Adler and Cannon, 2014. Whether this will result in an effective pushback against Obama's expansive views of his powers remains to be seen as of late 2014.

[158] As a result of Obama's November 20, 2014, Executive Order. See Ehrenfreund, 2014.

EPA imposition of GEF in the face of red and energy state opposition to its energy policies. The legal issue is whether the President can have EPA rewrite the Clean Air Act to allow them to implement his GEF "solution." The Obama Administration's actions in this and other areas raise similar issues concerning disregard for the laws they are required to "faithfully execute" in favor of independent Executive Branch actions to selectively enforce and reinterpret the laws to pursue their favored policies.

As of late 2014 the courts have largely agreed to what the Administration has proposed in the energy/climate area, although some decisions in June 2014 suggested that they might be having possible second thoughts. Congress is gridlocked since neither party controls 60 votes in the Senate to force action let alone 67 to overcome a Presidential veto; Obama is openly trying to move into the resulting political vacuum created by asserting a greatly expanded role for the Executive Branch. The resulting struggle may last for the remainder of Obama's term in office.

In brief, the Obama Administration's approach to climate is part of what I view as a pattern of attempted unilateral action by the Obama Administration in areas where there are disagreements between the Administration and congressional Republicans. Obama appears to believe that he knows best what the country needs and that he is free to reinterpret the laws passed by Congress to institute his policies. The Endangerment Finding was an early precursor of this policy, but clearly consistent with it since it involved expansion of EPA's powers by adding another major alleged pollutant for which it claimed responsibility. If the Obama Administration actually manages to implement its Climate Action Plan as well as its immigration and health insurance policies, there is nothing to stop this or any later Administration from assuming dictatorial power using the same techniques in other areas or even in all areas. I did not expect that what started out as invalid science pushed at the Presidential level would be part of a much broader challenge to the US legal system and even the Constitution. But it is.

So what started out as an obsession with reducing fossil fuel energy production and use has ended up creating what is likely to be major political and legal battles over the Clean Air Act and the Constitutional powers of the Executive and Legislative Branches. Even if EPA's proposed regulations represented good economics and good science, which they do not, I believe that they need to be defeated in order to reduce the increasing threat posed by an imperial Presidency that deliberately ignores Congress in an effort to unilaterally implement their favored policies over the objections of Congress. In my view the laws should be written by Congress, not an increasingly all-powerful Chief Executive. I would characterize Obama's behavior as dictatorial. History shows how quickly and easily democracy can be lost if not vigorously defended. The time has come to defend it.

EPA's proposed CO_2 regulations on existing power plants can only be defeated by Presidential action to withdraw them, by Congressional action, or by the courts. Otherwise they will take effect over the next few years.

In addition to the major shift away from bipartisanship on environmental policies, the Obama Administration appears determined to overthrow the cooperative federalism with the states enshrined in the major environmental acts that EPA is supposed to implement. Although there were occasional Federal Implementation Plans (FIPs), the most extreme action that EPA can generally take to force a state to follow EPA's mandates, imposed by EPA on states during the Clinton and George W. Bush Administrations, there were about 50 during Obama's first term.[159] Now there promise to be many more in order to force red states to implement GEF/EWD for existing power plants and potentially other areas. But there are many other examples of EPA's power grabs from the states.[160]

It now appears that there will be even greater confrontations between EPA and red states in 2015-2016. The Administration's

---

[159] Yeatman, 2014c.
[160] Yeatman, 2013.

proposed GHG emissions controls on existing power plants appear likely to set new records for Federal intrusion into electric power regulations, long the province of the states, as well as environmental ones. This may well result not only in a historic showdown over EPA's powers between the three branches of the Federal Government but also between the Executive Branch and red states since they may refuse to implement what the Administration proposes, somewhat similar to their actions on health insurance. As of very late 2014 opposition to the proposed EPA power plant rules from red and energy states appears to be mushrooming with some conservative organizations urging states to refuse to cooperate with EPA on their power plant rules. It appears unlikely that EPA can force states to pass legislation implementing GEF, but they may try to withhold highway funds and promise Federal subsidies as inducements.

The Obama Administration and their environmental supporters appear to be willing to gamble everything (except their reelection in 2012, which led to a pronounced slowdown in their agenda) on their climate proposals. Democrats lost control of the House of Representatives in 2010 in considerable part over the Administration's cap and trade bill because key Democratic Representatives who voted for the Cap and Trade bill were defeated on the basis of their votes, particularly in states that would have been hurt by the bill if it had been enacted.

In the 2014 elections the Democrats lost control over the Senate as well, at least partly due to climate issues. Obama will now face strong opposition in Congress to his GWD/EWD regulations as well as his actions on immigration and health insurance. Whether and how Obama can defend his climate and other go-it-alone policies from determined attempts by Republicans in Congress to block them remains to be seen. He has a veto and the Republicans have non-veto-proof majorities in both houses of Congress as of 2015. The new Senate Majority Leader is from Kentucky and strongly opposes EPA's attempts to control CO_2 emissions.

Politicians surprisingly seem to underestimate the significance of climate policy on their election prospects. Australia offers a stark example. After deliberately downplaying climate in his 2012 reelection bid, Obama gambled the effectiveness of the last two years of his tenure on active support of GWD/GEF/EWD-inspired EPA regulations. I continue to be amazed that Obama has staked so much on imposing his policies at the expense of everything else.

I believe that the fundamental need in environmental regulations is that they should be based on sound science and economics as well as the law as written, not the law as some might wish it to have been written. They also need to take into account the interest groups involved in order to develop regulations that can reasonably be accepted by the various stakeholders while satisfying the needs of the public as a whole. And they need to ascertain whether there may be unintended consequences that will be undesirable.

I believe that the underlying task of an environmental regulatory agency such as EPA is to sort out a middle ground that takes all these aspects in mind. The most serious problems are likely to result if a regulatory agency approaches its task with an ideological bias as to what it should do. This can yield regulations that are based on faulty science, bad economics, or bad law, or yield undesirable indirect effects. And the results of such ideological approaches can be disastrous.

The best known example of a disastrous EPA regulation was the ban on DDT in 1972. Even though EPA carefully gathered the relevant scientific information, it paid no attention to these findings in reaching its decision. The principal reason for the disaster, however, lies with not taking into account the indirect results of its action. A good case can be made that the indirect (and certainly unintended) result was the death of perhaps 50 million people, particularly children, who would otherwise have lived, in less developed countries.[161] An international campaign to wipe out malaria had made considerable headway in the 1960s based on the

---

[161] Discover the Networks.org, 2009.

low cost and unique properties of DDT in repelling as well as killing mosquitoes carrying the malaria parasite.[162] I was aware of this program because of my research on US aid to India, where the malaria eradication program was one of their success stories at that time. The impacts of the ban on US use were limited due to the eradication of malaria in the US by the early 1960s, so some may believe that a temporary ban might been justified in the US to avoid any possible risks at least until these risks could be better understood. It is now clear that DDT's impacts on birds, particularly raptors, did not justify the ban even in the US.[163] But what was either unforeseen or ignored was that the US ban led to actions by the UN and many less developed nations which had the effect of crippling promising malaria control programs in less developed countries, with disastrous results.

A better approach would have been to avoid a ban until and unless a thorough scientific case could be built for it. Many environmental organizations supported the ban both in the US and in less developed countries and some still do. I find this to be a very ugly legacy of the environmental movement, and is one more reason that I cannot support radical environmentalism. My main point here, however, is that it is only by careful, unbiased scientific and economic investigation and careful consideration of the indirect effects that disasters of this sort can be avoided.

Whatever benefits EPA has brought to the US and the rest of the environmental world, and there certainly have been some important ones, as I witnessed in Pasadena, there have been some important instances where EPA has been too responsive to current environmental fads. These include banning DDT without adequate consideration of its effects on DDT's unique effectiveness in controlling malaria, and adoption of an Endangerment Finding concerning emissions of carbon dioxide without doing any independent analysis. In the first case, the problem arose because of

---

[162] Roberts and Tren, 2010.
[163] *Ibid.*

the direct intervention by the EPA Administrator, who paid little or no attention to the scientific evidence gathered by the Agency. It appears that the Administrator was primarily responding to pressure from environmental organizations and a popular fad resulting from a book by Rachel Carson.

Unfortunately, I believe that the GWD/GEF/EWD-inspired actions of EPA are on the way to becoming the second worst EPA decision, this time because EPA was apparently told by the White House to proceed rapidly to regulate CO_2 regardless of the scientific evidence and even if the schedule imposed did not allow time to independently consider the science and economics of the decision. Unfortunately, EPA had little no choice but to go along, but should at least have had the ability to insist on a careful analysis of the science in my view.

The problem is how to create an administrative framework under which careful, unbiased investigations can be carried out and listened to in preparing regulations. EPA as it is currently structured is clearly not that institution. This has been particularly evident during the Obama Administration. EPA was not set up in such a way to encourage impartial and unbiased regulations since it reports directly to the President, and if the President is a radical environmentalist, as Obama is, EPA will have a strong green bias. The principal topic of this section is whether there is an alternative administrative arrangement that could yield better regulations.

The problem with the current EPA structure is that it is overly responsive to political pressure from the party/president in power or from public fads rather than to careful science and economics. Everyone at EPA who had followed the 2008 presidential campaign knew that an Endangerment Finding would have White House approval and would go through if Obama were elected. I was the only employee, however, who was willing to state in writing that it would not be based on good science and economics. This illustrates the vulnerability of EPA to political pressure since it was unwilling to give any serious consideration to evidence that was contrary to political decisions concerning a scientific issue. In the case of DDT,

EPA at least gathered the contrary scientific evidence, but the then Administrator chose to ignore it in favor of his preferred outcome.

In the case of the GHG Endangerment Finding EPA chose not to even gather it or to make an independent judgment. It simply reproduced what others, particularly the UN IPCC, had claimed in order to come to a pre-determined conclusion favored by the President. And there was no particular environmental reason to hurry the effort since years of effort (at best) would be required to achieve the necessary worldwide agreements to significantly change CO_2 emissions in the unlikely event that it could be done at all. Much more important in my view was to first build a solid case for (or against) the need for such controls and to sort out why EPA thought that the critics' claims might be scientifically wrong. Failure to do so ultimately led to opposition by most or all Republican congressmen and senators, which in turn led to the defeat of the Administration's proposed Cap and Trade legislation, and to some of the Democratic Party losses in the House of Representatives in 2010. Instead, the Administration decided that speed was more important than scientific objectivity or serious consideration of real public and expert comments from all sides. Perhaps they even understood how weak the science was and how careful review could only damage their case.

The Obama Administration now wants to push its radical environmentalism and "liberal" interpretations of environmental laws even further by ignoring the Congressional Republicans and even Congress as a whole and pushing every legal lever they can find to accomplish their goals, including resurrecting parts of the Waxman-Markey bill by regulatory means. This includes claiming that EPA can unilaterally make changes in the Clean Air Act to overcome the fact that the Clean Air Act as written cannot realistically be applied to existing power plant emissions of CO_2. Given the impasse in Congress, the Supreme Court will be effectively determining environmental law rather than Congress.

There are two obvious ways to avoid similar problems in the future. One would be to require affirmative Congressional approval of all major proposed regulations. The other would be to change the structure of EPA and perhaps other Federal environmental regulatory agencies so as to make them far more independent of the party in power.

Congress now has veto power over final regulations but subject to a presidential veto; opponents have not had enough votes to exercise this veto on environmental regulations and hence it has not resulted in EPA taking major account of anything other than the President's wishes. There have been several votes on proposed environmental regulations but none of these veto actions have been approved by the Senate. And now Obama has proclaimed that he will do what he wants environmentally independent of Congress. One way to change EPA's incentives greatly would be by making approval of major regulations conditional on Congressional approval. The House of Representatives has already proposed how this might be done in their REINS (Regulations from the Executive in Need of Scrutiny, H.R. 367 in the 113th Congress) legislation.[164]

The other way to prevent such actions by EPA in the future is to make EPA more administratively independent of the party in power. One problem with that is that neither party might end up taking political responsibility at the presidential level for environmental regulations and hence be held politically accountable for their actions. This political accountability did not prove to be a major factor in the 2012 Presidential election since climate was almost completely ignored in favor of other issues, particularly economic ones, so the accountability argument for direct control of EPA is currently very weak in my view.

One possibility might be to create a three or five-member commission, perhaps with staggered terms to run EPA with the chairperson from the party in power and both parties equally represented in the other members. Other less controversial regulatory agencies are governed using such an approach. The Endangerment Finding suggests that good science and economics

---

[164] To be discussed in Chapter 12.

are not possible under the current approach with a single Administrator appointed by the President if the President has strong pre-conceived environmental views. In the Obama Administration the skeptics were simply ignored and told to send in comments, which were never given a hearing or even serious consideration.

As outlined in Chapter 5, I was quite amazed at the actions of the Obama Administration in pushing their GWD agenda so hard in the face of the obvious political backlash that would occur in response to their actions, particularly from areas dependent on the coal industry, a favorite target of radical energy environmentalists and the Obama Administration. The treatment of the oil industry was not as harsh but resulted in considerable resentment as oil rigs moved out of the Gulf of Mexico for more hospitable regulatory regimes and as the Administration's Keystone XL Pipeline approval was repeatedly denied or postponed. The first response came in the 2010 Congressional election, when the Democrats lost their majority in the House in part because approximately 20 Democrats who voted for Cap and Trade were replaced by Republicans.[165] So I wondered why the Obama Administration persisted in such actions, which ultimately might have lost Obama his reelection in 2012. My conclusion was that Obama believed that his political skills could overcome any ill effects of his massive regulatory program and that his green revolution was a vital part of the legacy that he hopes to leave behind. He did not want to lose control over the House of Representatives, which has made it much harder for him to pursue his broader agenda, but he may have felt that it was a necessary price to pay for his legacy and would not result in a loss of the Presidency in 2012. If so, he was correct about the outcome of the 2012 election and apparently continues to believe he can implement his EPA agenda despite increasingly powerful Republican opposition.

EPA's experience with implementing GEF shows how easily regulatory agencies can be corrupted by politicians implementing

---

[165] Sensenbrenner, 2012.

the views of a relative minority directed towards one particular end without regard to its objective merits and without adequate safeguards to prevent the interest groups from using an alleged "crisis" of their own making to propel themselves into positions of even greater power and influence using the power of the Federal Government. The CIC has succeeded in "buying" the support of many members of the academic and bureaucratic elite; it has only been the common sense of the average person and a number of Republican senators who saw little or no evidence to support CAGW that prevented the Administration's green agenda from being fully implemented through the Cap and Trade bill in 2009-10.

When regulatory agencies are captured by either the groups being regulated or by those most favoring the regulations the whole enterprise is in danger of either being impotent or tyrannical. Keeping a balance such that neither side gains the upper hand and regulations are based on solid science and economics is the only way that regulation can avoid these outcomes. With some major exceptions such as DDT, this is what EPA and other environmental regulatory agencies largely managed to do until the Obama Administration. Their job was to tailor regulations that would accomplish what was needed at the lowest reasonable cost to the economy and affected industries.

Fundamentally, however, bureaucracies respond to political power, not science. Political power supplies the budgets, people, promotions, and increased salary levels that feed the bureaucracy. Going against it is not regarded as helpful. So it was not hard for the Obama Administration to turn the EPA bureaucracy into a supporter of their GEF ends. This is unfortunate because it is critical in my view for EPA to respond to good science, not political determinations of what the science "should" say. The problem is how to create an environmental regulatory regime that is centered on good science, economics, and law, not political expediency. This is not easily done. EPA's present structure has failed a major test of this during the Obama Administration.

It is not yet known exactly how the climate scare will end. It is clear that the GWD supporters' preferred outcome, approval by the US Congress as an alleged step towards international approval of carbon emissions reductions, did not happen in the 111th, 112th, and 113th Congresses. It is equally clear that approval of global mandatory carbon emissions are equally unlikely because of the lack of support by the United States Congress, the demands by the less developed countries that they receive payment for their efforts/compensation for alleged past "damages" from CO_2 emissions from the developed countries, and the fact that agreement by less developed countries to reduce CO_2 emissions is not in their self-interest. The main uncertainty is how the Obama Administration's intention to implement carbon emission controls through EPA action may turn out. The Obama Administration is trying to persuade the less developed world that there will be both substantial and increasing reductions in CO_2 emissions in future years and US payments to UN funds intended for the less developed countries. The less developed countries have made it clear that they will not decrease CO_2 emissions unless paid very large sums that the Western nations may not offer. It is conceivable, of course, that the whole movement will die as the current global temperatures continue to plateau or even drop into the next predicted low in the 2030s (to be discussed in Chapter 9). But that may be too much to hope for given the religious fervor of the alarmists and the continuing efforts of the CIC.

Enlarging GWD by including EWD, as Obama did in his June 25, 2013 climate speech, has made Obama himself appear to be willing to forego even the semblance of scientific credibility since there is currently no real evidence that 2012's unusual weather was caused by or influenced by CO_2 levels or that US or even world reductions in CO_2 emissions would have any effect on future weather. An interesting question is why the CIC now supports EWD. The apparent answer is that EWD is being used by the CIC and prominent Democratic politicians[166] to claim that any unusually adverse weather event is due to increasing CO_2 levels and can be avoided/diminished by reducing CO_2 emissions. And unlike global warming and climate change since 1998, there have been and will always be unusually adverse weather somewhere in the world that the CIC can try to use to support government-imposed CO_2 emissions controls. It does not seem to matter whether there is no valid evidence for EWD science or CAGW or climate change since 1998; they simply assert that this is what the science says. It also does not seem to matter whether government can actually do anything about any of these alleged climate/weather problems either. Anyone who denies any of these claims is a "denier," as in holocaust denier, or a "flat earther."

As early as 2009 as a result of the Obama Administration's policy, the skeptics had declared full scale war on the Administration's climate policy and brought the Congressional Republicans with them. The congressional Republicans, however, either lacked the votes or will, depending on one's viewpoint, to halt the Obama GWD/GEF/EWD onslaught, which apparently will continue until at least the end of Obama's term unless effective action is taken by Congress.

The Obama Administration apparently decided that it could use its control over EPA to do what it could not get Congress to give them in the way of a cap and trade bill. This effort has been more successful than I initially expected even after the loss of the House of Representatives in the 2010 election, but the Administration had apparently counted votes in the Senate and realized that no one except the courts could stop them. No one had ever attempted to risk a conflict with Congress over energy/climate policy before, and it represented somewhat of a risk as to whether they had counted heads correctly. And it ran the risk of alienating

---

[166] See, for example, a speech made by Senator Barbara Boxer, Chairman of the Senate Environment and Public Works Committee, as reported in Morano, 2013.

much of the business community and ultimately individual voters if energy prices increase and very arduous regulations that slows further recovery from the Great Recession are implemented. But as of late 2014 their gamble appears to be on track from a GEF viewpoint. A congressional confrontation over climate as well as immigration and health insurance may come in 2015 as a result of the Republican gains in the 2014 mid-term elections. In both cases, however, as in climate, there are court challenges, which could prove decisive.

The reason that so many Democratic senators have not been willing to stop EPA regulations may be that despite the fact that EPA is simply doing what the Administration wants, the senators may fear that a vote against EPA regulations would be used by opponents to claim that the senator was opposed to improved environmental quality. In other words, EPA has been able to put up enough of a pretense of scientific and environmental objectivity that Democratic senators fear attacks for being anti-environmental even though it is their constituents who will pay the much higher energy and other bills and receive nothing in return.

There appear to have been similar Executive Branch efforts to expand Presidential power by not adhering to the language in the new health insurance law and various immigration laws. These attempts to circumvent Congressional intent and review represent a new and serious escalation of the increasingly partisan political warfare between the parties in Washington—just the opposite of what Obama promised. The President instead uses his power to do what he perceives to be in his interests up to the point where he is stopped by Congress or the courts. I am concerned that this further attempt to extend Executive Branch power will further exacerbate the increasingly polarized politics and deadlock in Washington.

Science is different than ideology and politics. And it is vital that environmental policy be guided by science and economics and not by ideology or politics. The difference is that science is (or at least should be) based on comparisons with the real world in a way which is reproducible and verifiable. The problem with using

211

ideology or politics to decide environmental policy is that neither one has any real probability of yielding outcomes that will ultimately prove to be scientifically correct and therefore useful. And this can lead to great disasters because pollution control will end up having unintended side effects for no purpose and potentially huge resources will be devoted to solving non-problems or problems that cannot be solved by government and not to solving problems that really would be helped by government intervention. Further, if solutions to environmental problems are not based on science, they are less likely to be effective in solving the problem because they will end up attacking the wrong problem or the right problem in the wrong way.

Because the environment is inherently extremely complex, there is a substantial risk that the science involved in environmental protection will not provide adequate guidance as to what should be done without adding the effects of ideology or politics to the mix. To allow either ideology or politics to be used to analyze or solve environmental problems just makes the problems that much harder and the solutions more likely to fail. The global warming scare is a textbook example of what can happen when politics and ideology guide environmental policy rather than science. If all the intellectual resources devoted to the global warming debate had instead been devoted to preventing a deep water oil well blowout they might actually have managed to avoid the blowout.

Although the great warming/climate change/extreme weather scare has no real basis in science and should long since have been over based on the abundant science available discrediting it, it is not over because it is being pushed with every resource available to the Obama Administration, their CIC partners, and by some Western European countries. This includes large grants to climate researchers which have been and are only given to those professing true belief. So it is apparently not yet time for those receiving so much governmental support to jump ship despite the overwhelming scientific evidence for doing so. The stakes involved in the outcome of this public policy issue are growing for everyone in the United

States. Unless action is taken to remove anti-fossil fuel energy regulations and prevent the new ones proposed for the Obama Administration's second term, the tsunami of counterproductive environmental regulations I describe will engulf every aspect of our lives involving at least energy production and use. This is just what the CIC wants and I oppose.

Although it has been evident since the collapse of its favored cap and trade bill[167] and the Copenhagen summit, the Obama Administration has now proposed regulations that it will unilaterally impose to allegedly accomplish its climate objectives of reducing CO_2 emissions regardless of whether the rest of the world does so or not and regardless of the fact that the US has made greater "progress" towards reducing CO_2 emissions than any other country since 2006 even though the US did not ratify the Kyoto Protocol and has no obligation to do anything. This would replace the practically impossible problem of reaching an effective worldwide agreement by instead putting the US at a disadvantage in competing economically with most of the rest of the world that does not impose such regulations. This would result in making US products even more internationally uncompetitive than they already are, drive even more manufacturing to other countries, such as those in Asia that do not have such regulations, and reduce jobs for Americans even further. To see what might happen, all one needs to do is to look at Western Europe, which is much further down the radical energy policy road. Although they have additional problems due to their ill-advised decision to create a common currency before creating a common fiscal policy, another cause of their continuing economic problems is high and increasing energy costs due to their adoption of a green energy policy and reluctance/refusal to allow shale oil and gas development.

---

[167] The Waxman-Markey bill from the 111th Congress.

Current Balance of Forces

The current balance of forces appears to be on the alarmists' side, but much less so than during the first six years of the Obama Administration. As of early 2015 the CIC is supported by most if not all environmental organizations, many liberal foundations, most of the mainstream media, the primary climate journals, the national academies of science, the UN and EU climate efforts, many European governments, President Obama, EPA and other Executive Branch agencies, and one prominent candidate for President in 2016 (Hillary Clinton), and has ample funding from taxpayers and environmentally-oriented foundations. The skeptics have a strong but not overwhelming presence on the web, some support from the conservative media, widespread support from Congressional Republicans (including strong opposition to alarmist science by one declared 2016 presidential candidate—Ted Cruz), the leadership of the US House of Representatives and Senate, the Canadian (starting in 2011) and Australian (starting in 2013) Governments, and dedicated efforts by volunteers and retired professionals, but are starved for funding. This is contrary to the myth created by the alarmists that the skeptics are richly funded by environmentally "dirty" industries. Based on the more than five years I have spent in the skeptic world since retiring from EPA with a few peeks at the comparatively lavishly funded alarmist world, the alarmist-inspired myth concerning skeptic funding is simply that—a self-serving myth designed to build support among alarmists and provide a simple excuse to ignore whatever the skeptics have to say.

Anyone looking at this balance of forces should find it easy to forecast the most likely outcome. Events, however, have nevertheless been moving in the opposite direction in the last few years. The failure of the 2009 Copenhagen Conference and the proposed US cap and trade bill set back the GWD agenda substantially both in the US and elsewhere in the developed world. As discussed in Chapter 6, the Australian Green and Labor Parties overplayed their hand on climate by enacting a very visible and disliked carbon tax in 2011 and were ousted from power in September, 2013 in major part as a result. The Australian and Canadian Governments say they now want to form a climate "realist" group of nations. Several major European governments including Germany and Spain have greatly reduced their subsidies for wind and solar. Both Britain and Germany face increasing electricity availability problems and rapidly escalating energy prices as a result of their GWD policies.

But the ultimate decision on whether GWD/GEF will become a permanent reality in the developed world is likely to be made here in the US. Unless the US adopts it, other countries that do will be at a competitive disadvantage and will soon have to abandon it, as appears to be starting to happen in Western Europe. The US is far behind a number of Western European countries in adopting the GWD agenda and unlike many parts of Western Europe has organized political opposition in the Republican Party and a number of conservative groups. But with the 2014 proposed EPA regulations on power plants, which propose to implement GWD/GEF/EWD by regulatory stealth, the Obama Administration has probably put forward its "best" strategy in terms of maximizing the chances of success in achieving its objectives. If they can move rapidly enough they apparently hope to get these key regulations in place before the 2016 election and present skeptics with a *fait accompli* in spite of the Republican majority in both House and the Senate starting in 2015. The legal case for these regulations is very weak (as discussed in Chapter 6), however, so this is far from certain. Removing these regulations once in place would probably require either a major change of heart by the Supreme Court, a veto-proof Republican majority in the House and Senate, or an anti-GEF President in 2017. These are far from likely, but possible.

As discussed in Chapter 6, EPA published its new proposed regulations for limiting GHG emissions from new and existing power plants in 2014. Although EPA can directly set standards for new power plants, the states have ultimate authority for existing plants and may try to exercise it at least in red states. There are a number of major legal problems, so the drawback of the Obama strategy is that it relies on the very slow EPA regulatory process and major legal uncertainties. I expect that EPA will push the process as hard as possible, but it is always slow at best.

Perhaps the main reasons for hope is that time is on the skeptics' side. The alarmists have painted an increasingly dire and outlandish picture of the future without their proposed "solution," but somehow none of these undesirable outcomes ever seem to happen and may gradually appear ever less likely to happen to the non-expert, particularly after at least 17 years with no increase in global temperatures. These no-show apocalyptic outcomes include almost all if not all the undesirable outcomes they have prophesied. The recent CIC emphasis on EWD, however, may make it possible for the CIC to claim a new apocalypse every time a new extreme weather event occurs despite their decreasing frequency in recent years in the US. How much longer will the alarmists be able to get away with selling these apocalyptic outcomes when all the evidence is on the other side? The Australian Labor Government made the mistake of making the very large real costs of GWD evident to voters and lost power. Obama proposes to implement the GWD/GEF by regulation in hopes that no one other than a few readers of the *Federal Register* will notice. The problem is that the longer it takes the more likely it is that citizens will figure out that the prophesied GWD/GEF apocalypses are imaginary while the costs are large and very real and will significantly raise their household costs. They may even figure out that a little warming is good, not bad. So the long term outlook for defeat of GWD/GEF is good, but the next few years may be unpleasant and very nasty in this fight.

Proper Role of EPA

EPA can fulfill many roles. Although it purports to act entirely on the basis of science and occasionally economics, it reports directly to the President, and as a result responds to the President's political or even scientific dictates rather than face the consequences. Although there have sometimes been regulations that did not accurately reflect the science and more often the economics of a controlling a particular pollutant, EPA prior to 2009 pursued a relatively balanced approach towards using scientific and sometimes economic analysis with some notable exceptions such as the DDT ban and possibly the Montreal Protocol. With the start of the Obama Administration, however, there was at best only a pretense of doing either one. The strong radical environmental views of the incoming Administration swept through the EPA bureaucracy overnight as it became apparent that those not showing an adequate degree of greenness were not long for holding their current jobs. An obvious question is why bother to have an EPA under these circumstances? Why not just farm the work out to environmental organizations rather than pay a bureaucracy to carry out their wishes since the bureaucracy's views were going to be disregarded anyway? This may appear extreme, but in at least one prominent case, the regulation of GHGs from power plants, EPA has largely followed the ideas and blueprints for action formulated by an environmental organization rather than their own analysis and insights.[168]

So a President that came into office promising to heal the growing divide between Republicans and Democrats undertook actions within months of taking office that resulted in EPA being politicized too since he apparently felt there was no other way to implement his GWD views. And the career bureaucrats at EPA rapidly decided that it was best for them to help rather than raise the many inconvenient scientific and economic issues involved in implementing GWD through regulations. Personally, I don't blame them much since they essentially have few protections against

---

[168] In 2012 the Natural Resources Defense Council provided a proposed roadmap for the Obama EPA to regulate CO_2 emissions from existing power plants. According to Davenport, 2014, EPA used it as a blueprint for its June 2014 proposed existing power plant regulations.

adverse personnel actions by the Administration in power. Executive pay depends in part directly on the political appointees so independence of the civil service is largely a myth.

The problem partly results from the way EPA was created—as a government agency like any other reporting directly to the President. Nixon, who created EPA from disparate pieces scattered around the Executive Branch, probably did it this way to avoid the need to obtain Congressional approval for a different organization and possibly because he wanted to directly control what EPA did. But the net result was that there were no real safeguards built in to prevent EPA from doing the Administration's bidding or to insure that valid science and good economics were used. There is the possibility of judicial review, but court review generally involves administrative rather than substantive issues since courts usually do not review the science used by a regulatory agency. EPA is not currently a neutral arbiter as I believe it should be; it is rather an environmental advocacy Agency doing precisely what the President and the environmental organizations say should be done. Many other regulatory agencies have built in some safeguards to insure some independence from political interference. EPA has none. Illustrations of what this leads to are described in several chapters in this book.

On the one hand EPA can decide on the best way to implement an ideology such as radical environmentalism. On the other hand it can avoid any ideological bent and instead look for ways to improve public health and environmental welfare where there is strong scientific support for doing so and a strong economic case that society as whole would benefit from such a change. The Obama EPA is an example of the first approach. In fact, it is safe to say that EPA has been fully captured by the radical environmentalists and largely carries out their wishes. The second approach is my strong preference. Unfortunately, this ideal has never existed and may never do so, primarily because EPA has never paid enough attention to or given economics the important role that it needs to have and because of the current controlling role of politics in EPA decisions. It is nevertheless possible, however, for the second approach to be an ideal that EPA strives to attain.

As I see it, EPA's role should not be to implement anyone's agenda but rather to search for and implement solutions to environmental problems that are possible politically, economically advantageous, legally justifiable, and, for which there is clear scientific justification. This requires a good deal of independence, responsibility, and integrity. The last thing it needs is an ideological agenda which makes it impossible to achieve the careful balancing of economics, politics, law, and science needed to make the system work and to achieve its environmental objectives. Unfortunately, that is what it is currently required to do by the Obama Administration. The result is conflict, Congressional votes to overrule regulations, and either a real life experiment in radical environmentalism in action or a complete paralysis of the system. Radical environmentalism is where EPA is now headed.

Unfortunately, the strains on the system imposed by the Obama Administration's ideological bent may not be the last time that EPA will experience these or similar threats to their political, scientific, and economic independence. EPA is not set up in a way that would help to insure any of these goals. The Administrator serves at the direct pleasure of the President subject to the laws and funds approved by Congress. EPA has now taken to rewriting the laws and assuming the funds will be appropriated. Chapter 12 contains my suggestions as to how EPA could be restructured so as to both solve its current ideological problems and serve a useful purpose in the face of future circumstances.

One of the fundamental problems is that there is public confusion concerning the role of EPA. This probably arises from the carefully cultivated appearance of objective weighing of scientific evidence that EPA usually goes through in formulating regulations and perhaps its history of taking mostly fairly reasonable decisions. The reality, however, is shown by its actions thus far with regard to climate change, where EPA exercised almost no independent judgment but instead depended on the IPCC AR4[169] and various US Government reports reflecting its views. The best explanation of its actions was that it was doing exactly what it had been told to do by the White House. Yet influential Democrats such as Nancy Pelosi have continued to argue that it would be wrong to overrule EPA and that it would be better to follow the decisions made by EPA "professionals." As of the end of 2013, all of the votes on overturning specific EPA regulations during the Obama Administration have been unsuccessful in the Senate. One explanation for this is that the senators fear a backlash from voters if they oppose an EPA regulation because of EPA's environmental and scientific "halo" effect. This has enabled the Obama Administration to have it both ways—by actually making the decisions as to what and how EPA will regulate in the White House while portraying them as independent EPA decisions based on an independent scientific review. There is no doubt in my mind as to where at least the climate decisions have been made during the Obama Administration.

In fairness, the beginnings of the anti-energy use efforts by Federal regulations started during prior administrations as far back as the Carter Administration; some even began as initiatives by Congress. But the Obama Administration has seized on any and every regulatory approach that might be used to promote their radical environmental and particularly their GEF/EWD view of the world. The ban on incandescent light bulbs, which is already being implemented, illustrates what is happening. In brief, Congress and the Obama Administration are telling Americans what to buy and how to use what they buy instead of letting the market and ultimately the individual decide what to buy and how to use it in cases where there is no market failure that would justify such intervention. This has been tried before, of course, but mainly in Eastern Europe, with disastrous economic and freedom of choice consequences. Even worse, there is no scientific or economic basis

---

[169] United Nations Intergovernmental Panel on Climate Change, 2007.

for what the Obama Administration is trying to do with respect to energy use and climate. They have admitted that their efforts are based on the UN IPCC reports rather any effort by the EPA to do an independent assessment, as they are required to do by laws such as the Clean Air and the Federal Data Quality Acts. They are attempting to implement a radical environmental agenda including the GWD/GEF promoted by environmentalists around the world without even checking the scientific validity of the IPCC reports.

Implications for the United States System for Federal Funding of Scientific Research

Since World War II Federal funding of scientific research has expanded exponentially both in dollars and in fields of research. In some part this was a result of the "success" of the Manhattan Project where academic researchers solved a set of technical problems which in the end allowed the US to bring World War II to an earlier end. But it resulted in the availability of Federal tax dollars for research not always directly relevant to immediate national security concerns.

My concern is not the interesting question of whether such an expansion of Federal funding was the best policy but rather the obvious complete failure of the climate research programs funded largely by the Federal Government over a period of more than 20 years. Well over $39 billion has been spent for Federal research based on the CAGW hypothesis with nothing to show for it other than misleading research and bad policy advice. This enormous failure/scandal strongly suggests that there is something basically wrong with the system of Federal support for scientific research developed since World War II. Even though many agencies and departments have been involved in the funding of this $39 billion the result has been the same: failure. None of these agencies have devoted significant resources to alternatives to the CAGW hypothesis or proved the case for the CAGW hypothesis. Vast amounts have been spent on building computer models that prove nothing other than that computer modeling is expensive and of limited use in confirming or rejecting scientific hypotheses concerning chaotic, multivariate systems such as climate, as will be discussed in Chapter 9.

The question is how such an outcome from a major multi-decade Federal research program could have happened and what can be done to keep it from happening again and in other areas. It appears to be a systemic failure over many years, many agencies, and many Administrations. Why has the United States experienced such a failure in one of its supposed area of strength—scientific research? One obvious answer is that government is not much better at funding scientific research than it is in trying to determine industrial policy. A vast enterprise has grown up under which researchers submit proposals under a variety of programs but which usually end up being evaluated by their "peers." The problem at least in this case is that the "peers" are the researchers that have been pursuing bad science for several decades. They only approved proposals by other researchers who wanted to support the same CAGW hypothesis. And all this was exacerbated by an international organization, the UN IPCC, which endorsed and promoted this bad science and was substantially financed by the US taxpayers too.

One approach that would surely work would be to end Federal funding for research not directly related to military necessity. Some argue that this would be a good general outcome.[170] This is most unlikely to happen, of course, particularly given the urgency of some expensive research such as better understanding the non-CAGW aspects of climate, but the alternatives would be difficult to administer. Requiring say that a significant percentage of research grants be allocated to alternative approaches to any research effort would give rise to numerous problems and obvious waste if the alternatives were no better than the current main line of research, but may be better than doing nothing. Leaving the Federal research effort in its current form simply invites more monumental failures.

---

[170] Kealey, 2013.

9
Lessons Learned about Climate Science

People underestimate the power of models. Observational evidence is not very useful, adding,...*our approach is not entirely empirical.*
—John Mitchell, principal research scientist at the UK Met Office, 2011[171]

ANOTHER ONE OF the CIC base institutions is the science establishment, which Eisenhower clearly had in mind in his Farewell Address. Climate alarmists have tried very hard to justify their proposed GWD/GEF/EWD policies and huge proposed expenditures by governments and ratepayers by claiming that they are based on science, or even that science "demands" action. So it is important to examine these claims. Basically, the CIC has attempted to take a slightly plausible hypothesis, that human use of fossil fuels to generate energy has resulted in increased emissions of CO_2 and increased atmospheric CO_2 levels, which they alleged would result first in catastrophic global warming, then in climate change, and most recently in increased extreme weather. Slight

---

[171] Haapala, 2011.

plausibility is not scientific proof, however, and use of the scientific method shows that the hypothesis is invalid scientifically. This is fundamental since the whole alarmist enterprise collapses in this case.

Why the Scientific Basis for CAGW Is Invalid

The IPCC in particular and the CIC in general believe that there will be catastrophic anthropogenic global warming (CAGW), which means that global temperatures will increase catastrophically due to CO_2 emissions caused by human activities. Skeptics agree that there may be some increases but they will be minor and will be predominantly due to natural forces rather than human activity.

Why the scientific basis for CAGW is invalid will be discussed by first discussing the role of the UN IPCC, then the application of the scientific method to CAGW, then omitted variable problem/fraud, then an examination of two fundamental but dubious assumptions underlying CAGW, and finally a broader examination of the issues.

Role of the IPCC

One of the key components of the CIC is the UN IPCC. It describes itself as a scientific body, but I regard it as the main scientific component of the propaganda arm of the CIC. The IPCC has attempted to build a scientific case for CAGW and the IPCC's "solution"—governmentally mandated CO_2 emissions reductions— not to independently assess the science and the "solution." Their hypothesis, if it were valid, is significant for modern civilization because CO_2 is a byproduct of higher income economies as well as essential for most life on Earth so this sets up a possible conflict between economic development and catastrophic global warming/climate change and makes it doubly important that their conclusions be examined with great care.

In September 2013 the IPCC issued a draft of the fifth version of its assessment for Working Group I (physical sciences). This is the most critical part of their overall new report characterizing "the risk of human-induced climate change, its potential impacts, and options for adaptation and mitigation" and is the most recent in a series which they have issued every 5-6 years since 1990.

The IPCC's argument for significant effects of human-caused emissions of CO_2 on climate have shifted significantly between different reports. They have never admitted that earlier explanations were wrong, but keep inventing new ones. A brief summary has been prepared by Fred Singer.[172] Singer's conclusion is that "the IPCC has never succeeded in demonstrating that climate change is significantly affected by human activities—and in particular, by the emission of greenhouse gases."

One of the important things to note here is that these assessments are not intended to be assessments of the larger and more relevant topic of what science says about the causes of climate change but only of human-induced climate change. Unfortunately, the IPCC and other parts of the CIC have been very successful in getting the public to think that they have done the larger assessment rather than only the human-induced component. In my view their reports are nothing more than overly extensive, sophisticated catalogs of all the evidence they can find to support their predetermined CAGW hypothesis.

Thus the IPCC was not established to determine whether global temperatures were rising or why, but rather to determine the extent and influence of only one possible influence that they presumed to be significant. They have made no attempt to apply the scientific method to determine the validity of their presumed hypothesis or to explore obvious other factors strongly correlated with changes in global temperatures. A scientific assessment that presents only one side of an issue and omits any discussion of the results of using the SM is of little real value and a potential threat to determining a balanced scientific case for policy, but the IPCC has nevertheless been amazingly effective in convincing many

---

[172] Singer, 2013a.

individuals, governments, and organizations directly or indirectly to accept the IPCC's CAGW hypothesis and proposed "solution." This is the heart of much of the scientific climate debate—whether the IPCC's conclusions should be used to determine climate policy. The simple answer is that they should not because they examine only one side of the issue. They should also not be used by USEPA or any other country to determine climate policy, but they have been.

As a result of my earlier work on assessment documents for EPA during the 1970s and early 1980s, I expressed my problems with EPA's air pollutant assessment efforts well prior to the Technical Support Document for their Endangerment Finding for GHG Emissions. I was critical of many of these earlier assessments prepared for the Office of Air Programs by the Office of Research and Development (ORD), but for different reasons. I argued rather that they were overly long and lacked focus on the key issues. Unfortunately, it was often easier to include everything concerning a pollutant, even if only tangentially relevant, than to pick out the key issues and the evidence concerning them. But I regarded most of these earlier assessments as reasonably balanced on the whole in their approach.

The same problems I saw with the ORD assessments appear in the IPCC assessments. They are too long and do not focus on the key issues. But the more important flaw of the IPCC reports is that they are designed only to present the case for human-induced climate change. I strongly believe that using such one-sided assessments as the basis for EPA or any other policy is simply wrong. It appears likely that EPA used the IPCC reports and related US reports based on their reports for their Technical Support Document because they supported Presidentially-determined science, were handy and widely known, the White House wanted immediate action, and the legal need for careful, balanced assessment was regarded as low. In other words, the lawyers did not believe that EPA would be likely to be successfully challenged for not doing its own careful assessment despite the obvious biases of the IPCC reports. This legal judgment appears to have been correct since the Supreme Court has directly or indirectly supported EPA's actions in issuing the Endangerment Finding.. But that does not make it a responsible judgment on EPA's part in my view. As I argued at the time, the most important decision EPA has ever made deserved and needed an independent, broader, balanced assessment, regardless of what the IPCC had done. Since the IPCC has not conducted a balanced and therefore suitable assessment, it was even more incumbent on EPA to prepare one in my view.

In addition, it is important to understand that the IPCC is basically a political/propaganda organization rather than a scientific one. Although their underlying scientific reports are drafted by scientists (and all too often by environmental activists and graduate students, as we now know), the scientists are selected by national governments often on ideological grounds rather than scientific merit[173] and their critical Summaries for Policymakers are edited and approved by representatives of member governments, who may or may not be scientists, in closed meetings. The underlying reports are then changed to reflect the changes made in the Summaries. *In other words, the scientists and the conclusions are determined by member governments and both the summaries and the main reports are then changed to suit.*

This is as if I, as the EPA staff sponsor of numerous pollutant assessments prepared for EPA by the US National Academy of Sciences, which I was for a number of years, had insisted that I be allowed to determine the makeup of the NAS committees charged with writing each assessment, then edited and changed the assessment report conclusions which EPA had sponsored, and finally claimed that the NAS reports represented the views of NAS. This, of course, never happened because NAS would not have allowed it and I would not have wanted it to happen since I wanted the reports to reflect NAS' views, not EPA's. But this is just what happens in the preparation of each IPCC report, particularly the

---

[173] Tol, 2014.

Summary for Policymakers, which is the only section of each report that is widely read. Thus, the principal conclusions of the IPCC reports, particularly the final Summaries for Policy Makers, are determined by representatives of the same governments who later claim that they represent the dictates of science and the scientific basis for their GWD/GEF policies. The choice of scientists and national representatives to IPCC meetings are often heavily influenced by the environmental agencies of each government that have the most to gain by advocating GWD/GEF policies. In other words, national governments determine who writes the reports, and the major conclusions the reports reach, and then claim that it represents what the science says. This is unacceptable in my view. The IPCC reports are nothing more than summaries of information and arguments that support what the governments and their environmental agencies want them to say. Fundamentally either they represent what the science says or what the governments say. They cannot be both. Until they represent what the science says they should play no role in determining climate policy and should no longer be supported by US and other taxpayers.

The IPCC reports have become increasingly alarmist over time despite an ever increasing gap between predicted and actual global temperatures and the discovery of numerous other problems. In their first report they actually mentioned the possibility that natural forces could be changing the climate. By their 2007 report they claimed that there was a 90% certainty that humans are the "dominant cause" of global warming since the 1950s. The 2013 WG1 report upped this by claiming that there is a 95% certainty (see Figure C-7 for a graphic illustration of how IPCC certainty has increased as the disparity between their models and measured global temperatures has increased). Even if true, the AR5 report no longer reports a preferred ratio of temperature change for a doubling of CO_2 and in fact increases the range for this critical variable on the lower end.[174] So the 95% may have little impact on

---

[174] Crok, 2013.

projected warming. These percentages are the personal opinions of a few of the authors and reviewers of one section and are not based on any verifiable or reproducible statistical analyses. It is not a scientific conclusion but rather a political one apparently made for its propaganda effects. Meantime the scientific evidence against CAGW continues to multiply, including the 16 or more year hiatus in global warming since 1998. So although many scientists are involved, the IPCC reports are basically a political statement rather than a scientific one. Yet alarmists universally claim the opposite.

A vital role in this whole process is played by other UN groups, the mainstream media, and the rest of the CIC propaganda effort, which attempt to claim that the IPCC reports are actually scientific assessments rather than attempts to justify CAGW and GWD/GEF. The most important of these groups is the UN Framework Convention on Climate Change (UNFCCC). At its initial meeting in 1992 they declared that human emissions of CO_2 were causing significant and dangerous climate change. This viewpoint had no factual basis then or now; it was rather the IPCC's role to present whatever evidence they could find for these conclusions.

Since then the UNFCCC has conducted annual meetings to try to pressure countries into reducing their CO_2 emissions. Each meeting is wrapped in a propaganda blitz exaggerating the significance of the IPCC's findings and implying that they extend far beyond its much more limited mandate. The Executive Secretary of the UNFCCC is appointed by the UN Secretary General, which implies that the upper levels of the UN are in support of this slight-of-hand. The IPCC reports and the annual UNFCCC meetings are then reported in an approving manner by another vital part of the CIC, the mainstream media, in the world's media markets. They are under even fewer constraints as to the accuracy of their statements, and all too often state that CAGW is proven and not subject to questioning like all scientific hypotheses. Another CIC component, the environmental organizations, then declares that everyone on Earth simply must abide by their commandments concerning the generation and use of energy. Many Western national governments, including the Obama Administration, then join in and echo all of the above.

In the days before there was an internet, this sophisticated and massive complex of coordinating groups would long since have sold their product (government-imposed CO_2 emissions reduction) to a world that had no other widely available source of information. When the leaders of the CIC determined that global warming was too minor to be sold to the public, they rebranded their effort as preventing climate change. But climate always changes, and they have found it difficult to show much change and what was bad about what has changed. So recently they again rebranded the effort as preventing extreme weather. After all there is sure to be extreme weather somewhere on the planet! And each time the rest of the CIC obediently fell into line, all with no questions being raised by the mainstream press.

Fortunately, there are now alternative sources of information—the internet and a few generally conservative publications. Whether people will use this source before their lives are altered by placing unnecessary and counterproductive government restrictions on their energy availability and use, and therefore their economic well-being, is the remaining question. The CIC apparently still believes they have a chance at succeeding or Western governments would not have approved the new Summary for Policymakers for the AR5 WG1 report despite the overwhelming evidence on the other side.

Critical Role that Should be Played by the Scientific Method

According to the scientific method, a scientific hypothesis must be tested by comparing real world data to the consequences of any hypothesis. This is how Albert Einstein was able to persuade the world that his ideas on relativity had merit despite his challenge to the then generally accepted Newtonian view of the world. Scientists kept proposing real world tests of the consequences of his hypothesis in hopes of disproving it but each test confirmed the validity of the consequence being tested. After a number of these tests, the opposition conceded that his hypothesis had validity.[175] An earlier but similar process resulted in the acceptance of Newtonian mechanics and other hypotheses which gradually assumed the status of theories, only to be modified by Einstein's new hypothesis.

If the comparison of a consequence with real world observations is not confirmed, the hypothesis should be rejected. There are only two alternatives from a scientific viewpoint when this happens: Discard or modify the hypothesis or discover an error in the data used to reject it. From a scientific viewpoint, it is totally irrelevant how many public officials or scientific organizations—or how prominent they may be—support a particular hypothesis. Even public support by the US National Academy of Sciences, the President of the United States, or the Pope is irrelevant to the scientific validity of a hypothesis. Those who continue to proclaim the merits of an invalid hypothesis are simply showing their ignorance of the scientific method or their willful intention to misrepresent the science.

In order to show a scientific basis for GWD/GEF, alarmists must show that CAGW satisfies the scientific method (or, to put it more accurately, that the hypothesis cannot reasonably be disproved). Nothing else qualifies. Listing numerous arguments for a hypothesis, which is what the IPCC has done, does not qualify. Alleged indications of a warming Arctic or warmer summers or rising sea levels prove nothing except that they may or may not be due to climate variability and natural climate cycles. The same thing holds for allegedly melting Antarctic ice shelves, or melting mountain glaciers, or slowly rising sea levels.

The reality is that despite expenditures by the US Government in excess of $39 billion over several decades for climate research

---

[175] For a description of this extended process see, for example, Crelinsten, 2006.

and $43 million more for IPCC assessments,[176] the IPCC and the CIC never attempted to apply the scientific method to their CAGW hypothesis to show that it has scientific merit. This would have required an effort to disprove the hypothesis, not prove it. But that was apparently the last thing they wanted to do. It is impossible to "prove" that a scientific hypothesis is valid—only that it is invalid. Only by repeated unsuccessful efforts to disprove a hypothesis is it possible to say that it may have some validity. This is a very important point: Since the IPCC has never tried to disprove the CAGW hypothesis, they have never proved it since only by attempting to disprove a hypothesis can it gradually be proved. So even if the IPCC were correct in every scientific statement they have made (hardly the case, of course), *the fact that they have never tried to disprove the hypothesis means that they could not have proved it.*

Instead, after five reports over 23 years the IPCC has attempted the impossible—to prove that their hypothesis is correct by constructing a changing set of arguments for it. The IPCC's "evidence" almost always concerns alleged increases in temperatures compared to some earlier period and sophisticated, ultra expensive computer models showing vastly greater warming than subsequent actual data shows for the same period. Neither of these prove anything scientifically, and even suggest the weakness of their case.

The models relied on so heavily by the UN simply show what the model builders believe would happen if the hypothesis and all their other assumptions were correct. The model results are interesting, but irrelevant in deciding whether their CAGW hypothesis should be accepted or rejected. This is because they do not compare the implications of the hypothesis with real world data other than past temperature data which the models have been modified to emulate.

As will be discussed shortly, CAGW is an invalid hypothesis and must remain so until it can be shown that no reasonable effort to disprove it is unfavorable to CAGW. And until it becomes a

---

[176] Haapala, 2013.

scientific hypothesis validated using the scientific method there is no scientific basis for spending even one dollar to solve the alleged problems posed by it, or even calling it science. But many nations of the world, particularly in Western Europe, have spent vast sums and continue to do so (see Chapter 11 for some estimates) on the basis of this invalid hypothesis. Obama has now proposed that the US do the same thing despite the economic disaster that has resulted in Western Europe (as discussed in Chapter 6).

Why EWD Science Is Invalid According to the Scientific Method

Starting in about 2012 the CIC has placed much greater emphasis on EWD rather than GWD or GCCD. They have not disavowed GWD or GCCD, but increasingly ignored them in preference for EWD. Despite assertions to the contrary by many alarmists, including President Obama, there is simply no evidence that extreme weather events are becoming more frequent or severe as atmospheric CO_2 levels rise or that man-made CO_2 emissions have anything to do with these events. A supporter of CAGW, Roger Pielke, Jr., even testified in late 2013 as follows:[177]

- *There exists exceedingly little scientific support for claims found in the media and political debate that hurricanes, tornadoes, floods and drought have increased in frequency or intensity on climate timescales either in the United States or globally.*

---

[177] Pielke, Jr. 2013. . It is important to note that Pielke, Jr., has taken an alarmist viewpoint with regard to CAGW, so his very different views concerning EWD science are particularly striking. Yet the Obama Administration is trying to sell their power plant CO_2 reduction campaign on the basis of EWD, and, since June, 2014, as noted later in this Chapter, on health "co-benefits" from closing coal power plants. See also Idso, 2014 and Hockey Schtick, 2011 for additional studies.

- *Similarly, on climate timescales it is incorrect to link the increasing costs of disasters with the emission of greenhouse gases.*

- *These conclusions are supported by a broad scientific consensus, including that recently reported by the Intergovernmental Panel on Climate Change (IPCC) in its fifth assessment report[178] ... as well as in its recent special report on extreme events.[179]*

Pielke, Jr. thus notes that even the UN IPCC does not support EWD!

This is unfortunately a case where the CIC appears to be trying to respond to extreme recent weather events for public relations purposes without even the pretense of scientific support. We can all agree that it would be nice if humans could avoid the results of extreme weather. The CIC's general and uncritical adoption of extreme weather as the problem they seek to solve on the basis of little or no scientific support suggests both their unscrupulousness and their desperation about the inadequacy of CAGW and GWD to achieve the scientific and political acceptance they seek. There is no evidence that increasing CO_2 emissions increases extreme weather events or that reducing such emissions, assuming that could be done by governments, would reduce such events. But as of late 2014, most CIC spokespersons talk almost exclusively about extreme weather and the urgent need to reduce it by reducing CO_2 emissions.

One way to examine the validity of EWD science is to make the inference that since atmospheric CO_2 levels are rising, then extreme weather should also be rising if EWD science were a valid scientific hypothesis. But as Pielke, Jr., points out, available observations dispute this and the Administration has not launched a

---

[178] UN IPCC, 2013; see also Watts, 2013a.
[179] UN IPCC, 2012.

credible counter effort to make their EWD science case as of laate 2014. In other words, the Obama Administration has thrown caution to the winds on EWD science and is now joining the CIC in using the *ad hominem* attacks and epithets long used by other parts of the CIC at the highest level. They appear to be making up the EWD "science" as they go along to support their EWD-based program and hoping that official pronouncements will drown out any responses that skeptics make.

The choice of EWD as the new goal of GEF is curious in another respect. Deaths and the death rate related to extreme weather have declined by 93 and 98%, respectively, over the last 85 years, in significant part because of the increasing use of fossil fuels.[180] Fossil fuels have made possible fertilizer and pesticides, which have contributed to high food stockpiles which can be moved to countries in need by fossil fuel-powered conveyances. Extreme weather prediction has been made much more accurate as a result of satellites launched by fossil fuels. So rather than the use of fossil fuels causing extreme weather, they have helped to make humans much less vulnerable to extreme weather. Using less fossil fuels as the CIC advocates is thus likely to make the effects of extreme weather worse, not better, particularly since governments cannot reduce extreme weather—only mitigate the effects of it when it inevitably occurs.

The case for the validity of EWD science will not be discussed further in this Chapter since it is so transparently invalid. Rather, the rest of this Chapter will concern the earlier CAGW hypothesis. This has received much less attention by the Obama Administration and the rest of the CIC in the last few years, but they have not disowned it either. And the EPA Endangerment Finding is based on it. This Chapter will do this by first examining the scientific validity of the major claims, then the advantages of alternative explanations, then alternative means for implementing climate change controls, then the issue of science transparency and integrity, and finally the

---

[180] Goklany, 2011.

question of whether American science has been seriously compromised by the CAGW debate.

Why CAGW Is Invalid According to the Scientific Method

There have been a number of tests of various aspects of the CAGW hypothesis.[181] Everyone I know of has failed. Only one failure is enough to invalidate the hypothesis. In other words, if the IPCC had applied the SM, they should have long ago concluded that the CAGW hypothesis was invalid. This has become an increasing problem for the IPCC as an increasing number of studies have become available. By the time of their Working Group I AR5 report in 2013, the only way they could avoid invalidating their models using observational evidence was to include outdated studies with substantial shortcomings, obfuscating what they had done, and inadequately informing policymakers on this critical issue in their Summary for Policymakers.[182]

Although sophisticated applications of the scientific method have shown this for some time, it is now possible for anyone to largely verify this by looking at any major global temperature series and seeing the temperature "hiatus" that has been going on since at least 1998. Atmospheric CO_2 levels have continued to increase during this period but temperatures have not (see Figure C-1). It was the drop in temperatures in late 2007 that originally led me to suspect that CAGW was wrong. It is a reasonable inference from CAGW that as long as atmospheric CO_2 levels increase, temperatures should also increase. But this is simply not the case since at least 1998 based on all major world temperature datasets. The alarmists claim that 17 years is too short a time to invalidate their hypothesis. I argue that their CAGW hypothesis should have

---

[181] Carlin, 2011 discusses several tests; others discuss particular ones, including: Garfinkel et al., 2014; Miskolczi, 2014; Lindzen and Choi, 2011; Douglass et al., 2007; Douglass and Christy, 2008; Hammer, 2013; and DiPuccio, 2009.

[182] Lewis and Crok, 2014

included some mechanism to explain/predict the at least 17 year hiatus if they wanted the world to believe their explanation. This simple test fails based on current data. The IPCC models and reports have continued to project rapid increases, and are leaving reality ever further behind.

Various far out explanations have been offered by alarmists, of course, but the answer is all too obvious—the CAGW hypothesis is simply wrong, must be discarded as a scientific hypothesis since its consequences do not correspond with real world observations, and must be replaced by other hypotheses that do, just as the scientific method calls for. Yet the CIC still wants the world to spend trillions of dollars reducing CO_2 emissions on the fear that Mother Nature will eventually validate their scientifically invalid hypothesis and the consistently too warm computer models it is based on. This is ridiculous and constitutes one of the greatest scientific scandals of recent decades. The rational alternative would be to do nothing and await the future scientific information they claim will validate their hypothesis, if it ever comes. Those who insist on proclaiming that invalid hypotheses are science after they have been disproved are not scientists but cultists. Given the possibility of economically superior geoengineering alternatives discussed later in this Chapter, there is no cost to waiting as long as geoengineering options are not ruled out, which GWD supporters usually try to do in any way they can.

Any policy decision based on CAGW is highly likely to be wrong and any funds spent on the basis of a failed hypothesis are very likely to be wasted, including more CAGW-based research, windmills, solar plants, and corn ethanol in this case. The CIC drumbeat of alarmist, mainstream media, UN, and western government propaganda continues just as before, of course, but it is just that—propaganda—despite the support of numerous national academies of sciences and professional scientific societies. Mother Nature has spoken. Only ideology/religion support CAGW. But the IPCC has nevertheless issued its WG1 AR5 report singing much the same old tune (although with a few more caveats), largely ignoring any attempt to account for the growing disparity between their projections and reality (see Figure C-7), and not applying the scientific method.

How much longer will those funding the IPCC, including the US taxpayers, continue to waste money on CAGW and GWD/GEF/EWD rather than recognizing that they are pursuing a failed hypothesis? A strong hint is provided in a report that representatives of the governments sponsoring the IPCC who met in Stockholm in September, 2013 agreed to downplay any discussion of the temperature "hiatus" in the Summary for Policymakers for Working Group I.[183] This suggests to me that they have no intention of ending the IPCC or the GWD/GEF scenario anytime soon, but have rather decided to continue the CAGW false alarm/hoax/scam for at least another five years until the next report. Deciding to stop digging a deeper hole is not easy when the hole is very large and the embarrassment for those involved would be overwhelming. It is much easier to keep digging and let someone else refill the hole down the line. Governments seem to be particularly prone to such approaches.

But is it not possible that if we waited just one more year that global temperatures would soar up to the ever increasing IPCC model levels? Yes, but one prominent *alarmist* IPCC lead author put the chances of this at 2% in 2013[184] based on the probability that the models could result in a 15 year "hiatus," but he said that if temperatures continued unchanged for another 5 years they "will need to acknowledge that something is substantially wrong with our climate models."[185] As this is written in early 2015, 2 additional years have gone by with the "hiatus" still going strong. In other words, this lead author thought that the world should continue to waste as much as $1 billion per day for another 3 years on less than

---

[183] Booker, 2013,

[184] Kramer, 2013. This is a report of a *Spiegel* interview with Hans Van Storch as reported in WattsUpWithThat.com

[185] *Ibid.*

a 2% chance that the IPCC models are correct. As a long term student of markets, I would not put any of my money on this faint possibility, however, and I would not recommend that governments do so either, especially given the previous history of CAGW SM failures, which should have resulted in abandoning CAGW long ago. And as will be seen later in this chapter, there are reasons to expect temperatures will either continue the current "hiatus" or fall until the early 2030s, and even then likely increases may not be distinguishable until the early 2040s. By then even the alarmists will have either given up or died out. But the funds already spent cannot be recaptured in almost all cases. New funds spent on GWD/GEF in the meantime will disappear down the ever growing GWD/GEF hole, just like all the previous ones. So look forward to even more abandoned wind turbines and solar panels in need of maintenance (as illustrated in Figures 10-1 through 10-3 at the end of Chapter 10).

As will be discussed later in this chapter under alternative natural explanations for climate change, there actually is an alternative[186] to the AGW hypothesis which has satisfied every test so far done to verify it. And it reaches exactly the opposite conclusions to that of the IPCC—that is, that Earth's temperature is independent of CO_2 increases (human-caused or otherwise). That does not mean that this hypothesis is correct, of course, but it makes continued expenditures on GWD/GEF appear even more absurd until and unless the new hypothesis is proved invalid and CAGW is miraculously resurrected from the dead. Is it better to spend resources on a hypothesis that has been proven invalid or on one that passes all the tests done so far? I think the answer is obvious. An interesting footnote, is that the author of this new hypothesis was told by his employer, NOAA, that he could not spend additional time on his hypothesis,[187] so he left. So much for Federal Government openness to new science.

---

[186] Miskolczi, 2014; see also earlier publications by the same author.
[187] Marohasy, 2009.

Why the IPCC's Conclusions Do Not Satisfy the SM Even if Every Projection Were Correct

The IPCC's 2007 report notes: "Most of the observed increase in globally averaged temperatures since the mid-20th century is very likely due to the observed increase in anthropogenic greenhouse gas concentrations." In 2013 the IPCC Working Group I Summary for Policymakers claims that "it is extremely likely that human influence has been the dominant cause of the observed warming since the mid-20th Century." The IPCC has no empirical proof that human carbon emissions are the main cause of planetary warming (despite their alleged 90 or 95% certainty); their primary "proof" is that they claim that their scientists cannot think of another explanation. This is a form of deductive logic, called the principle of exclusion, which is not a valid approach under the SM because it uses deductive rather than inductive logic. The problem with the principle of exclusion is that there may be other explanations which the IPCC scientists have not considered or may not have even been discovered as yet. In fact, there are, which I will discuss shortly and where necessary in more detail in Appendix C.2. Even Darwin had a problem with the principal of exclusion, but ultimately realized his error.[188] So in brief, the IPCC's argument is unacceptable as proof under the SM, regardless of whether their conclusions turn out to be correct or not. Yet the alarmists claim that CAGW is "settled science" and that skeptics are "deniers" or "flat earthers." It is not settled science; it is not even science. In fact, there is nothing more anti-scientific than the idea of "settled science." CAGW is an invalid hypothesis and thus ideology/religion dressed up to look like science. And there is clearly another possible explanation—natural climate variability—which they have not shown to be wrong despite an alleged diligent search.

In order to show that global warming is human-caused, alarmists would need to show at least that it exceeds natural

---

[188] Macrae, 2011.

variability. Since as of early 2015, there has been no statistically significant warming from 1998 using all major indices (including the surface thermometers so favored by the CIC), only minor warming in 1997-98, and no observable net warming in satellite temperatures from 1979 to 1997,[189] this certainly has not been shown; in fact, the natural variability explanation becomes ever more likely the longer this temperature "hiatus" continues. As outlined above, CAGW has not happened and appears unlikely to happen. The climate models at best show what their authors believe to be the physical relationships between various atmospheric phenomena. They have never been validated and have consistently substantially overestimated projected future temperatures. Computer programs make it easier to calculate the results of making various assumptions, but say nothing about scientific validity.

Because CO_2 is a trace gas in Earth's atmosphere, and a weak greenhouse gas, it can only have a significant greenhouse effect on climate if it can be shown that it triggers a strong positive feedback from a much more abundant gas with much stronger greenhouse properties. The IPCC argues that such a feedback exists with water vapor, the major greenhouse gas in the atmosphere which does have much stronger greenhouse properties, but has never presented much observational evidence that the feedback is strong or even positive. Without such a feedback, the CAGW hypothesis is not viable. So one way to apply the scientific method to the CAGW hypothesis is to see whether there is observable evidence for a number of implications of the strong water vapor feedback hypothesis. Acceptance of the hypothesis implies that at least four observations should be present.[190]

The best known of these observations is that there should be a tropical hotspot in the Upper Troposphere. No such tropical

---

[189] See Appendix A, Figure 2-13, and Appendix Figure C-12.
[190] Carlin, 2011. See also Appendix C, which provides some added discussion.

hotspot has been identified and the other three tests are also negative.[191] No hotspot means no strong feedback effect between changes in atmospheric CO_2 and changes in water vapor, so CO_2 levels are not of much concern. The IPCC agrees that the hotspot should exist, but no one has managed to find one despite extensive efforts and data. This key component of CAGW also fails the other three SM tests.[192]

Only one failure is enough to find that CAGW is invalid scientifically. Yet the CIC acts as if the CAGW hypothesis were alive and well. It is not. It is a failed hypothesis, nothing more. People are welcome to hold religious beliefs/ideologies if they so wish, but they should not claim that they are based on science if they cannot be validated using the SM. Nor should they use a failed hypothesis as an argument to remake Western energy production and use at enormous cost in the name of science. To add insult to injury the 2013 Working Group 1 Summary for Policymakers omits discussion found in the previous report of upper troposphere temperatures while mentioning other layers after having the upper troposphere problem specifically pointed out to them in an earlier draft.[193] Such omissions imply that the IPCC is not engaged in preparing impartial assessments of the science but rather making the best case it can for a viewpoint already held. And the CIC widely calls skeptics "deniers!" At the very least this omission serves as strong evidence that the IPCC assessment reports are not objective and are best characterized as verbose, expensive propaganda exercises unwittingly funded by the taxpayers of the world.

Similar findings (although with much less emphasis on the SM) concerning the scientifically unsupported findings of the IPCC reports have been extensively documented by exhaustive assessments authored or edited by prominent skeptics,[194] These are

---

[191] *Ibid.*

[192] *Ibid.*

[193] Lloyd, 2013.

[194] Idso *et al.,* 2013; Idso and Singer, 2009; and Idso, Carter, and Singer,

known as the Nongovernmental International Panel on Climate Change (NIPCC) reports. Another study[195] reviews seven other published studies and concludes that:

> *The seven papers discussed use different methods to critique global warming but are all based on empirical data and are in rough agreement that any increase in global average temperature due to a doubling of CO_2 is more likely to be about half a degree than the 3.26 degrees determined by the IPCC [AR4, Box 10.2]....*
>
> *The global warming models amplify CO_2's effect by 3 – 7 fold, but no matter how you measure it [outgoing long wave radiation, cloud changes, optical depth, historical temperatures, vertical heating patterns in the atmosphere] the real measurements contradict the models and their assumptions about the feedbacks appear to be unconnected with real data.*

In addition to all the other failed tests of the AGW hypothesis already discussed in this section, there is still another one. The AGW hypothesis essentially argues that increasing atmospheric CO_2 acts like putting a blanket over the atmosphere which decreases Earth's emissions of longwave radiation into space (hence the name greenhouse gas). If such radiation decreases, global temperatures increase since temperatures reflect the balance of incoming and outgoing radiation. Hence a reasonable test of the AGW hypothesis is that outgoing longwave radiation should decrease with rising atmospheric CO_2 levels. NOAA has determined, however, that such radiation has been increasing during the last 30 years[196] during which time atmospheric CO_2 has been increasing. This test is particularly damaging to the AGW hypothesis because it concerns

---

2011, 2013, and 2014.
[195] Cox and Nova, 2013.
[196] Hammer, 2013.

the very period that AGW supporters use to support their claims concerning global warming. Clearly the hypothesis is incorrect or the data are wrong. Since NOAA is not exactly an unbiased observer concerning climate issues, it appears highly unlikely that the data are wrong.

The evidence is simply overwhelming and settles the issue pending new data, but not in the way claimed by the CIC or, more recently, Obama. But until the CIC admits this, little progress can be made in climate science and there will be a continuing drain on scarce resources in pursuit of invalid science and uneconomic "renewables."

The essence of a useful scientific assessment is that is should even-handedly explore both the strong and the weak points of candidate hypotheses to explain the subject being assessed. As David Wojick has explained:[197]

> What is systematically omitted from the SPM [Summary for Policymakers] are precisely the uncertainties and positive counter evidence that might negate the human interference theory. Instead of assessing these objections, the Summary confidently asserts just those findings that support its case. In short, this is advocacy, not assessment.

Particularly important is the application of the scientific method to determine whether hypotheses considered can be disproved. The IPCC limits itself to trying to prove rather than disprove their hypothesis as scientific methodology requires. So the reader is left to do this job. This means that it is not an assessment at all—just a summary of the evidence for one (invalid) hypothesis. The USEPA did not make an independent assessment, as it has done for many other suspected pollutants, but instead used the IPCC reports and other US reports based on them as if they actually were valid assessments. So the inadequacies of the IPCC reports are also

---

[197] Wojick, 2001.

inadequacies of the USEPA TSD report. A more detailed summary of scientific issues concerning the CAGW can be found in Appendix C.1.

Discussing each of the many problems with the IPCC assessments is not possible here, so I will only mention two of them. In AR3 the IPCC prominently featured the well-known hockey stick developed by alarmist Michael Mann. AR3 failed to discuss the statistical problems later shown by Steve McIntyre and Ross McKitrick and published on McIntyre's Website, ClimateAudit.com. By AR4, the hockey stick had disappeared without a trace, just like the upper tropospheric hotspot disappeared in the AR5 WG1 Draft.

EPA's Attempt to Substitute Peer Review for the Scientific Method

In USEPA's view the UN IPCC reports and other assessments based on it were so satisfactory an assessment of current climate science that no independent EPA analysis was necessary, primarily because of the IPCC's "rigorous" policy on peer review. EPA cites this review policy as the reason it accepts these reports rather than others, such as the NIPCC reports.[198] It found that there is a sharp contrast, as follows: [199]

> The [NIPCC] organization does not appear to have established any procedures for author selection and provides no evidence that a transparent and open public or expert review was conducted. Thus, the NIPCC's approach stands in sharp contrast to the clear, transparent, and open procedures of the IPCC, CCSP, USGCRP, and NRC."

---

[198] Idso and Singer, 2009.

[199] In Response 1-12 to the public comments on the EPA proposed Endangerment Finding, US Environmental Protection Agency, 2009.

Recent reports show that as actually carried out the UN IPCC AR4 assessment was much less than rigorous in the application of its peer review guidelines, however.[200]

Lost in this exchange, however, is whether the yardsticks being used by the UN and the EPA are reasonable. Both organizations appear to assume that peer review is the important characteristic of valid science included in scientific assessment reports. I disagree. The important attribute is whether the hypotheses considered do or do not correspond with observable reality according to the application of the SM.[201] It is only this crucial correspondence that determines the scientific validity of a hypothesis, not how many or how distinguished the reviewers may be who agree with a given hypotheses. So although there is some discussion of the arguments raised by the NIPCC reports, no real effort appears to have been made to consider using the NIPCC reports at least in part on the basis of whether the report had "adequate" peer-review guidelines.

According to the EPA, only the IPCC and similar reports including such peer review meet EPA's "exacting" review standards. How accurate or how closely the NIPCC and other skeptic reports correspond with real world evidence appears not to be of any real importance to the EPA—just how comprehensive the stated review process was supposed to be. Yet when deviations from these standards in the IPCC reports are detailed, EPA maintains that the IPCC conclusions would not have been materially affected rather than admitting that their expressed confidence in the UN procedures was misplaced. This is also an argument that the substantive scientific merits of the non-IPCC assessments do matter, but only when the procedural aspects have not been comprehensively implemented. The reverse should be the case.

---

[200] Laframboise, 2010; Knappenberger, 2010,
[201] Carlin, 2011.

The Purposes of Peer Review

The basis for the underlying argument is what is fundamental to the scientific method: Correspondence with real world data or procedural review requirements. In examining this issue it is useful to recall the history of scientific peer review. It was basically introduced so as to decide which papers submitted to printed journals should be included (primarily for the purpose of saving then precious journal space) and to decide whether the science is well presented and moderately sound. The first purpose may have actually been useful in the days when journals were of limited size based on their printing and mailing costs. Peer review subsequently served a third purpose—to provide a basis for discriminating between the output of various authors/professors and thus providing a basis for conferring academic tenure on some but not on others.

The second purpose involving a determination of whether the science is well presented and moderately sound may serve some limited purpose but also introduces the possibility for mischief such as has occurred in climate science where a small group of scientists used it to prevent views contrary to theirs from being published in major journals (as shown by the Climategate Emails). But the first purpose is technologically obsolete since Web publication of added papers is very low cost and may be almost free. Use of Web-based journals has the added advantage that they are normally free to all users rather than limited to the select few who can afford often very expensive subscriptions. Peer review of papers for journal publication has some very important disadvantages, of which the most important is that it is sometimes used to prevent publication of non-conventional ideas that may have great merit.

So the extension of journal-based peer review to determining the scientific merit of competing hypotheses is a very important policy issue since it may lead to reducing the importance of comparisons of competing scientific hypotheses against real world data. This is exactly what appears to have happened in the case of the CAGW hypothesis of global warming. In fact, alarmists have widely cited better peer review as an important reason to support their hypothesis; according to the Climategate emails, leading warmist scientists actively conspired to prevent skeptic-oriented papers from being published in major climate-related journals.

The Fundamental Issue: How Should Scientific Hypotheses Be Judged?

All this highlights the fundamental issue of whether scientific hypotheses should be judged on the basis of whether they have appeared in peer-reviewed journal publications or on the basis of correspondence with observed real world data. I believe very strongly that it is the latter rather than the former that should be used. One important reason is that peer-review is subject to the same "group think" that science should seek to avoid in order to be objective and useful. And that is exactly what has happened in the case of the CAGW hypothesis. Despite the absence of any relevant real world data comparisons[202] to support their case, warmists try to use the widespread support (the so-called "consensus") among sympathetic scientists for their hypothesis to argue that it should be accepted. There are enough global warming supporters among climate scientists so that with a little careful selection favorable peer reviews can be obtained for any desired alarmist hypothesis. Hence such views can pass the peer review standard whether a hypothesis really stands up to comparisons with real world data or not.

Obviously if this were the standard, we would still believe that the Earth is flat and that the Earth is the center of the universe, to mention just two widely supported hypotheses disproved by their lack of correspondence with real world data.

It is very unfortunate and may even prove disastrous that EPA and other environmental regulatory institutions appear to have made peer review procedures of much more importance than

---

[202] Carlin, 2011.

correspondence with real world data. Scientific assessments need to determine the correspondence between hypotheses on the basis of real world data, not relative "peer review" procedures. This needs to be corrected before immense damage is done to our crucial criteria for judging scientific hypotheses and to our economy as a result of using faulty science for public policy purposes.

Omitted Variable Problem / Fraud

The CAGW hypothesis not only fails SM tests. It also ignores alternative explanations that have vastly stronger associations (correlations) with global temperature records but this is generally not reported by the IPCC.[203] Instead, the IPCC claims that they cannot think of any other explanation for their alleged warming anomaly besides human activity. Perhaps they might run a few correlation analyses to get some ideas. Normally scientists run statistical tests and put their effort into analyzing the factors that appear to be most closely associated (correlated) with the variable of interest, in this case global temperatures. There are at least 25 published studies that suggest that solar variability or solar magnetic field variability are closely correlated with global temperatures.[204] There are also strong associations between global temperatures and ocean oscillations. Now correlation does not prove causation, but lack of correlation or low correlation strongly suggests lack of causation. The IPCC's main excuse for omitting total solar irradiance (TSI) is that it does not vary nearly enough to account for global temperature variability. This is true, but does not justify dropping solar variability or solar magnetic variability from consideration since there appear to be indirect effects that are much larger and there may be others that they do not yet understand. Svensmark's hypothesis, to be discussed later in this chapter, is one such indirect effect.

---

[203] See Appendix C.1.
[204] Rawls, 2012

If the IPCC had actually carried out a correlation analysis and failed to include a solar or solar magnetic variable or an ocean oscillation variable, they would be guilty of what is called an omitted variable problem or, if done on purpose (which I suspect it has been) fraud, because it omits the most important variables affecting climate change. An omitted variable problem/fraud occurs when an explanatory variable is omitted from a statistical analysis and its explanatory power is misdirected to another variable that is correlated with it.

Three variables, solar or solar magnetic activity, ocean oscillations, and atmospheric CO_2 on the other, are correlated with temperatures since the Little Ice Age (LIA). By leaving out the solar/solar magnetic and ocean oscillation variables, their strong explanatory power is shifted to the atmospheric CO_2 variable. This omitted variable problem is examined in almost every introductory statistics class that has ever been taught, but seems to elude the IPCC to this date, which suggests fraud rather than a problem. This highlights the fundamental issue concerning warming in the 20th Century: Are temperature changes due to increases in atmospheric CO_2 levels or in changes in solar/solar magnetic activity or ocean oscillations?

One obvious question is why the IPCC does not present a formal statistical analysis of the variables affecting climate change using data from a wide variety of time periods? Simple such analyses show that the strongest correlations are with ocean oscillations and solar activity. The correlation with CO_2 concentrations is weak in the years before 1998 and negative since then.[205] Over longer periods the relationships would probably be even worse. From a commonsense viewpoint, the question is whether the large variations in Earth's climate are due to changes in a trace gas whose greenhouse effects reduce logarithmically as concentrations rise or in the source of our heat and light, the sun, or oscillations long observed in the oceans that take up most of our planet? The fact that

---

[205] See Appendix C.1.

the IPCC ignores these omitted variables is not likely to be accidental in my view.

Even if the IPCC may not be technically guilty of omitted variable problem/fraud, Al Gore and everyone else who has compared CO_2 changes to global temperatures without mentioning that solar/solar magnetic variability or ocean oscillations both have higher correlations is guilty. Both solar variability and ocean oscillations will be discussed later in this Chapter and in Appendix C.2.

Three Major Assumptions Underlying CAGW "Science"

But there are still more problems with IPCC science beyond not asking the relevant question, being primarily a political rather than a scientific organization, ignoring the scientific method, and possibly having omitted variable problems/fraud. The first is that they assume that increases in atmospheric CO_2 due to human-caused emissions are having an important greenhouse effect on temperatures. The second is that they assume elaborate climate computer models (called atmospheric General Circulation Models or GCMs) can take into account all the natural factors affecting climate well enough to be able to determine that any remaining anomaly can be attributed to human actions. They then use this assumption as the basis for seriously proposing altering the world's energy generation and use systems at enormous cost to the world's taxpayers and ratepayers so as to allegedly prevent this anomaly from becoming catastrophic, which they believe it will. The third is that CO_2 levels determine global temperatures, not vice versa. Since neither of the first two assumptions is valid, and the third is doubtful, CAGW "science" starts out with three more major strikes against it.

Assumption: CO_2 Is an Atmospheric Thermostat

The first assumption, that minor changes in the current atmospheric CO_2 levels have a major greenhouse effect, is far from credible. The

251

general idea is that CO_2 in the atmosphere operates much like a thermostat in your house. According to the IPCC's hypothesis, if atmospheric CO_2 levels are increased by human-caused emissions, this results in higher levels of water vapor in the atmosphere and the higher levels of both CO_2 and water vapor have a greenhouse effect which decreases outgoing radiation from the Earth and thus results in higher heat retention and higher temperatures here on Earth.

One of the problems with this hypothesis is that CO_2 constitutes 0.04 (4/100ths) of 1% of the atmosphere and has less than 0.1% (one-tenth of 1%) of the warming effect of all atmospheric gases. This compares with the major GHG, water vapor, which is a much stronger GHG and much more common and contributes over 90% of the greenhouse effect. Even worse for the CO_2 greenhouse effect is the fact that the greenhouse effects of increasing CO_2 decrease rapidly with increasing CO_2 concentration, so that most of the CO_2 greenhouse effect occurs at much lower than current levels and only a very minuscule amount at current levels. This can be illustrated by several charts (see Figures C-9 and C-10). Finally it is important to realize that human-caused emissions of CO_2 are just a minor part of overall CO_2 emissions into the atmosphere. And there are two major CO_2 sinks removing CO_2—plants and oceans. The CIC would like people to think that all human caused CO_2 emissions stay in the atmosphere for long periods of time, but this nonsense. The higher atmospheric CO_2 goes, the more plants and oceans absorb.

So CO_2 is a minor GHG. It also happens to be essential for life on Earth and has been at far higher levels when life on Earth was developing. So although CO_2 may be the best human-influenced "culprit" the IPCC has been able to find, it is a remarkably unlikely choice as the predominant factor influencing global temperatures given its very minor and steadily decreasing marginal greenhouse effect at current concentrations. Like most climate skeptics, I agree that increasing CO_2 levels may slightly increase temperatures over what they would otherwise have been (although even this is far from certain), but strongly disagree that such increases will be catastrophic. Catastrophic increases depend crucially on feedback effects alleged by alarmists of changes in atmospheric CO_2 levels on water vapor concentrations, and are discussed in more detail in Appendix C.1.

The attempt by the Obama Administration to call CO_2 "carbon pollution" may be good propaganda on their part but has no relationship to reality since CO_2 is not a pollutant. It is a very small but absolutely vital component of the atmosphere that makes possible life on Earth, stimulates plant growth, and has only minor effects on temperatures at current or likely future atmospheric CO_2 levels.

Assumption: Global Climate Models Are Useful for Predicting Future Climate Changes

The second assumption, that GCMs built by alarmists accurately model past and future natural climate variability, is even weaker. It presumably came about because one of the two parent organizations of the IPCC, the World Meteorological Organization, composed of the national meteorological governmental organizations of member countries, apparently believe that the models these organizations use to predict weather could be adapted to predicting climate change. The meteorological models are able to make useful predictions up to about 10 days in advance, but not much beyond that. In fact, the UN IPCC AR3 actually made the following admission:[206]

> In climate research and modeling, we should recognize that we are dealing with a coupled non-linear chaotic system, and therefore that long-term prediction of future climate states is not possible.

---

[206] United Nations Intergovernmental Panel on Climate Change, 2001, Section 14.2.2.2.

So the IPCC agrees that climate is a "coupled, non-linear chaotic system" and "therefore that long term prediction of future climate states is not possible." I regard this official statement by the IPCC as devastating but entirely appropriate. The climate system is chaotic and multivariate. So although climate is deterministic it is not determinable.[207]

What this means in plain English is that the US Government has wasted tens of billions of dollars building climate models that cannot begin to approximate the real climate and then compounded the error by asserting that the model results prove the necessity to spend trillions more remaking the energy economy of the US and the world. The IPCC makes similar claims. The models prove nothing of the sort and appear unlikely to be able to approximate the actual climate even with decades of further expensive efforts at the taxpayers' expense. But there is much much more leading to similar conclusions concerning the knowledge that IPPC scientists had of the inadequacies of the models they use.[208] So why does anyone place any confidence in their "projections" (as they correctly insist on calling their model outputs despite their use as predictions when it comes to making policy recommendations) based on such models? Without the invalid model results the entire GWD ideology collapses in terms of having even the appearance of a scientific basis.

Yet the CIC in general and the IPCC in particular claim that they can forecast global temperatures for decades or even centuries in advance. The models simply show the results of making a series of assumptions by their authors as to how the physical parameters influencing climate interact. The models have consistently greatly overestimated global temperatures over the life of the IPCC.[209] Yet the IPCC has the audacity to claim that these models show ever increasing certainty as to the human impacts on climate change at

---

[207] Brown, 2014.
[208] Ball, 2014. See also Caprara, 2014, and Coyoteblog.com, 2014.
[209] See Monckton, et al., 2015, for an explanation as to why they run hot.

the same time that there is an ever increasing divergence between the models and real world temperature measurements (see Figure C-7).

One of the many assumptions in their models is that increases in CO_2 increase upper troposphere water vapor, as just discussed. Measurements, however, show that upper troposphere water vapor has decreased (thus allowing more heat to escape into space) as CO_2 has increased.[210] So not only is the greenhouse effect of increased CO_2 minor but the effect on water vapor appears to be negative, not positive as the IPCC claims. The effect of the CO_2 "thermostat" is so small that it can and should be ignored. Yet the CIC claims just the opposite and argues that humans must change their pattern of energy use on the basis of model assumptions that do not correspond with real world observations. And when skeptics point out this and other inconsistencies between the models and real world measurements they are told that CAGW represents "settled science" and that the skeptics are in the pay of "evil" polluters. How much more scandalous can a scientific scandal get? The essence of science is comparisons between hypotheses and physical reality, and has nothing to do with "consensus."

One of the many other weaknesses of the models is that they do not represent cloud physics well,[211] particularly in the case of cumulus clouds.[212] In general, changes in cloud cover have a much greater influence on the warming effects of the sun compared to changes in CO_2. Models that cannot deal accurately with the major effects of changes in clouds cannot usefully say much about the comparatively minor effects of CO_2. One particularly important determinant of cloud cover is the concentration of aerosols. One recent paper argues that the IPCC underestimates global cooling

---

[210] Gregory, 2013.
[211] For two recent publications on this issue, see Chen *et al.*, 2013, and Koren *et al.*, 2014.
[212] Koren et al, 2014.

from man-made aerosols/clouds by a factor of 27 times.[213] Until this problem is solved, model results should not be used in climate policymaking,[214] and probably not when it has been resolved either because of their many other problems.

Anyone can verify the strong reflective effects of clouds by standing outside when low clouds suddenly obscure the sun. Changes in CO_2 levels, on the other hand, require sophisticated instruments and cannot be directly observed without them.

Leaving aside the overwhelming case that CAGW does not satisfy the scientific method, it can be argued that the CIC's use of models essentially amounts to scientists trying to tell nature how it should work rather than trying to tease out of nature by patient experiments how nature actually works. The GCM models assume that global temperatures are primarily determined by CO_2 levels because that is what the builders believe is the case. But it is not the case.

The question of whether numerical models can add to scientific knowledge dates back to Greek and medieval science, as described by Ken Haapala as follows:[215]

> [This ancient science] embraced concepts such as the earth centered solar system of Aristotle, the Ptolemy scientists, and their models.
>
> Contrary to popular belief, initially, the intense criticism of Galileo came from his colleagues, Aristotelian scientists at the University Padua, and elsewhere. The criticism by the leadership of the Catholic Church came later. Galileo advocated that scientific knowledge comes through observation, not authority or convenient models. He outraged his colleagues by empirically testing the assumptions of Aristotelian science and found them wanting.

---

[213] Ibid.
[214] Hockey Schtick, 2014.
[215] Haapala, 2011.

Lessons Learned about Climate Science

He advocated that mathematics, because it is precise, is the proper language of science, not Latin, and he wrote in the vulgar script of Italian.

Newton synthesized the findings of Galileo, Kepler, and others into his theories of planetary motion, which required significant mathematical calculations. The precision of the mathematics in describing the movements of the planets was daunting. Incorrectly interpreting this mathematical precision as imparting scientific knowledge, many subsequent scientists claimed that scientific knowledge could be acquired by mathematics alone, such as geometry. Of course, subsequent observations, variations in observed planetary motions (leading to relativity theory) and erratic behavior of particles (leading to quantum mechanics), demonstrated that mathematics and numerical models are tools for understanding, and not sources of knowledge. Results must be confirmed by observations. Unfortunately, many 21st century scientists...appear to embrace beliefs held in the late 19th century that mathematical procedures can create knowledge. Such scientists appear ready to accept the results of their models over empirical observations, a concept rejected in the 20th century.

This recent, numerical-model based science contradicts the principles of modern empirical science - that empirical observations are the final determinants of validity of scientific theory and models. For example, the strength of the standard model in physics comes not from its mathematical complexity, but from repeated and exhaustive testing of its assumptions and high precision between the results of the model and observations (experiments).

So, in brief, what the GWD/GEF advocates have characterized as "settled science" is many centuries out of date. As of 2013 the CIC has used over $39 billion of taxpayer dollars to produce medieval science, which they attempt to sell on the basis of "consensus" opinion in science—exactly the same problem faced by Galileo in arguing his case against his academic brethren.

Similarly, the arguments made by the CIC to defend their case have been around for many centuries. Many if not most of them were described by Aristotle[216] over 2,300 years ago as logical fallacies. So what we have is medieval science defended by logical fallacies.

An illustration of their fallacious thinking is the following. GWD followers have long used the consensus or headcount fallacy identified by Aristotle. Just because it is alleged that many people say they believe a thing to be so, that is no evidence that many people say it, still less that they believe it, still less that it is so. The mere fact of a consensus—if there were one—tells us nothing whatsoever about whether the proposition to which the consensus supposedly agrees is true or false. Usually, the CIC cites climate scientists or experts as the basis for the consensus viewpoint. This is still another of Aristotle's fallacies: the reputation or appeal-to-authority fallacy. Merely because a group has a reputation or a particular expertise, it may not deserve it; even if it deserves it, it may not be acting in accordance with it; and, even if it is, it may be wrong

Science is not like politics. There is a "correct" answer—the answer which corresponds to all the relevant observational evidence. The process of determining this answer is called the scientific method. Unfortunately, many of the relevant American scientific societies, the Obama Administration (including the Environmental Protection Agency and the President himself), and the National Academy of Sciences have used a different standard but have tried to characterize it as "science." So what we have is the best medieval science that the taxpayer's money can buy. The American public deserves and must insist on better since neither the Obama Administration nor the American science establishment will.

---

[216] Aristotle, 2007.

Unfortunately, as in Galileo's time, this will not be easy to bring about given the magnitude of the forces arrayed against it, which represent the powers that be in both government and science.

As in Galileo's case, it really does not matter who the proponents of CAGW "science" are or what their backgrounds or status may be; the only relevant information is that provided by application of the scientific method. Just because President Obama has announced his wholehearted support for CAGW science does not carry any scientific weight even though it almost certainly means many billions more wasted on climate modelers, windmill builders, and solar installers. It does not require a PhD in climatology to reach these conclusions. All it requires is relevant data and application of the scientific method. Unfortunately, the CIC refuses to discuss the scientific issues raised by skeptics and claims that all their opponents are either unqualified or in the pay of special interests or both; furthermore, the public has been badly served by the advocacy of the Obama Administration for the GWD/GEF as well as the mass media in their reporting of and strong support for them.

The broader picture is that some less than well-known academics seized an opportunity to make themselves media stars, squelch any other academics who dared to oppose their ideas as equivalent to holocaust deniers or in the pay of polluters or unqualified or some combination of these, denied their opposition the opportunity to publish in major science journals, subverted the peer review process, tried to substitute "consensus" for the scientific method, withheld or "lost" their research data as a means of subverting freedom of information laws, became advocates rather than scientists, were richly rewarded with research grants which they made sure were not available to the heretics, toured the world to speak at CAGW conferences at various exotic locations, and even encouraged pseudo-psychological papers smearing their opponents. How do I know this: Climategate, news reports over many years, and my own experience.

I believe that this unfortunate state of affairs developed in part because of the self-interest of these climate-related academics who understood that the easiest way to get their important government funding and public attention was to support the GWD/GEF cause. This was the case even before the Obama Administration but is even more the case now since the last thing the Obama Administration is likely to do is to fund climate skeptics, who the President has publicly characterized as members of a "Flat Earth Society" and the Secretary of the Interior has stated are not welcome in her Department.

The George W. Bush Administration was actually very skeptical of GWD but avoided ample opportunities to question CAGW "science," which they allowed to flourish despite a prohibition on implementing the GWD "solution." This was the lesson from the abortive EPA Air Office draft TSD for an Endangerment Finding. This effort could not be tolerated in the Administration's view and had to be squelched, which it was, but very quietly. So the growing CAGW cancer was left to the next Administration to deal with—which they did by totally accepting both the "science" and the proposed "solution" of the GWD and even enlarging it to include GEF and EWD. All this might have been avoided or at least blunted by encouraging a public debate on CAGW, but this never occurred during the George W. Bush Administration. They made life easier for themselves, but emboldened and infuriated the alarmists and laid the groundwork for the Obama Administration's adoption of GWD/GEF.

The major underlying problem in my view is the availability of government research funds for only one side of the scientific debate. Two solutions are to end all government research support for climate issues or to guarantee that all sides in a scientific debate receive relatively equal funding.

Assumption: Global Temperatures Are Primarily Determined by CO_2 Levels

The third major crucial assumption made by the IPCC is that global temperatures are primarily determined by CO_2 levels. Another possibility is that CO_2 levels are determined partly or predominantly by temperatures. Obviously if this is the case, changing CO_2 emissions will have little effect on temperatures. New research suggests that it is global temperatures that primarily determine CO_2 levels. One piece of evidence is that at the end of ice ages temperatures have risen before CO_2 levels, thus implicating temperature changes as the cause of CO_2 changes. There is also evidence that changes in temperatures occur prior to changes in CO_2 levels in the very recent past.[217] Events in the future (CO_2 changes) cannot influence events in past (temperature changes).

In more recent research Murry Salby reports that he has been able to explain recent changes in CO_2 levels on the basis of temperature changes alone.[218] The reason may be that water absorbs less CO_2 at higher temperatures. Hence if temperatures rise, more CO_2 will leave Earth's vast oceans and increase the concentration of CO_2 in the atmosphere and vice versa. Thus while CAGW supporters claim that humans are to "blame" for atmospheric CO_2 increases and hence for alleged global warming, this new research implies that temperature changes are caused largely by other effects and that it is the temperature changes themselves that primarily determine CO_2 levels. Temperatures have been trending up since the end of the Little Ice Age in the early 1800s, so under this hypothesis it would appear that CO_2 levels should have too.

The alarmists have assumed that increases in CO_2 since the Industrial Revolution are due to human-caused fossil fuel use, and that this justifies their ultra-expensive proposed curbs on fossil fuel use. Their models all assume that global temperatures track predicted CO_2 levels. Pehr Bjornbom, a Swedish climate scientist says that he has been able to replicate Salby's conclusion that

---

[217] MacCrae, undated.

[218] Hockey Schtick, 2013a. Note that Salby's work had not been published as of the time this book was written, so may be subject to revision when and if it is. See also Darwall (2014a) for a broader, more historical account of Salby's research.

temperature drives CO_2.[219] But Salby's conclusions are far from a proven hypothesis at this point. If his new research holds up to further scrutiny (it has only been available in presentation format for about 18 months as of late 2014), it would leave the CAGW hypothesis and the GWD "solution" without even a fig leaf. In other words, the alarmists and the IPCC may have gotten their fundamental assumed relationship between temperature and CO_2 backwards. If the science is wrong, policy based on it will also be wrong. Salby's evidence for temperatures largely controlling net CO_2 emissions is outlined in Appendix C.1.

One of the key technical factors in climate policy is the climate sensitivity to changes in CO_2 (technically the change in global temperatures for a doubling of atmospheric CO_2 levels). This is what the AR5 Working Group 1 report says about it:[220]

> *The assessed literature suggests that the range of climate sensitivities and transient responses covered by CMIP3/5 cannot be narrowed significantly by constraining the models with observations of the mean climate and variability, consistent with the difficulty of constraining the cloud feedbacks from observations.*

My interpretation is that the authors are saying they have no idea what climate sensitivity is. This is reinforced by their decision to expand their range of uncertainty for this critical parameter from that shown in their previous AR4 report while at the same time increasing their alleged certainty that climate abnormalities are due to human activities. The upper bound on climate sensitivity appears to have greatly decreased based on new 2015 research to the point that climate alarmism can no longer be justified even using the IPCC approach to doing so.[221]

---

[219] Hockey Schtick, 2013b.
[220] Section 9.7.3.3
[221] See discussion near the end of Appendix C.1.

Lessons Learned about Climate Science

Two of the important scientific questions posed by CAGW are to establish whether weather/climate events have been abnormal and whether any abnormalities are caused by human activities rather than natural effects. This is a very difficult thing to do even under the best of circumstances because of the great variability of both weather and climate, the limited data available for periods more than 30 years ago and especially more than a few hundred years ago, and the substantial changes in climate exhibited by the Earth over millions and billions of years. The GWD/GEF supporters have taken the unlikely approach of describing what they believe to be abnormalities and their human causes without examining the range of variation from the historical record. Yet they have the audacity to then claim their 90/95% certainty that the alleged abnormalities are due to human activities because they claim that they cannot think of any other explanation for minor temperature increases in the late 20th Century. They do this even though they admit that many important factors influencing climate change are poorly understood. So how can they be 90/95% certain that warming is due to one possible factor when many others are poorly understood? They cannot.

One should not be surprised if this approach has led to controversy. It originated with the basic charge of the IPCC, which was to examine only the effects of human activities on climate, and ended with the user governments dictating the outcome. This has naturally led to numerous twists and turns to try to maintain the favored hypothesis against increasingly non-supportativemi evidence from the real world.

Given that the IPCC did not test their CAGW hypothesis using the SM, a more logical and much more believable approach would have been to first establish what natural variations in global temperatures have been and if possible the reasons for these variations, then discuss the general trends in climate over the relevant past, then examine against this background whether recent changes have been abnormal, and only then look for human or other causes of these possible abnormalities. Their limitation to the human causes means that the IPCC has only examined a minor portion of the problem and effectively presumed the answer before they started. Until this basic flaw is corrected, the IPCC's reports should be regarded as gathering evidence on one side of an issue but not a serious assessment of the basic scientific issues or as the basis for any action. Since EPA's TSD and Endangerment Finding are based directly or indirectly on the IPCC reports, the same comments apply to it.

In general, the IPCC's case is remarkably weak for the tens of billions of taxpayer dollars that has been spent on it and the underlying research on which it is based. But then after five tries with little or no demonstrable improvements in scientific certainty, it seems more likely that they have made the best case they could for their assigned hypothesis and that building a solid case was impossible to begin with since the CAGW hypothesis is invalid. The outcome was pre-determined by their charter; their job was to fill in the details as best they could, but it was an impossible assignment. The unfortunate thing is that anyone believed their conclusions, but many believers maintain that they do.

So even in the very unlikely case that CAGW should later be proved correct in its IPCC formulation, I would argue that wisdom is on the side of currently doing nothing but gathering and hopefully objectively analyzing much more comprehensive information by a much more neutral organization than the IPCC and waiting to take action when and if it should be proved that action is needed and that something useful could be done for a cost that would make it worthwhile compared to adaptation or other control measures (all highly unlikely). As will be argued at the end of the Chapter, there appear to be vastly more effective and efficient means for solving climate change problems than that proposed by GWD/GEF proponents. Fortunately, these better means could be implemented very rapidly once we work them out and are sure that something is actually needed, unlike GWD/GEF, which requires detailed advance knowledge decades in advance that we simply do not have.

So there is happily no need for current action to reduce CO_2

emissions. The GWD/GEF proposal for "corrective" action is even weaker than their analysis of the need for it. It is doubtful that decreasing CO_2 emissions will have much effect on the climate. And it is quite clear since the failed Copenhagen conference, there is virtually no chance that the nations of the world can agree on effective, enforceable international agreements remotely like what the GWD/GEF proponents have proposed or that their "solution" would achieve what its proponents claim even if they did. GWD/GEF actions by individual countries would be useless and hurt the countries involved by reducing their competitive advantage compared to other countries that did not undertake such actions.

Natural Climate Variability

It is vital that global climate change be considered in the context of what is known about climate variability in the geological record. Only in this way can a judgment be reached as to whether the IPCC's conclusions concerning how unprecedented Earth's climate has been in the 20th Century is reasonable. In the 4.5 billion years since Earth's creation, there have been huge and continuing changes. CO_2 levels and temperatures have changed greatly with little apparent relationship. As Patrick Moore has testified, "the fact that [Earth] had both higher temperatures and an ice age at a time when CO_2 emissions were 10 times higher than they are today fundamentally contradicts the certainty that human-caused CO_2 emissions are the main cause of global warming."[222] Even in the last 3 million years or so climate has undergone enormous changes of huge importance to life on Earth and until the last 90 years or so these changes have clearly not been significantly influenced by human activities.

The broad picture is that over the last 3 million years the Earth has varied cyclically between long ice house periods called ice ages and much shorter, warmer climate like today, called interglacial

---

[222] Moore, 2014.

periods. The broad trend during this period has been towards ever colder temperatures. For the first 2 million years of this period, the ice age cycles were about 41,000 years each. Over the last million years, however, global temperatures have exhibited about 100,000 year cycles and shown downward instability about 10,000 years after the start of each interglacial periods and upward instability about 90,000 years after the start of each ice age.

A fascinating question is why this repeated pattern exists? Why are the highest average global temperatures fairly stable on the upside during interglacial periods? Why do not temperatures just continue ever higher as the CAGW models claim they will as CO_2 levels increase? Why are they comparatively stable on the downside after an ice age takes hold? What causes the upward and downward instability between these two periods? Until we understand such basic questions we may not be able learn how climate works and certainly not control it.

The relative stability of interglacial periods on the upside is discussed in the next subsection. Little seems to be known why there appears to be a floor on the downside of an ice age. Much more appears to be understood concerning the instabilities between interglacial periods and the ice ages. The best evidence appears to be that Earth's temperatures have been low enough during the last 3 million years so that relatively minor changes in the incidence of sunlight on Arctic land areas caused by known variations in the Earth's orbital parameters appear to have made the difference as to whether old ice survives the annual Arctic summer melts and therefore gradually builds into continental glaciers.

If this critical annual melting does not occur for long enough a continental glacier forms, which results in a positive feedback in terms of even colder temperatures in part because of the ice albedo effect caused by the fact that ice and snow reflect more sunlight than the earth and water surfaces that they replaced. With further build-up, the continental glaciers begin to expand and move southward, particularly over what is now Canada and the Northern US, resulting in further positive reinforcement from the ice albedo effect. As they move south, the sunlight striking the Earth in the newly covered areas becomes stronger and even more is reflected away from the Earth's surface. It is not until sunlight on the far northern land area becomes strong enough to melt more ice than forms each year on a consistent basis that the ice albedo effect goes into reverse, the glaciers finally recede, the ice age comes to an end, and the next interglacial period starts.

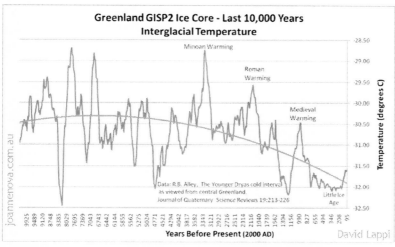

Figure 9-1: Holocene Temperatures from GISP Ice Cores at Top of Greenland Ice Sheet[223]

Although not exactly the same in each ice age cycle, the ice age and interglacial temperatures for each cycle have been somewhat similar. The interglacial periods have been characterized by a temperature peak followed by irregular decreases until there is a drop into the next ice age, with temperatures 6 to 8 degrees C lower. Some researchers believe that the drop into new ice ages has been quite sudden and takes place in a decade or two. Our current

---

[223] Nova, 2010, which in turn is based on data from Alley, 2000, except for the upward line at the very end, which was added by David Lappi because the Greenland database comes to an end in 1855.

interglacial period, called the Holocene, has so far exactly followed this pattern and started about 11,000 years ago, so it is likely to be nearing its end and shows it (see Figure 9-1).

Although the alarmists argue that all this has been changed by their rediscovery of the long-known greenhouse effect, there is currently no reason to think that this pattern has changed, and certainly no reason to fear the outcome if it has recently changed in the warming direction as GWD adherents claim. What we do need to fear is a new Ice Age or even a new Little Ice Age such as Earth experienced in the latter part of the last Millennium, not the possibility of avoiding one. So there is currently no reason to become concerned about adverse abnormalities in climate involving increases in temperatures at this point in the highly unlikely event that they should occur. Major such increases from the latter stages of an interglacial period have not occurred over the last five glacial cycles and are highly unlikely now. GWD proponents argue that increased greenhouse gas emissions as a result of human activities will result in alarming increases in global temperatures, but have so far not found any scientifically valid basis for this view.

If there were reasons to think that global temperatures were going to increase above the Minoan high or even the previous Medieval Warming high (about 1,000 years ago) at this point in the cycle this would indeed be abnormal and worthy of note, but this has not happened and there is no indication that it will. And if either of these events should happen we should welcome them since it might indicate that the Earth might be getting out of its current ice age cycles, which would certainly be the most welcome environmental development in almost three million years. What appears to be happening instead is that the continuing series of lower highs since the Minoan high is continuing (see Figure 9-1), just as it has in every previous ice age over the last million years for which we have data. Why anyone should fear a possible end to one of the worst environmental disasters imaginable—a new ice age—is beyond comprehension and suggests that those doing so have no comprehension of recent geologic history or that they deliberately want to ignore the lessons of this history. The climate models used by the CIC account for none of this. Climate oscillates; the models do not. The models show no Little Ice Age (LIA) nor do they explain why it happened.

So Is There an Abnormality?

Global temperatures have been irregularly declining for at least 3,000 years based on Greenland ice core data similarly to what has occurred near the end of previous interglacial periods. The increases during the Twentieth Century have been well within normal bounds over the last 800,000 years for which we have ice core data from Antarctica. What is worrisome is not these minor and exaggerated increases highlighted by Gore and other alarmists but rather the LIA beginning about 1350 or so and not ending until the early 1800s. The increases since then can reasonably be viewed as a normal rebound after the LIA, and without them we might already have started the descent into a new Ice Age. We should welcome every temperature increase that comes along, not fear it or propose changing the energy-using basis for modern society in a fruitless effort to avoid them.

The key question is what fraction of the observed impacts of climate change the IPCC identifies is human-caused and how much is natural. After examining seven of the more relevant and intelligible IPCC claims, Andrews found that "there is no good evidence linking human activities to any of the observed impacts of climate change listed in the Summary for Policymakers of the WG2 report."[224] So there is no good evidence of a human-caused abnormality—just a continuation of the historical record over the last almost million years for which we have ice core data.

You may be surprised to learn that I sincerely hope that the CAGW hypothesis concerning large increases in global temperatures were correct since such an increase would suggest

---

[224] Andrews, 2014.

that Earth may miraculously escape the next ice age—and all humans need to do to bring this about is to continue emitting CO_2 in a business as usual fashion. Were it only so! Unfortunately, the reality is that this wonderful and much to be desired outcome is highly unlikely, and we need to assume the worst anyway—descent into a new ice age sometime in the next few thousand years, and possibly as early as the next few centuries.

The "Solution"

If there is no adverse abnormality, there is no need for a "solution" for the effects of human activities on climate/extreme weather changes. Although the "problem" to be averted keeps being redefined, the GWD/GEF "solution" never changes and is for humans to cooperate in reducing their CO_2 emissions through internationally-agreed regulation of CO_2 emissions. Such cooperation is clearly not going to happen in an effective way under current circumstances even if there were actually a need for it. Even in the unlikely event that it did, the result would be a catastrophic reduction in energy use taking civilization back to energy use levels last seen a century or more ago if the 80% decreases usually advocated by environmental groups and the failed 2009 US cap and trade bill are taken seriously.

If less developed countries did not undertake similar reductions (a most probable circumstance), there would be no way to achieve the proposed reductions worldwide by reductions by developed countries alone. US and world economic production would be greatly reduced as a result because of the world economy's dependence on reliable, inexpensive energy. This would take years of sacrifice and in the end would never happen in the US or other democratic countries where the losers could and almost certainly would object politically and sooner or later vote out of office politicians who attempted to carry it out. The primary victims would be the poor in both developed and less developed countries, who would bear most of the sacrifices. If decreases were made only in Western energy use, which is the best that alarmists

270

can hope realistically for, this decrease will be more than offset by the rapidly rising energy use in developing countries, particularly China and India. Even if Western taxpayers were to agree to large income transfers to the developing countries to compensate them for the West's alleged sins in emitting CO_2 in past years and to enable them to adapt to warmer temperatures, as the UN has proposed, the developing countries have not and in my view will never agree to ending their plans for improving their economies and their peoples' standards of living—nor should they.

In brief, the GWD/GEF "solution" is a fantasy that will not and should not happen.

Reality

It has been long recognized that weather is chaotic. Systems are chaotic when future behavior is fully determined by the initial conditions, but the systems are not predictable. In the case of weather predictions, they are generally regarded as having some accuracy for at most a week or two. Weather certainly cannot be predicted for years or decades or centuries in the future using the large computer models tried so far. Nevertheless, climate modelers claim that they can predict temperatures for decades and centuries in the future using very large computer models using basic physical relationships determined by the authors but leaving out major players such as the sun, the sun's magnetic fields, cosmic rays, the Earth's location in the universe, and important emergent phenomena (to be discussed later in this chapter), and poorly handling clouds.

This is nonsense, especially now that the basic assumptions concerning the causation of changes in CO_2 are in doubt. Those models used by the IPCC have proved to be overly alarmist compared to reality over the life of the IPCC. Even the temperature data used by the alarmists is very questionable. Since the CIC generally controls the surface temperature measurements and have increasingly taken to "adjusting" the data in such a way that it always seems to better support their case, it is now doubtful what it means.

The satellite temperature data only started in 1979 but appears to be very solid. It shows that the only major temperature increase during this period occurred in 1998 during a "super El Nino," and the permanent increase only amounted to $0.3°C$ (see Figure C-12). This one-time increase hardly appears to be anything that anyone should become concerned about. There is no reason to think that the current generation of climate models will improve in the future since their authors will likely still have the same exaggerated views as to the role of CO_2 emissions. So I question why US taxpayers continue to fund the IPCC and the model builders?

The alarmists' case is weak at best, and would be good news, not bad, if their AGW hypothesis actually results in modest temperature increases. A new ice age would particularly ravage the northern countries, including the northern US, based on previous ice ages. It would be the end of the world as humans have known it in the northern latitudes for over 10,000 years during the entire development of modern civilization. To argue, as alarmists do, that moderate warming temperatures due to the greenhouse effect would be a calamity that must be prevented is simply absurd.

In reality, climate is a very complicated system that humans have only a limited understanding of and has a great many major variables, not just CO_2 levels. A full understanding of these variables has eluded humans for hundreds of years; if anything, knowledge may have decreased in recent years because of the unwarranted emphasis placed on one relatively unimportant variable—CO_2—by many researchers. Until more researchers and politicians are willing to admit how little we know, little progress can be expected.

A splendid example of this is provided by the previously mentioned more than 17 year "hiatus" of global warming since at least 1998 on top of the much lower warming prior to 1998 than predicted by the IPCC. The word "hiatus" suggests that warming will resume, so is not a good choice since whether there will be a resumption of the minor warming from the 1970s to 1998 is unknown. Although some alarmist scientists are willing to admit that they do not understand the reasons for this "hiatus," most continue to act as if this is no reason to change their consistently wrong predictions. Even worse, they claim that they know the exact percentage of reductions in CO_2 emissions needed to achieve a desired reduction in global warming/climate change/extreme weather even though they have been wrong even about whether increasing CO_2 levels has resulted in global warming since at least 1998.

At the very least GWD/GEF supporters should be less dogmatic about how much they know about climate unless they want to appear even more ridiculous concerning energy/climate policy than they already are. The idea that they know how much of a decrease in CO_2 emissions will produce a specified change in global warming/climate change/extreme weather and that a global agreement can be reached to achieve that decrease are simply fantasies—and potentially very expensive ones at that. The alarmists argue that only model builders know enough to have an input into climate policy, even though the models they have built have been consistently too hot.

It would be far better, of course, if the alarmists recognized the obvious—that they have no scientific basis for their policy prescriptions—only an ideology/religion they are trying to sell on the basis of invalid science. Almost all the climate research funding has been for CAGW-supportive topics for several decades. A number of more promising ideas have been developed but generally ignored by the climate science establishment. A substantial part of the remainder of this chapter will briefly outline some of these research ideas. In some cases they are presented as ideas that are worthy of further research if funds were to be made available for more useful research. They also suggest how limited the IPCC analysis has been.

Numerous Indications of the Dubious Nature of CAGW

The average concerned citizen may have some difficulty sorting out the scientific arguments made by each side, although the consistent

273

gross overestimation of projected global temperatures by the IPCC during its entire existence and the at least 17 year pause in global warming should make even the most science-phobic citizens wonder. But it is much easier for such citizens to judge the tactics used by the CIC.

One of the major ironies of the whole CAGW story is that the major initial political support for CAGW was allegedly because Margaret Thatcher was having problems with the British coal miners and thought that CAGW might help her in this regard.[225] So the initial political support came from the far right of the political spectrum, and was apparently primarily politically motivated. In more recent years, however, GWD/GEF has now come to be consistently supported by the liberal, usually labor-supported parties, in many Western countries even though it is their supporters who are most likely to be adversely affected if GWD/GEF were implemented (to be discussed in Chapter 11). So perhaps the left also has a political motivation?

One of the most obvious indications of the dubious nature of GWD/GEF is the repeated attempts by the movement leaders to redefine what it is that they claim they are saving the world from while always keeping the "solution" unchanged (reduction of CO_2 emissions by government fiat), suggesting that it is tis and only this "solution" that is their real objective. For many years they claimed that global warming was occurring and would be catastrophic in its effects. Despite numerous attempts to retroactively change the basic ground-based temperature data, apparently to make it better fit their model-based CAGW hypothesis[226] (the alarmists largely control the data through their control of most national meteorological offices), there has been no significant global

---

[225] Personal conversation with Christopher Monckton, a former advisor to Margaret Thatcher.

[226] See, for example, Delingpole, 2014a; Delingpole and Eastwood, 2014; RealScience, 2014, Delingpole, 2015, Homewood, 2015, and Delingpole, 2009.

warming trend since at least 1998 using either ground or satellite-based data. For this or other reasons, they then claimed they were concerned that climate change, a much more nebulous and less easily determined phenomenon, was the problem they were attempting to solve. In 2012 they appeared to switch to extreme weather events, perhaps because several such events occurred during 2012 and it was becoming increasingly evident to the average person that there was no significant current climate change or global warming.

In 2013 there was an apparent change in rhetoric—saving the world from "carbon pollution"—to quote President Obama.[227] As discussed in Chapter 6, this has the feel of a propaganda effort to convince the public that the Administration's GWD/GEF campaign is simply an extension of EPA's traditional efforts to curb actual pollutants.

I find it curious that even though US carbon emissions fell 12% between 2005 and 2012 and were at their lowest level since 1994,[228] the CIC continues to advocate still more reductions despite no real evidence that the 12% has had any observable effects. A serious movement does not repeatedly and retroactively change the data and the goal posts, particularly when the previous goal posts are looking increasingly scientifically unsupportable, and the scientific basis of the new goal post has not been made and is even weaker than the previous one.

Rather than respond to the many and diverse scientific points made by skeptics concerning CAGW, the CIC has almost universally adopted the argument that there is a scientific "consensus" on CAGW and no one who holds a different viewpoint should be listened to. Those holding skeptical views are instead the subject of *ad hominem* attacks and epithets. In 2013-14 this characterized the response even by the President and other senior

---

[227] Obama, 2013.
[228] Gold, 2013. Based on estimates by the Energy Information Administration of the US Department of Energy.

US officials. This may explain the unusual emphasis among dubious CIC research efforts to show how large the consensus is alleged to be. This approach may have been effective in preventing skeptics from being heard or listened to but is unrelated to the scientific merits of the alarmist case.

If scientific problems are pointed out, the usual response is *ad hominem* smear attacks and epithets against the authors.[229] If this does not quiet the issue, the all too common CIC response is to deny rather than condemn an error they may have made. If that fails, an attempt is often made to justify what was done. If that fails, the final step is all too often to argue that the mistake was of no importance. In almost no cases have CIC supporters countered with serious technical responses to technical issues raised by skeptics. They have almost always adopted public relations responses, not scientific ones.

Like many ideologies/religions/cults, the CIC reserves its worst opprobrium for non-believers. People who switch from believers to non-believers or deviate only slightly from the approved ideology are particularly viciously attacked and often threatened. One recent example is Lennart Bengtsson who agreed to join the advisory board of a skeptic organization and then withdrew as a result of intense pressure such as withdrawal of co-authorships.[230] Another is Roger Pielke, Jr., who differs from the CIC primarily on EWD and was included in the 2015 post-Willie Soon attack[231] by some congressional Democrats.

Some of the reprehensible tactics employed by the CIC are to create and disseminate the myth/fabrication that skeptics should not

---

[229] A good recent example is the 2015 attack on Dr. Willie Soon, apparently prompted by his co-authorship of a scientific paper (Monckton, *et al.*, 2015). The attack started in the *New York Times* (Gillis and Schwartz, 2015), but quickly spread to many other CIC-oriented media outlets and even to some congressional Democrats (Goddard, 2015a).

[230] Robinson, 2014.

[231] Rust, 2015.

be listened to because they are the tools of and paid by environmentally irresponsible polluters, that skeptics are similar to holocaust deniers so that they should be universally called "deniers," and that skeptics should be compared to those that argued that the Earth is flat. Perhaps the worst of these is the repeated assertion that there is a quid pro quo between fossil fuel industry officials and skeptic scientists, for which no convincing evidence has ever been offered over more than 20 years.[232] Since CIC components have accepted numerous large grants from polluting industries, the CIC is on particularly weak grounds in their widespread insistence that such funding corrupts climate science and advocacy.[233] These tactics have worked to the CIC's advantage primarily because they control most of the mass media so are not subject to much investigative journalism that would reveal the blatant public relations reasons that the CIC has and still is[234] using these tactics.

The reality is that the skeptics are woefully underfunded (as discussed in Chapter 6), particularly compared to the CIC, and that the CIC is guilty of shamelessly using a term used in discussing the ghastly atrocities at Auschwitz and other Nazi extermination camps to score points in a science debate that they cannot win. The believers in a "flat Earth" are most appropriately compared to the CIC rather than to climate skeptics since in both cases they represented the establishment of the time. Those responsible for these reprehensible CIC tactics reflect badly on the CIC and its cause. The fact that the CIC engages in these tactics and avoids direct one-on-one discussions of the science should suggest to the non-scientifically inclined the extreme weakness of the GWD/GEF case.

Generally speaking, it has become common practice for the CIC to exaggerate confidence in its model results, seek to suppress the publication of inconsistent findings, misrepresent findings,

---

[232] Cook, 2014.
[233] Gosselin, 2015.
[234] See, for example, Gillis and Schwartz, 2015.

refuse to reveal methods and data, ignore opposing evidence, retroactively alter data, particularly temperature data, without strong or often any justification, attempt to discredit anyone who expresses views contrary to those held by alarmists, and make unsupported representations supporting CAGW and EWD which have no basis in science. If you have a weak scientific case, this may well be the most effective way to counter skeptics from a public relations standpoint, but should not raise much confidence among independent observers in those using these tactics.

Over many years the alarmists have made a wide variety of dire predictions concerning alleged climate catastrophes including substantially higher global temperatures, increased frequency of extreme weather events, sea level rise, ocean acidification, ice cap melting, etc., and denied the possibility that any related events could be due to natural variability. There is no good evidence linking human activities to any of the observed impacts of climate change,[235] but they have been unwilling to admit that they were wrong. In the meantime, observed reality has become increasingly separated from their predictions (see Figure C-7). All too often the alarmists have responded by increasing the level of alleged alarm and even to claiming that their predictions are even more certain to occur.

But there are many other suspicious things going on, some of which are summarized by the following questions posed by a climate skeptic[236] concerning CIC behavior:

- *Why is it that every one of the cockups and blunders we uncover in their papers always err towards a warmer global climate?*
- *Why do they persistently withhold the data on which their conclusions are based?*
- *Why do they, in their own words, hide behind Freedom*

---

[235] Andrews, 2014.
[236] ThePointman, 2013.

of Information laws, as a reason to keep such data hidden?

- *Why do they, in their own words again, hide behind Non-disclosure Agreements, as a reason to keep the data hidden?*

- *Why are they so vague about the exact methods used on the data to derive their results?*

- *Why do all their computer climate models run hot?*

- *Why have they consistently overestimated the climate's sensitivity to CO_2?*

- *Why don't they ever design experiments attempting to disprove their theories?*

- *Why do the Climategate emails reveal their deep private doubts about the science, which they've publically reassured everyone was settled?*

- *If the science was so solid, why'd one of their number feel they had to resort to identity theft to discredit the opposition?*

- *Why are they telling each other to delete emails to circumvent Freedom of Information requests?*

- *Why do they feel they've got to "redefine the peer review process" to prevent dissenting science papers being published?*

- *Why do they need to get science journal editors removed from their jobs because they dared to publish a dissenting paper?*

- *Why, after being the beneficiary of billions of dollars of research funding in the last two decades, haven't they by now proved their case beyond a reasonable doubt?*

- *Why is anyone who simply questions the science being equated with a holocaust denier?*

- *Why are they attempting to substitute science by consensus for scientific proof?*

- *Why do they say the world suffering six successive years of freezing winters is somehow caused by global warming?*

- *Why have global temperatures not risen in the best part of two decades while CO_2 levels have kept on rising?*

- *Why in the decade following 1990, were the number of ground temperature stations selected to calculate global temperature reduced from the available 14,000 to a mere 4,000?*

- *Why for Russia, with a land area over twice the size of the USA, are only a handful of southern ground-based thermometers selected to calculate its temperature?*

- *Why has the Moscow-based Institute of Economic Analysis (IEA) been saying for years that the average temperatures calculated for Russia are quite simply wrong?*

- *Why did it fall to a skeptic volunteer force organised by Anthony Watts to regrade the data integrity of the alarmist's few cherry picked temperature stations?*

- *Why is there an unexplained divergence between the global temperature derived from satellite observations and their ground based measurement?*

- *Why did it take them nearly fifteen years to finally concede that the global temperature had not risen in all that time?*

- *Why at the end of nearly every one of those fifteen years was it loudly proclaimed to be the hottest one on record?*

- *Why did the experts we're supposed to trust not feel the need to correct such scientifically inaccurate claims?*

- *Why, in the absence of the heat predicted by their theories, do they suddenly assert it must be by some mysterious mechanism hiding undiscovered at the bottom of the world's oceans?*

- *Why isn't the Argos network of ocean monitoring buoys showing any such warming?*

- *Why, in the absence of any global warming, did they switch the threat to climate change?*

- *Why do we see supposedly objective scientists acting like catastrophists, haranguing us to do as they say or we're all going to die?*

- *Why can't we get a straight answer to those simple questions rather than abuse, outright propaganda or simply being ignored?*

- *Why can't these supermen of miraculously settled science ever say they just don't know?*

An interesting added suspicious set of circumstances concerns the CIC's allegations that global warming will lead to ocean "acidification"[237] due to the absorption of increased CO_2 by the oceans. It appears that this concern was based in substantial part on modeling results rather than the relevant historical data by researchers at the National Oceanic and Atmospheric Administration (NOAA).[238] When the historical data is used, no decreased alkalinity is found.[239] Seems like a familiar story!

In recent years the CIC has become ever more blatant in making scientifically unjustified arguments in support of their campaign. Leighton Steward has compiled a few of them together with his comments on them:[240]

---

[237] There is no real possibility that the oceans will become acid. The CIC claim is apparently that they will become less alkaline, which alarmists try to call "acidification."

[238] Watts, 2014b.

[239] *Ibid.*

[240] Steward, 2014.

- *The rate and magnitude of recent warming is unprecedented.* This is absolutely false. Peer reviewed studies, including the journal Climate Dynamics, recently concluded that average global temperatures stopped warming 15 years ago. Looking farther back, there have been many periods of rapid warming before man's measurable release of CO_2.

- *The number and intensity of major hurricanes and tornadoes is rising.* The 2013 Atlantic hurricane season was the first Atlantic hurricane season since 1994 to end with no known major hurricanes. Data published by Florida State University indicates global cyclonic intensity has been trending down for 20 years.

- *Droughts and floods are more frequent and intense.* Again false. According to 106 peer reviewed global drought and 47 global flood studies, this is not true.

- *Forest fires and acreage destroyed have intensified.* The National Interagency Fire Center statistics of total wild land fires and acres destroyed from 1960-2012 concludes that there is no evidence to support this claim.

- *The rate of sea level rise is increasing.* Global statistics refute this claim. Sea level is continuing its rate of rising seven inches per century, unrelated to human contributions to global warming. There are some local areas where sea level is either rising or falling but no global increasing rate of sea level rise.

None of these questions and inaccuracies prove that GWD/GEF science is wrong, but they raise suspicions about the trustworthiness of the CIC on climate issues, including a number of US Government agencies. And if citizens accept the alarmists' ideology on the basis of the alarmists' trustworthiness, this appears to be misplaced. If they had a strong scientific case, the alarmists would be likely to use it to question the skeptics' science instead of calling them ugly names, questioning their integrity and qualifications, and claiming that scientific issues should be settled by consensus, which they repeatedly claim favors their case by an overwhelming percentage (often 97%) on the basis of flawed methodology (as discussed in Chapter 6).

Conspiracy, Fraud, False Alarm, Hoax, or Scam

There is little question that CAGW and EWD science do not constitute valid scientific hypotheses and that this has been or should have been evident for a number of years in the case of CAGW. The next question is whether the principal proponents of CAGW and EWD science knew this or whether they purposely pushed CAGW and EWD science knowing this or even whether they did so in order to make money. In the first case, their activity can be characterized as a false alarm. In the second, it would qualify as a hoax. In the third it would be a scam.

For all the reasons discussed in this Chapter, there is no doubt in my mind that their activity qualifies at least as a false alarm. But there is considerable evidence that it is a hoax as well. The major evidence is the Climategate emails, which were written by a group of climate researchers central to the CIC science enterprise. Instead of concentrating on the science they were supposedly studying, many of their emails concerned how to circumvent providing data requested by other researchers, keep skeptics from publishing in major journals, and doctor data to make it appear to support CAGW. They certainly did not sound as if they were primarily conscientious scientists doing their job of seeking scientific truth. When talking among themselves (they thought privately), the Climategate scientists sounded much more like partisans pushing their ideologies through every means at their disposal, regardless of its propriety, not pursuing scientific truth.

If they truly believed in their scientific findings, I see no rational reason why they would discuss any of these ways to circumvent the system. If skeptic researchers wanted the data, why not give it to them? If the data did not fit the hypothesis, why not change the hypothesis rather than hide the data? If others wanted to publish papers supporting a different hypothesis, why not encourage them and then develop their own papers presenting their hypotheses? At the very least Climategate clearly indicated that activities of the CRU and its supporters lacked scientific integrity. Much more likely is that they undertook their less than forthright actions because they did not really believe that their favored hypotheses could carry the day on their own merits. This is further suggested by repeated efforts by surface temperature data managers in various parts of the world to change temperature data in ways that made it better fit the GWD hypothesis. All this suggests that many alarmists did not really believe their hypotheses, which means they were engaged in a deliberate hoax. It also suggests scientific fraud since alarmist scientists have made repeated efforts to misrepresent their findings as supporting alarmist hypotheses rather than allowing the data to determine their hypotheses, as the SM requires.

I further suspect that many of the core CIC scientific proponents did not believe in their basic hypotheses because of the basic nature of the IPCC reports. Most scientists understand the scientific method and the role it plays in determining scientific validity. So why have they not used it rather than avoiding it? There is no other way to build a valid scientific case for their hypotheses. Hypotheses must be challenged and shown to be consistent with available data. Most scientists understand the use of statistical analysis to assist in determining the correlation between various hypothetical causes of an effect. So why have they largely avoided using it? There is only one likely answer: They knew what these techniques would yield and did not want the answers they would get using them. If so, this puts their activities in the hoax category. Further, most of the CIC's public relations tactics such as avoiding discussions of CAGW with skeptics appear to be most easily explainable as an attempt to avoid a losing argument.

There can be little doubt that in many cases CIC jobs depended on the outcome of the climate debate since the institutions employing them would almost all lose if the CIC should lose the scientific debate. And since all but a few of the researchers obtained funding from government agencies which largely only gave funding to true believers, it is safe to say that the scientists involved in the CIC cause had a direct financial incentive. So although it is impossible to prove, there is a basis for the suspicion that CAGW and more recently EWD are scams since many if not most of those involved expected to gain financially by supporting scientific hypotheses they did not really believe.

Some aspects of the CICs efforts are more clearly scams than others, particularly the promotion of ethanol for motor vehicle fuel. It is hard to argue that this is anything other than a scam to divert gasoline expenditures to corn farmers and agribusiness interests promoted by politicians beholden to these interests. Windmills and solar panels are not far behind, although it is possible that some supporters actually believe the propaganda for these economically and environmentally undesirable technologies. Not far behind these is outlawing incandescent light bulbs, which clearly helps light bulb manufacturers, does little for energy conservation, and increases the risk of exposure to mercury, a toxic substance that EPA is currently trying to very strictly (over) regulate.

Another question is whether the CIC is engaged in a conspiracy. There can be little doubt that there is extensive coordination among the major CIC components. Not only are there frequent international and national conferences where the leading players get together, but the deliberate changes in the party line, such as the shifts from GWD to GEF to EWD, suggest very careful coordination and probably a conspiracy.

General Conclusions about the CAGW Scientific Assumptions

The unfortunate situation is that the IPCC and alarmists in general simply try to ignore attempts to point out that their CAGW hypothesis fails the SM standard used by the rest of science. The EWD science hypothesis is not even supported by the IPCC as of 2014, but is by President Obama. Often alarmists simply ignore what the skeptics have said, but if they respond at all they usually argue that those who have pointed this out are either unqualified to reach these conclusions or are being paid by evil polluters of various sorts. If pushed hard enough, one of their many websites will put forth scientific obfuscation that they claim proves the skeptic's claims wrong. In some of the most critical issues, such as the missing upper tropospheric hotspot, they will write and publish implausible papers to support their alleged case. Then, perhaps worst of all, the alarmists (including Obama) will claim that there is an overwhelming consensus for CAGW or other alarmist "science" among scientists, despite the fact that science is not based on consensus but on comparisons between hypotheses and observational reality.

This situation would be unthinkable in most any other scientific field, and the perpetrators would be labeled as academic pariahs and receive little or no further research funds. But they get away with it in this field because of an apparently unending supply of government research funds for those supporting CAGW and more recently the EWD science hypothesis and little or none for those opposing them. This supply of government research funds for an invalid hypothesis is the primary reason CAGW has developed into the worst scientific scandal of the current generation. The spoils go to those who support invalid science rather than to those who support valid science. This has been going on through successive US administrations for more than 20 years, but has been particularly flagrant during the Obama Administration, which publicly supports CAGW and even EWD science regardless of what the science says.

The results of this behavior have not only included the enormous resulting waste of resources on GWD/GEF "investments" but also the lesson that it offers for future corruption of science at the hands of government for political reasons or financial gain and for the building of additional industrial complexes to promote other scientifically invalid hypotheses which may also be exploited for private gain. As outlined in Chapter 6, this scandal appears likely to continue until or unless the funding stops and those politicians supporting this behavior are voted out of office.

A Much More Robust Natural Explanation for Climate Change

It is easy to poke holes in the CAGW science. It is an edifice built with numerous architects and too many model builders using an overly simplified approach to a very complex problem, and it shows. But it is sometimes difficult to persuade others of this, no matter how scientifically invalid the hypothesis is, without presenting a better alternative. Unfortunately, it is also dangerous for anyone who attempts to present such alternatives because the science involved is not only not settled, it is still rapidly developing. Only two things are clear—that CAGW and EWD science are invalid and we still have very little understanding of Earth's climate.

I will present a number of ideas, most of which have been published in peer-reviewed technical journals but not all of which are generally accepted even by the skeptic community, and most of which are dismissed by alarmists as not worth considering with the claim that the IPCC has already spelled out the "settled" science that explains climate change. *Future research may well prove some parts or even major portions of these partial alternatives to be wrong. But they are demonstrably far better than CAGW in terms of explaining the available evidence.*

It is widely acknowledged that the weather is chaotic in nature. Climate change does, however, have a number of

287

regularities. But the regularities are sufficiently complicated that humans have had a hard time understanding them. Since prior to 1940 humans had brought about the emission of very little CO_2, climate change prior to then was probably primarily naturally caused under AGW theory. The CAGW approach offers very little assistance in learning what these natural causes might be since it does not explain most of the observed data. Some have alleged that the increasing temperatures at the end of the last ice age might be related to increases in CO_2. The problem with that is that the temperature increases started before the CO_2 increases, so the CO_2 increases were more likely to be a result of the temperature increases. So the major research task is to understand what the pre-1940 natural causes of climate change might be. Once we understand that, we can see if the post-1940 climate behaved in a similar manner.

As will shortly be discussed, it is generally agreed that the successive ice ages that Earth has been subjected to for the last million years are primarily determined by the 100,000 year Milankovitch cycles, which determine how much sunlight reaches the northern latitudes. A number of alarmists believe that their rediscovery of the effects of increasing GHGs means that the effects of the cycles are now a minor effect dwarfed by the warming effects of GHGs, but for all the reasons just explained in this chapter, I disagree. The disagreements become even sharper as to whether there are shorter-term cycles and what they are determined by. Yet it is evident from looking at temperature measurements/reconstructions that there are important shorter-term cycles as well including a 60 year cycle and possibly others as well. So there has been some regularity in Earth's climate changes.

If, as argued above, CAGW must be abandoned as a scientific hypothesis for lack of correspondence with observed reality, an important question is where to start to construct a viable science of climate change. This is primarily a matter of scientific curiosity, but if such an alternative could be identified, it might help to persuade people to abandon CAGW. If, contrary to the IPCC's conclusions,

Earth's climate is primarily influenced by natural forces, these forces can be usefully characterized either as astronomical influences from outside Earth or as natural forces on Earth or in its atmosphere and oceans.

Astronomical influences are likely to be cyclical—usually perfectly cyclical—since cycles are what characterize the astronomical world. Night follows day; a solar year has 365 days, etc. So there is an obvious place to look for such influences—cyclic, repetitive patterns in factors that might affect climate, particularly in the sun, which provides most of our heat and light.

Non-astronomical natural phenomena affecting Earth's climate from Earth or its atmosphere are much more varied and comprehensive than often realized. There is reason to believe that these phenomena in some cases act as a climate control system for Earth and that this system may greatly reduce the minor influence of CO_2 on global temperatures. This apparent control system may act much like a throttle and governor does on a car and much less in a precise cyclical manner characteristic of astronomical forces. Some of these Earthly phenomena appear to respond to thresholds in temperatures and other climate variables.

The phenomena that make up the control system go by the unusual name of emergent phenomena, which are complex ordered phenomena that arise from interactions between a number of simple elements and are self-organized. Examples of emergent climate-related phenomena include ocean oscillations as well as thunderstorms, tornadoes, clouds, hurricanes, dust devils,[241] the South Asian monsoons, and the thermohaline circulation (including the Gulf Stream) in the major oceans. A number of these phenomena are well known individually, particularly hurricanes/typhoons and tornadoes, but it is not generally realized what an important role they appear to play in climate as a whole.

Climate-related emergent phenomena are often based on phase-changes by water and generally appear to result in making

---

[241] Eschenbach, 2013.

surface temperatures more uniform. Much of this is based on how water acts when it transitions between its solid (ice), liquid (water), and gaseous (water vapor) states, when it absorbs large amounts of heat. When various thresholds for temperatures and other variables are crossed, climate systems react in very different ways from what they did before the thresholds were reached. Each response varies according to current local conditions so any cycles involved are generally somewhat irregular rather than precise. And finally, emergent phenomena exhibit overshoot, so that once the relevant threshold is exceeded the resulting phenomena may continue even after the initial threshold has been re-crossed in the opposite direction. Because of these characteristics such phenomena are very difficult if not impossible to model accurately. This may explain in part why climate models have proved so inaccurate.

The surprising thing is that since most or almost all Federally-funded climate research over the last 20 years has concerned CAGW or GWD, very little has concerned either astronomical cycles or emergent phenomena despite their evident potential importance. Particularly in very recent years some interesting and much more useful research has been done by a few independent, often unfunded researchers, most of whom are climate skeptics.

Most of the rest of this chapter will sketch some of that research that I find particularly interesting. In some cases, such as the influence of the Milankovitch Cycles on ice ages, the science is quite persuasive; in others more research is clearly needed. At the very least this section will suggest that the IPCC was less than objective in deciding with allegedly more than 90, now 95%, certainty that recent minor temperature increases could not be explained by natural causes. In order to reach such conclusions it is necessary to rule out the effects of all possible astronomical cycles and climate-related emergent phenomena on Earth. It is very clear that the IPCC has not done this. They claim that their GCM models take all major natural factors into account. But they deliberately ignore a number of possibilities. A number of these will be discussed in this section.

Lessons Learned about Climate Science

The evidence is overwhelming that the CAGW hypothesis explains very little except where it has been forced to do so by adjusting various model parameters to make the output fit one particular set of temperature data, usually for the late 20th Century. Alternative hypotheses, on the other hand, can be found that may explain many of the observations in relevant areas of climate science despite the IPCC's claim that they cannot think of any natural explanation. Besides the fact that the crucial CAGW hypothesis does not satisfy the scientific method, the very broad correspondence between observed data and an alternative set of hypotheses convinces me that some alternative hypotheses are much to be preferred even though they should be subjected to the same SM criteria. This section will briefly explain a set of these alternative hypotheses and explain why they do so much better a job in explaining the observed data. Further research by those willing to consider non-CAGW alternatives may result in new hypotheses that will do an even better job of explaining climate observations than the hypotheses to be presented in this section. But we already know enough to say that the climate problem is more complicated than can be explained by the overly simplistic CAGW hypothesis.

The surprising thing is how much an interested person with little background in climate science can learn for him or herself by examining available research. In fact, in my view a lack of a climate model-building background is a significant advantage because such persons will not be burdened by all the preconceptions that CAGW-anointed climate model builders have previously brought to the problem and incorporated in their elaborate but largely useless models.

One obvious conclusion is that humans understand surprisingly little about climate change despite many years of effort and the potential importance of the knowledge to humans on Earth, particularly as we approach the end of the warm Holocene. Although we have a good understanding of what causes ice ages, we understand much less about why climate fluctuates in other respects, as it does. The first goal might be to understand at least the general causes of the marked fluctuations during the Holocene. Why did the Little Ice Age occur? Why were there significant variations during the LIA? What caused the temperature highs during the Holocene? Until we understand the reasons for these fluctuations we cannot say that we understand climate even during the Holocene let alone previous interglacial and ice age periods. Claims by researchers that they understand parts of the overall climate regime (like CAGW) should be discounted unless they can show that they have an overall understanding that explains the broader swings at least during the Holocene. Since CAGW supporters have a difficult time explaining the late 20th Century and cannot explain the temperature plateau so far in the 21st Century, CAGW clearly does not meet this standard.

One of the most obvious things about our Earth is that it is a blue planet because so much of the surface is made up of water. Accordingly it is the physical characteristics of water that play an important role in determining how Earth responds to variations in climate-related natural forces affecting it. Water freezes at a certain temperature and forms a solid (ice), leading sometimes to the formation of glaciers and sea ice. If ice lasts through successive summers on land, it builds up and may in time become a glacier. Continental glaciers are common during ice ages. Water vapor, created predominantly by the evaporation of sea water, is the most important greenhouse gas in the atmosphere and is an important determinant of how the atmosphere responds to incoming radiation from the sun (through variations in cloud cover) and outgoing radiation from the Earth (through the greenhouse effect). The oceans are the principal heat reservoir, not the atmosphere, which holds comparatively very little. There is increasing evidence that clouds regulate atmospheric temperatures over water, particularly in the tropics.[242] Water makes Earth habitable; minor gases such as CO_2 play a vital role in life on Earth but have only a comparatively small role in climate. Computer models that assume otherwise are

---

[242] Eschenbach, 2013b.

just that, models. There is a constant interaction between Earth's vast oceans and the water vapor in the atmosphere, which makes possible vital precipitation in many parts of the Earth and maintains the balance of water vapor and some other atmospheric constituents such as CO_2 in the atmosphere. Emergent climate phenomena are often water related.

Despite the oceans of propaganda many of us have been bombarded with designed to persuade us that global climate has been significantly affected by human-caused influences, I have been unable to confirm that any of the alleged catastrophic outcomes are likely or even probable. The only effect of human activity on warming/climate that clearly plays a significant role in climate is the urban heat island (UHI) effect, which is clearly man-made since urban areas were created by humans. UHIs appear to exist around most large urban areas, which are generally warmer than the surrounding areas because of the heat absorbing characteristics of urban areas compared to rural areas. Whether man's influences extend beyond these UHIs is unknown and unproven, but probable, but also probably minor.

It should be noted, however, that land-based temperature measurements are not very reliable as indicators of global temperatures because the measurement sites are far from uniformly distributed, changes in the immediate surroundings of the measurement instruments themselves due to urbanization,[243] and widely reported attempts to tamper with the data by CAGW-inclined officials to make the data better support their hypotheses. The continuing urbanization of the US and much of the world appears likely to have resulted in a substantial increase in land-based temperature measurements in urban areas unrelated to changes in global temperatures. If the areas surrounding a thermometer are altered, the readings will usually be affected unless careful attention is paid to strict siting guidelines, which have not been adhered to

---

[243] Watts *et al.*, 2012.

very well in the US and maybe even less elsewhere.[244] Further, those responsible for these measurements have been shown to have made inappropriate and even unexplained adjustments to the thermometer readings and the selection of those used (which they generally control), most of which seem to result in exaggerating recent temperature rise.[245] These two factors appear to account for a significant portion of the increases reported from land-based temperature measurements in the third quarter of the 20th Century. Most of these problems can be avoided by using satellite temperature data, which GWD adherents and even NASA analysts rarely mention, let alone use, perhaps because they have no control over the data used and because it has a different trend that does not support their pre-existing viewpoint.

This section will attempt to explain some of the main causes of climate change, which I interpret to mean changes in global atmospheric temperatures. If, as argued above, CO_2 has little effect on global temperatures at current atmospheric CO_2 levels, the question is what does influence changes in global temperatures? Some good clues as to where to look can be found from correlation studies. These suggest that there appear to be at least five major influences: Emergent phenomena and ocean oscillations, large volcanic eruptions, changes in low cloud densities due to variations in the sun's magnetic field, incoming cosmic rays, and cyclical variability of the sun's irradiance, and changes in the sunlight incident on the northern latitudes due to Earth's orbit and orbital properties. Each of these major influences will be discussed in this part of Chapter 9.

Unlike the IPCC, I believe that climate change is a very complicated phenomenon with many causes, most of which are natural. Although the IPCC recognizes a few of the many natural causes of climate change, particularly volcanism and the Milankovich cycles, they believe that there is no significant

---

[244] *Ibid.*

[245] DuHamel, 2012. See footnote 226 for additional citations.

influence from ocean oscillations or celestial bodies outside of the Earth. And they appear to have ignored emergent phenomena. Prior to the IPCC, the predominant opinion was that Earth's climatic changes were primarily due to variations in the sun. Despite their best efforts to hide the obvious, one of the major problems with the IPCC's analysis is that they choose to ignore the contributions of emergent phenomena and most astronomical influences just discussed. Solar variations are too small in their view to have a significant impact on climate. There is now strong evidence not only that they are wrong but rather that these other influences are the major (but not the only) ones. The UN IPCC claims that they cannot think of any natural cause for recent land-based temperature increases, so in their view climate change must be due to AGW, but of course this cannot be correct given the negative findings concerning water vapor feedback using the scientific method. It should also be noted that this particular argument of theirs has no place in science; just because they cannot think of a natural cause does not mean that it does not exist—just that they have chosen not to believe that there might be one. What the IPCC is doing is excluding natural explanations even though there may be natural explanations that the IPCC does not yet understand or chooses to reject. But if there is a natural explanation, one of the IPCC's principal but invalid arguments disappears anyway. This section will sketch some possible such natural explanations.

A number of efforts have tried to relate global temperatures with various cycles. These include assumed internal cycles unique to the Earth, the orbital characteristics of the Earth around the sun, changes in cosmic ray fluxes, the sun's magnetic fields, sunspots, and planetary influences on the sun. It is not possible to explain each of these efforts or the results. I have rather picked out a few of them which I find particularly interesting. Only improved data and further research is likely to show which approach best explains the data.

Earlier Views

Prior to the advent of the IPCC and interest in the effects of increasing CO_2, the predominant view appears to have been that variations in global temperatures over periods less than 100,000 years were primarily due to solar variability since the Sun is Earth's major source of heat and light.[246] A number of researchers have studied this over the years, and they have found some apparent relationships between sunspot cycles and global temperatures. Some have even developed a hypothesis to explain this apparent relationship.[247]

This hypothesis is roughly as follows. Solar variability has been studied for at least 400 years. The general conclusion prior to 1990 was that the Sun is the major driver but there was little agreement as to the exact mechanism. But starting in 1990, the IPCC instead attributed warming to GHGs/humans. In 1998, however, Svensmark suggested a mechanism for indirect solar variability effects.[248] Now many CAGW skeptics cite solar variability as the major cause and basis for their skepticism. In recent years there has been a furious debate/war on this issue. There has been some new research in recent years, however, some of which will be summarized in the following sections.

Predominant Views Prior to 1990[249]

- *"Earth's temperature often seems to correlate directly with solar activity: when this activity is high the Earth is warm"*
- *"During the famous 'Little Ice Age' during the 17th Century, the climate was notably cooler....This*

---

[246] Hoyt and Schatten, 1997
[247] Prominently, Svensmark, 1998
[248] *Ibid.*
[249] Based on Hoyt and Schatten, 1997.

correlated with the Maunder Minimum on the sun, an interval of few sunspots and aurorae"

- *"In the 11th and 12th centuries, a "Medieval Maximum" in solar activity corresponded to the "Medieval Optimum" in climate"*

- *"The 20th century has been marked by generally increasing levels of solar activity"*

Possible Natural Cause: Ocean Oscillations

Probably the clearest cause of climate changes other than the Milankovich Cycles are ocean oscillations, which are one of the natural variations generally ignored by the IPCC. Large oceans, particularly the Pacific, have long been known to exhibit such oscillations. Inspection of satellite temperature data available since 1979 strongly suggests that global temperatures are not primarily influenced by gradually increasing CO_2 levels but rather are associated with periodic major ocean oscillations, particularly the 3-5 year El Nino Southern Oscillation (ENSO) and the 60 year Pacific Decadal Oscillation (PDO) found in the largest ocean, the Pacific (see Figures C-12 through C-14). The North Atlantic Oscillation (NAO) also appears to play an important role and is related in its cycles to the PDO. As discussed in Appendix C, the ENSO is composed of a natural succession of El Ninos and La Ninas,[250] which have been associated with global temperatures changes of about 0.3°C above and below the global average temperature since 1979, respectively, but were evenly balanced between 1979 and 1997. The former has the effect of taking heat out of the ocean and putting it into the atmosphere. The latter does the opposite.

The PDO acts like an "envelope" of ENSOs and has both positive and negative phases that result in a disproportionate number of El Ninos and La Ninas, respectively, and hence increases

---

[250] For a detailed, illustrated introduction to ENSO, see Tisdale, 2012 and 2004.

and decreases global temperatures over its cycle. The PDO cycle appears to be approximately 60 years in length composed of about 30 year positive and negative phases each. The PDO cycle may be related to temperature records over the last 150 years (see Figure 9-2). Since 1979 when satellite temperature data began, the PDO was in a positive phase until the very early 2000s, and has now entered a negative phase (see Figure C-13), as have temperatures,[251] as shown in Figure 9-2.

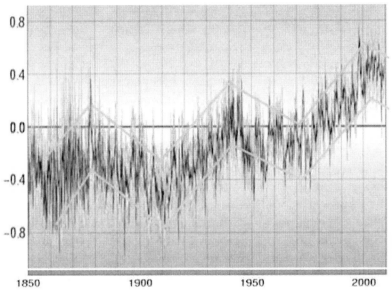

Figure 9-2: How Global Temperatures Since 1850 Fit into a 60 Year Cycle[252]

The only non-normal ENSO increase in global temperatures during that period was about 0.3°C in 1998 as a result of a "super El Nino" which was out of phase with ENSO. Despite this, it shared most of

---

[251] Assuming that flat temperatures are included in the negative phase. A rationale for including flat temperatures can be found in Appendix C.2.
[252] Cheetham, 2012.

ENSO's characteristics and may be associated with the positive PDO during this period. The increased heat in the atmosphere resulting from the Super El Nino appears to have remained in the atmosphere until at least early 2015, perhaps because there has been no super La Nina to take it out. There was no steady increase in global temperatures as one might expect if temperatures were responding to the gradual rise in CO_2 levels in the atmosphere, or as portrayed in most temperature graphs of the late 20th Century built by GWD adherents. This observation suggests that CO_2 increases are unlikely to be the cause of the minor warming during 1979-2014, all or most of which occurred in a one year period during 1997-8 based on satellite temperature measurements. As discussed earlier in this Chapter, this is one of many reasons I strongly question the IPCC hypothesis that CO_2 changes determine temperatures rather than the other way around. *The satellite global temperature charts simply have no apparent correspondence to the changes in atmospheric CO_2, but a strong resemblance to ENSO.*

A very interesting question is exactly what determines the timing of the ENSO (and perhaps the PDO) oscillations. One possibility is that it is determined by a natural system that responds to unusually warm water in certain parts of the Pacific Ocean. ENSO appears to be an emergent phenomenon that redistributes heat in the Pacific Ocean both east and west and north and south and between the oceans and the atmosphere with the result that geographical temperature differences are reduced. The PDO may also be an emergent phenomenon that stabilizes sea temperatures by adding additional or stronger El Ninos or La Ninas during alternate 30 year periods that result in a warmer or colder atmosphere, respectively. What brings about the PDO is uncertain, but one possibility is that it responds directly or indirectly to changes in solar activity. What is clear is that we cannot determine what determines global temperatures without understanding both of these phenomena. To ignore both of them as if they did not exist, as the IPCC generally does, is guaranteed to throw away potentially valuable clues to the larger puzzle that is climate science.

On the scientific side, the CIC did get one thing right—that global temperatures would rise in the late 1900s. I doubt that they understood why—the natural 60 year Pacific Decadal Oscillation cycle—but at least with a little help from data manipulation they could and did present a case that global warming was taking place, which it was, although so little that it could and should have been ignored as unimportant. Since then, however, nature in the form of the downward 30 year segment of the 60 year cycle has taken hold, and they have no reasonable explanation for it using their ideology. Some have tried to explain this development, of course, by arguing that the predicted heat has disappeared magically into the deep ocean[253] where there are few monitors or that the critical missing tropical hotspot magically jumps from one part of the globe to another in a way that makes it hard to pinpoint or measure. But the new IPCC AR5 report appears to try to ignore the hiatus in global temperatures.

Possible Natural Cause: Other Emergent Climate Phenomena

The Earth appears to have a very interesting and complicated temperature control system involving a number of non-linear emergent climate phenomena[254] in addition to those just discussed under ocean oscillations. Some of these phenomena appear to keep temperatures from exceeding levels near where the phenomena emerge in oceanic tropical areas, which receive a disproportionately large portion of the Earth's incoming sunlight. This control system appears to be based on temperature dependent emergent phenomena and not on climate forcing, the basis for most climate models. When various thresholds are crossed, the system reacts in very different ways from what it did before the thresholds were

---

[253] As of late 2014, the available data does not support this hypothesis during the period 2005-13. See Llovel et al., 2014.

[254] Eschenbach, 2010. See also, Eschenbach, 2013a.

reached. They arise and disappear and even move around spontaneously when conditions are right. If so, this explains in practical terms why large atmospheric models are not very useful for climate science since it would be very difficult to model the large variety and geographic specificity of emergent phenomena even if an effort were made to do so.

Very generally speaking, Willis Eschenbach, the principal proponent, argues that the climate system absorbs perhaps only 70% of the sunlight energy it receives. The other 30% or so does not make it in or is removed because of emergent phenomena. These include cumulus clouds and thunderstorms, particularly in oceanic tropical areas. These as well as some other climate-related emergent phenomena, such as ENSO and PDO, are Earth's real thermostats in my view, not CO_2. The sun's energy could vary significantly and temperatures in the oceanic tropical areas appear likely to change comparatively little. In fact it has. The sun has become much stronger since Earth's origin, giving rise to the Faint Young Sun Paradox[255] because Earth's temperature has not similarly increased.

The temperature moderating influences of climate-related emergent phenomena are found at both the macro and the micro levels. As just discussed, at the macro level, whenever the Pacific Ocean gets "too warm" on the surface, an El Nino/La Nina "heat pump" kicks in and removes some of the warm water, pumping it first west and thence towards the poles.[256] Unlike increases in atmospheric CO_2, ENSO is one of the few phenomena that can readily be seen to actually change global temperatures on a continuing basis simply by looking at satellite temperature graphs (see Figures C-12 and C-14).

At the micro level, when the sun comes up and temperatures start to rise during the day in tropical and other warm areas near water, water vapor is evaporated into the atmosphere, where it

---

[255] Sagan and Mullen, 1972.
[256] Eschenbach, 2013a.

forms cumulus clouds. These increase the reflection of incoming light away from the Earth and decrease the light actually received at the surface. This reduces the sunlight energy reaching the surface and can be viewed as the throttle of the climate heat engine.[257] As the heat and cumulus clouds further increase during the afternoon, some of the cumulus clouds are transformed into cumulonimbus or thunderstorm clouds. The columnar nature of these clouds act as vertical heat pipes carrying heat near the surface to much higher elevations, which moderates surface temperatures and results in increasing radiation to space since the rarified air at these altitudes offers less of an impediment to outgoing radiation. Eschenbach characterizes thunderstorms as the "governor" on the climate heat engine, and thus can be viewed as the thunderstorm thermostat.[258] So rather than CO_2 being the primary thermostat causing Earth's temperature to rise, as the IPCC maintains, it appears more likely that it is the timing of natural processes very common in the tropics that keep temperatures from increasing much in oceanic tropical areas.

These tropical micro emergent phenomena appear to be very important to understanding climate, but are generally not attributed by the IPCC as having a significant role in climate change. Since much of the sun's energy is received in oceanic tropical areas, these natural control systems or thermostats may have a substantial influence on global temperatures, primarily by stabilizing them and preventing temperatures from increasing substantially in the oceanic tropics. It may also help explain why atmospheric CO_2 appears to have so little influence on temperatures. Higher CO_2 levels may just mean that the micro emergent phenomena (cumulous clouds and later thunderstorms) appear slightly earlier in the day since the onset of these phenomena depends on temperature. These are the same cumulous clouds that the IPCC climate models are not able to model very well, as previously discussed. It also suggests that

---

[257] Eschenbach, 2010.
[258] *Ibid.*

volcanic eruptions may not have a substantial influence on temperatures in the oceanic tropical areas,[259] in this case by pushing the micro emergent phenomena a little later in the day. Presumably, however, there are smaller direct effects in regions and times of the year when there are few or no micro emergent phenomena or water. There are, however, other emergent climate phenomena such as dust devils, tornadoes, and the Indian monsoon on land that also play significant roles.

I find these phenomena of great interest because they provide natural explanations for the stability of temperatures near large bodies of water in the tropics over time.[260] I had always wondered why temperatures were comparatively uniform in most such areas that I visited.

I noticed similar cloud/thunderstorm emergent phenomena in the daily weather in coastal New Guinea when I was there and less frequently in the California Sierra during the summer. The Sierra are hundreds of miles from the ocean but have many lakes and thus a ready source of water vapor in summer. As John Muir pointed out, the infrequent thunderstorms result in lower temperatures after the storms in the late afternoons; this is just what happens most days in the oceanic tropics, which suggests that the requirements for the thunderstorm thermostat to operate are a source of water vapor and warm temperatures. I have never experienced a thunderstorm in a polar areas, presumably because they do not operate there because the thresholds are not crossed.

These micro emergent phenomena are very poorly understood, but need to be better understood if we are to understand climate change, or in this case the apparent relative climate stability. They would appear to greatly reduce the theoretical possibility that increasing CO_2 will substantially increase oceanic tropical temperatures above recent levels since CO_2 does

---

[259] Eschenbach, 2014.

[260] A very interesting chart in this respect showing the possible results of this phenomenon in the tropics can be found in Figure 7 of Tisdale, 2013.

not exert any direct influence on when clouds or thunderstorms form in tropical oceanic areas. From a scientific viewpoint, emergent phenomena as a whole cannot easily be analyzed in the same manner as systems without emergent phenomena and complicate model building greatly.[261] They may well be one of the reasons that climate model building is so disappointing.

An interesting question is how non-tropical and non-oceanic areas maintain their temperatures? These areas also receive sunlight—but much less in the case of non-tropical areas—as well as the movement of heat from tropical areas by air and water circulation. But are there other emergent phenomena that control these temperatures as well? The answer appears to be partially yes. The Earth's oceanic thermohaline circulation (which includes the Gulf Stream) plays an important part in this by redistributing significant heat from the tropics to the North Atlantic and North Pacific as well as the Southern Ocean. When it is interrupted, as has been hypothesized to have occurred during the Younger Dryas period at the end of the last ice age, Northern Hemisphere temperatures may have collapsed as a result,[262] in this case sending the Earth back into a new 1000 year mini-ice age.

It appears reasonable, however, that the northern latitudes should be much more variable in their temperatures since they are dependent on their much more meager sunlight and various pathways for equatorial heat to reach them by way of the thermohaline oceanic circulation and the atmosphere. The sunlight is itself more variable because of the Milankovich Cycles in the long run and marked variations in the incoming solar radiation found between latitudes in the northern latitudes. And the northern latitudes are indeed more variable temperature-wise than the oceanic tropical areas, as would be expected under a thunderstorm thermostat hypothesis that applies primarily to the tropics.

I believe that emergent phenomena do play an important role

---

[261] Eschenbach, 2013a.
[262] Broeker, 2003.

in preventing large increases in temperatures in oceanic tropical areas because of the relatively uniform temperatures and daily cloud cycles I have observed in the oceanic tropics and because they explain why interglacial periods seem to hit somewhat similar upper temperature limits. There must be a reason why global temperatures have not gone much above those experienced in recent interglacial periods. If so, there is even less reason to fear CAGW. If the thunderstorm thermostat hypothesis is correct and plays a significant role, CAGW is simply impossible (also not mentioned by the IPCC) in the oceanic tropics. So before wasting more money on CAGW it would appear important to understand the emergent phenomena.

A possibly related phenomena is that Earth's climate system may automatically adjust for increases in one greenhouse gas by reducing others, thus keeping the overall greenhouse effect relatively constant. All the IPCC climate models assume this possibility away by assuming that relative humidity is a constant while CO_2 changes. But there are extensive measurements showing that atmospheric relative humidity has been decreasing for a number of years while CO_2 levels have been increasing.[263] A different and more plausible but as yet unproven hypothesis[264] is that Earth's atmosphere keeps the greenhouse effect stable, not water vapor concentrations, the most important greenhouse gas. If so, that would explain why no tropical hot spot has been found[265] and why current increases in CO_2 levels are occurring at the same time as temperatures are remaining stable on a decadal scale.

Possible Natural Cause: Volcanism

Although doubts have been raised as just discussed as to whether the effects of substantial volcanic eruptions can really be seen on annual

---

[263] Gregory, 2009.
[264] Miskolczi, 2014; see also earlier publications by the same author.
[265] Gregory, 2009.

temperature charts for the late 20th Century,[266] it has long been believed that large volcanic eruptions that push significant quantities of aerosols into the stratosphere result in several years of lower temperatures. The reason is that the eruptions result in in emissions of sulfur dioxide, which is converted into sulfuric acid, which condenses rapidly in the stratosphere to form fine sulfate compounds. These aerosols increase the reflection of solar radiation back into space, which cools the lower atmosphere (troposphere).[267] Because they are in the stratosphere, they are only slowly removed, unlike particles in the troposphere, which are rapidly removed by rain. If the thunderstorm thermostat hypothesis is an important effect, the effects of volcanism may have been overestimated in the tropics, but may be more important at higher latitudes.

One possible illustration of the effects of large volcanic sulfur releases to the stratosphere in the past 7,000 years which occurred on the island of Lombok in what is now Indonesia in 1257.[268] The summer of 1258 was reported to be very cold in Europe[269] and there appears to have been a significant acceleration in the expansion of the ice cap on Baffin Island, the home for many of the continental ice age glaciers in North America about that time, as discussed in Appendix C and shown in Figure C-15. Although much smaller eruptions in the last few centuries appear not to have resulted in much change on global temperatures,[270] it is possible that such a large one as that in 1257 might have had effects on non-tropical oceanic areas such as Europe. One possibility is that major eruptions have small effects on climate in the oceanic tropics because of the micro emergent phenomena but may have greater effects in the northern latitudes which are largely not controlled by

---

[266] Eschenbach, 2014
[267] US Geological Survey, 2011.
[268] Lavigne *et al.*, 2013.
[269] Walsh, 2013.
[270] Eschenbach, 2014. See also Catto, 2014.

the cloud "throttle" and the thunderstorm "thermostat."

Possible Natural Cause: Transmission of Solar Variability to the Earth's Climate

Most statistical studies find a very strong correlation between temperatures and solar variability (see Appendix C.2). Hundreds of published, peer-reviewed studies demonstrate solar influences on climate.[271] The IPCC has argued that the variability of total solar irradiance (TSI) is too small to directly account for climate changes observed here on Earth. That does not preclude, however, other indirect effects of solar variability from causing climate changes or an amplification mechanism that we do not yet understand. Given the invalidity of the CO_2 explanation, the problem is to find these indirect effects or amplification mechanisms. A number of proposals have been made. The most prominent one is the cosmic ray hypothesis formulated by Henrik Svensmark.[272] Another proposal concerns the effects of changes in ultraviolet radiation from the sun, which has much larger variations than TSI, on climate.[273] Another possibility is that solar variability influences oceanic oscillations in such a way as to magnify the effects of changes in solar activity.

In 1998 Svensmark proposed one such indirect effect in his hypothesis that global climate is largely controlled by the presence of galactic cosmic rays (GCRs or CRs for short) from supernovae outside the solar system or their reaction products in the lower atmosphere, where they serve as condensation nuclei for water droplets that aggregate to form low clouds. When cosmic rays increase, more low clouds are formed, more sunlight is reflected by the clouds, and less sunlight reaches the surface of the Earth. This results in cooler global temperatures. When CRs decrease just the opposite happens. Most other effects such as changes in CO_2 levels

---

[271] HockeySchtick, 2015.
[272] Svensmark, 1998.
[273] Nova, 2015.

are comparatively minor in their effects on global temperatures in Svensmark's view. Research reported in 2011 by CERN[274] confirms experimentally the physical basis for such CR-cloud creation relationships. I find this research by Svensmark and CERN to be of particular interest because it explains how changes in solar activity can change global temperatures far in excess of what the IPCC says is possible based on changes in TSI.

Cosmic ray effects on Earth are determined by at least two factors, the cosmic ray fluxes that reach the solar system and the effects of the solar magnetic field in reducing the fluxes that impact the Earth. One of the most important primary changes in the level of incoming cosmic rays to the solar system occurs because of the movement of the solar system within the Milky Way Galaxy. When the solar system is located in an arm of the Galaxy, CR fluxes are much higher than when it is located between two of the arms because of the increased proximity to the CR's source, supernovae (see Figures C-15 and C-16). The solar system is currently located in an arm so global temperatures are presumably comparatively cold compared to when it was between two arms of the Galaxy about 50 million years ago, when temperatures were much higher. It is logical that supernova explosions relatively near the solar system would also result in more CRs reaching the Earth, and there is some evidence that this may be the case.

Although the effect is much smaller than the changes in CR fluxes to the solar system just discussed, changes in the sun's magnetic field also influence the number of CRs that impact the Earth. A more active sun leads to a stronger field and fewer CRs reaching Earth's atmosphere (see Figure C-18). This would explain why Earth's temperatures are higher when the sun is more active, as it has been in the 20th Century compared with the Little Ice Age, when it was less active. And it would explain the long-held observation that the climate during the Little Ice Age was cloudy, rainy, and cold and there were few sunspots. Under this hypothesis

---

[274] Kirkby *et al.*, 2011.

lack of solar activity led to few sunspots and lots of cosmic rays and lots of low-level clouds and cold. There have long been observations that temperatures in Antarctica move in a trend opposite to those elsewhere at least in the short run. This also follows from the Svensmark hypothesis since more clouds in Antarctica result in less reflectivity since more of the sunlight reflects off the lower reflectivity clouds rather than the higher reflectivity snow and ice on the ground. So changes in CR fluxes affect Antarctic temperatures in ways that can be explained by the hypothesis. This difference is not explained by the CAGW hypothesis.

Possible Natural Cause: Cyclical Character of Sun's Variability

The next question is what determines solar variations in TSI, sunspots, and other indicators of the sun's activity that influence global temperatures. According to the Svensmark hypothesis, solar activity affects its magnetic field, which affects the CR fluxes reaching Earth, which affects low cloud density, which appears to help determine global temperatures. If climate changes are not primarily due to human emissions of GHGs, then Earth's climate is influenced by totally random changes resulting from the chaotic aspects of weather, by emergent climate phenomena, or by somewhat regular cyclic changes often observed in heavenly bodies, or all three.

Although newer and much less widely accepted even by skeptics, Nicola Scafetta has recently proposed that changes in the sun's activity are based on cycles determined by the interaction of the sun's internal dynamo with the tidal effects of motions of the larger planets, Jupiter and Saturn. The results of this interaction, which he bases on sunspot number records, are cycles of 61, 115, 130, and 983 years he says. Using these cycles, Scafetta believes he has been able to explain longer term temperature changes down to the multi-decadal level during the Holocene, including various complex minima during the Little Ice Age. This is quite remarkable given the rather unusual pattern of temperatures observed during the LIA and is the level of understanding we need to see in any useful explanation of climate change. The PDO cycle of about 60 years may be related to an approximate 60 year cycle that appears to be supported by temperature records over the last 150 years (see Figure 9-2). Inspection of Holocene temperatures suggests that there may also be about a 1000 year cycle evident during recent millennia (see Figure 9-1), which may be related to Scafetta's 983 year cycle. His 115 and 130 year cycles may explain the rather jagged highs and lows during his proposed 983 year cycle (see Figure C-31). So his cycles have at least some plausibility in terms of the temperature records, unlike the IPCC's approach, which has a hard time even explaining the late 20th Century and is useless so far in the 21st Century.

The approximate 60 year cycle would explain the rise in temperatures from the 1910s to the 1940s and from the 1970s to the 2000s (and the fear of global warming), and the falling or flat temperatures from the 1940s to the 1970s (and the fear of a coming ice age) and since about the mid-2000s. Similarly, looking at temperatures over the Holocene (see Figure 9-1), the roughly 1000 year cycles in recent millennia would explain the longer term rise and fall of temperatures approximately based on a millennial time frame (*i.e.*, approximate highs roughly at about years 100, 1080, and 2060 AD). The combination of 60 and 1000 year cycles (as illustrated in Figure C-29) would explain the global temperature increases in the closing decades of the 20th Century since the two cycles would be pushing in the same direction and the relatively flat temperatures during the down-swings of the 60 year cycle (such as we are experiencing now) since the two cycles would be operating in opposition to each other. Together, this would make up the full 60 year cycle. The 60 year cycle would thus explain why alarmists voiced ice age fears in the 1970s and global warming fears in the 1980s and 1990s. It is probably the most evident of the cycles that have the major effect on climate according to Scafetta, but he proposes two others (115 and 130 years) which he believes account for climate changes as a whole during the Holocene in addition to his 60 and 983 year cycles.

The CAGW hypothesis, of course, cannot explain any of these apparent regularities. It appears more than doubtful that changes in human emissions of CO_2 would have any significant effect on such cycles even if they could be brought about by governmental action. Appendix C.2 contains a more detailed discussion and a broader comparison between the CAGW hypothesis and these cycles.

There appears to be a relation between the number of sunspots and global temperatures (more sunspots are associated with higher temperatures). There also may be an association between longer sunspot cycles and future lower temperatures.

Although little has been proven, Scafetta says that his hypothesis explains much of the climate data available over the Holocene, all naturally.[275] Scafetta's projections depend greatly on where we currently are in each cycle and the contribution of each of the four solar cycles he has hypothesized. He argues that the 61 year cycle peaked in 2006. He forecasts an intermediate low in the 2030s followed by a major high about 2060 for the 983 year cycle and in 2067 for the 61 year cycle. He believes that the low in the 2030s may be lower than usual for a 61 year low because one of his intermediate cycles (the 115 year cycle) will have a low at roughly the same time as the 61 year cycle.

The important thing at this stage of understanding the climate problem is not whether Scafetta's hypothesis proves to be accurate, although that would be a welcome bonus. Rather the important thing is that it shows the scope of what is needed to understand climate change. Unless a particular hypothesis or group of hypotheses lead to a broad understanding of how climate works and are validated using the scientific method we cannot really say that we understand climate. And until this happens, spending money on alleged "solutions" to climate "change" is very likely to result in waste, not worthwhile results that benefit anyone except those

---

[275] Scafetta, 2012.

advocating them.

Possible Natural Cause: Changes in Sunlight Incidence on Northern Latitudes Due to Variations in the Earth's Orbital Patterns

Variations in the receipt of the sun's energy on the northern latitudes (where most of the Earth's land is located) due to variations in the Earth's orbital parameters have long been hypothesized to have major influences on climate. The temperature cycles exhibited by the cyclical ice ages of approximately 100,000 years that Earth has experienced over the last million years have long been attributed to Earth's orbital patterns called the Milankovitch Cycles after their discoverer. But there were always discrepancies between ice volume and sunlight incidence near the Arctic Circle until Nigel Calder[276] and Gerald Roe[277] realized that changes in sunlight incidence only changes the rate of change in ice volume. Variations in sunlight incidence changes the rate at which ice accumulates/melts, but does not have an immediate proportional change in volume.

So what should be compared with sunlight incidence near the Arctic Circle is the time rate of change of the ice volume (which can be measured using ice core data), not ice volume. This change makes great sense to me because the most ice free Arctic seas are usually found near the end of August based on my experience sailing in them. This is two months after the highest sunlight incidence. In other words, sea ice continues to melt in July and August even though sunlight is falling. So the rate of melting is presumably decreasing on average during these months, but melting is still going on. The result of using the rate of change in ice volume is spectacular agreement between sunlight incidence and the rate of change of reconstructed ice volumes, with correlations well above

---

[276] Calder, 1974.
[277] Roe, 2006.

Lessons Learned about Climate Science

the 99% confidence level (see Figure C-33).[278] So with this reformulation of the effects of the Milankovitch Cycles there is every reason to have confidence in being able to predict future changes in ice volume by extending the precisely known sunlight incidence cycles seen over the last 400,000 years.

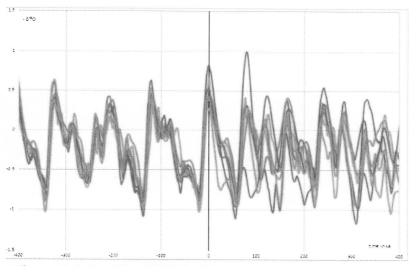

Figure 9-3: Predicted Global Temperature Proxy Due to Milankovich Cycles Based on 7 Different Models with Present Shown as 0[279]

Figure 9-3 summarizes the results of 7 models of global temperatures (using an oxygen 18 isotope proxy used in ice core analysis) built on this theory, one of which is by Roe. Every model agrees that the Milankovitch cycles predict a long descent from the current interglacial temperatures into a new ice age with all its horrors for humans, particularly those located in northern latitudes. Comparison with Figure 9-1 suggests that this descent started at least 3,000 years ago. The new lows made during the LIA appear to

---

[278] See also *ibid*.
[279] Iya, 2012.

me to be a precursor for what we have in store for us in coming millennia. This is the real risk, or should I say likelihood, posed by climate change, and it is not related directly to CO_2 levels or impacted by minor warming in the late 20th Century.

So not only have the alarmists gotten the problem wrong, they appear to have even gotten the sign wrong. We should cherish every bit of warming we can create (to the extent that we can), not try to reduce it as the CIC claims its "solution" would do. This is what happens when invalid science is used to guide public policy—we may get the policy wrong or even as in this case backwards. We should not be worrying about minor increases in open water in the Arctic. We should instead be looking for ways to increase areas of open arctic water and global temperatures in general, perhaps even including the possibility of *increasing* CO_2 emissions. And although serious cooling problems may be several centuries or even a few millennia in the future, it is not too early to start figuring out how to deal with them if we are to avoid the ultimate and most likely of environmental disasters—a new ice age.

According to David Archibald,[280] three of the last four interglacial periods have ended within 3,000 years of where we are now in the 100,000 year cycle based on the period of maximum temperature using Vostok ice core data from Antarctica (see Figure C-34). The other interglacial period ended much sooner. Past ice ages have resulted in huge glacial sheets coming down from Northeastern Canada across North America and from Northern Europe over Central Europe. These sheets have scoured everything in their way off the face of the Earth. Mankind's puny structures would not even slow them down. The result would be a mass human migration from the northern latitudes towards the tropics, the loss of all human-built structures in the northern latitudes, the widespread failure of many crops grown outside the tropics, and the loss of very large numbers of humans, animals, and non-cultivated plants. The only environmental catastrophe worse than a new ice

---

[280] Archibald, 2013.

age may be a supervolcanic eruption such as occurred about 70,000 years ago at Mount Toba in present-day Indonesia which may have resulted in a 6-10 year volcanic winter and a reduction in the worldwide human population to as little as 10,000.[281]

Many GWD adherents believe that the temperature changes due to the Milankovitch cycles are insignificant in their effects compared to their hypothesized CO_2-induced global warming effect.[282] I believe that this is unlikely to be the case and that the Milankovitch cycles are much more certain to occur and be consequential. Basically the CAGW adherents argue that because of their alleged positive feedback between CO_2 and water vapor, the effects of the 100,000 year cycle are so minor as not to be worth considering.[283] But as noted, this hypothesis does not stand the application of the scientific method so they really have no scientific basis for their viewpoint. I know of no reason to believe that these cycles have been repealed, and believe that there is every reason to think that the next ice age will start right on schedule over the next three millennia or so. As shown in Figure 9-1, Greenland icecap temperatures have declined fairly steadily since the Minoan high about 3300 years ago, with decreasing highs since then and a particularly extended period of cold (known as the Little Ice Age) during the most recent millennium. Temperature measurements from the Penny Ice Cap, a large ice cap on Baffin Island in the Canadian Arctic, tell a similar story.[284] Current temperatures are not unprecedented as the alarmists so often claim.

So there is little reason to think that the next ice age will not appear on schedule and that this may happen within a few thousand years. Unfortunately, there is little understanding as to exactly how it will start, surely an important question for humans in general and North Americans in particular, who have so much riding on the

---

[281] Ambrose, 1998.
[282] Revkin, 2012.
[283] Ibid.
[284] Zdanowicz et al., 2012, Figure 10.

outcome. This book will explore what I believe to be a reasonable hypothesis that they start with a significant increase in Northeastern Canada in multi-year ice and snow covered areas. Most snow and ice now melt during the Arctic summer. For the area that does not melt for the first time during the summer, the Earth's reflectivity increases during that summer relative to earlier summers when it did and there is increased likelihood of further increases in future years as relatively more solar radiation is reflected back into space from an even larger area which does not melt in the summer. The reason that ice and snow build-up is so critical is that ice and snow reflect a much greater proportion of incoming radiation from the sun than land or water, which they replace on the Earth's surface as temperatures fall. So as the ice and snow extend to larger areas, radiation reflectivity increases during the summer, snow and ice are less likely to melt during the summer, ice and snow builds up on the previous year's ice and snow, and temperatures further decrease, which further accelerates the ice build-up and the decrease in Northern Hemisphere temperatures with each passing year that there is not a full melting back to the original area. This is called the ice albedo effect and is an example of a positive climate feedback. More ice leads to even more cold and more ice.

In the early stages of an ice age it is reasonable to assume that a warm summer or summers come to the rescue and melt back the ice and snow to some smaller area. This would reduce the ice and snow feedback effect and postpone the development of a large and expanding continental glacier. This may be what may have started to happen in the 13th and 15th Centuries (see Appendix C). But history suggests that at some point the needed warm summer will not happen to appear at the critical period and the ice buildup will become serious and self-reinforcing. This seems more likely to happen as the sunlight incidence on the northern latitudes become less and less due to the Milankovitch cycles, when solar activity is weak, and volcanism is high.

Progressive increases in the non-melting area presumably may initially occur in one of two ways. One is as a result of very large volcanic eruptions which deposit very large quantities of volcanic material in the stratosphere which reflects more incoming radiation from the sun and which some agree results in a significant lowering of global temperatures. The other is that they start as a result of periods of unusual cold in the northern latitudes resulting from non-volcanic natural forces. Presumably ice ages can start with various combinations of these two natural mechanisms. This will be discussed further later in this Chapter.

The bottom line is that the Milankovitch Cycles and their effects on ice formation in the Northern latitudes is one of the best understood drivers of climate change. We ignore them at our peril. The implications are now obvious. Earth may have somewhat narrowly avoided a new ice age in the last millennium. Greenland ice cap temperatures have been falling since their high over 3,000 years ago. New cold periods may occur perhaps near each mid-millennial period for the next few hypothesized 1000 year cycles and one of these may result in a new ice age starting. Such a new ice age may not be likely to occur in the next few hundred years, but such an onset of a new ice age could prove catastrophic to countries located in the northern latitudes, especially Canada, Scandinavia, Northern Russia, and Northern United States. Because of the albedo feedback effect of ice formation, once a large mass of ice has built up in Northeastern Canada, it spreads until it covers much of North America with devastating results that are so well documented in the Great Lakes and many other lakes in Northern North America.[285] This would be consistent with at least a million years of Earth history. Climate alarmists are likely to argue that everything has been changed by their rediscovery of human-caused emissions of CO_2 in the last century. I argue that such emissions appear to be fairly minor since there is no significant feedback effect involving water vapor, and that under the circumstances we need more such emissions rather than less in order to decrease the strong likelihood of a new ice age. Alarmists argue that human-emitted CO_2 will

---

[285] Ruddiman, 2005.

317

remain in the atmosphere for very long periods; I am much more skeptical.[286] Just as the effects of decreases in CO_2 emissions are unlikely to significantly affect global warming, however, increases are equally unlikely to have much effect on preventing global cooling.

The principal advantage of the natural cycles and emergent phenomena explanations for climate change is that it appears to fit the data I know of. It explains the decline of Greenland ice sheet temperatures over the last 3000 years. It explains the step increase in global temperatures as a result of the very strong El Nino of 1998. It explains the multiple cold periods during the Little Ice Age. By contrast, the CAGW hypothesis explains the global case of warming in the 1975-2006 period using surface land data but nothing else. I do not think there is any serious competition between the explanatory power of these two hypotheses. That does not prove that the former is correct, just better. Yet the IPCC has shown no interest in exploring such hypotheses to date.

Combining Sunspot Numbers with Ocean Oscillations to Reproduce Global Temperatures

By combining the variations in the number of sunspots, one of the longer-running databases based on actual human observations, with ocean oscillations as explanatory variables, Dan Pangburn has managed to reproduce global temperatures with amazing accuracy since 1850 and with less certainty (due to less accurate temperature records) since 1700.[287] Significantly, he finds that adding a CO_2 variable does not increase the accuracy of his equations. This suggests that solar sunspot numbers and ocean oscillations are strongly correlated with global temperatures and CO_2 is not. Hence it might be productive to start any impartial investigation of the

---

[286] Carlin, 2011.

[287] Pangburn, 2013. Appendix C.2 contains a more detailed explanation of his research than is possible here.

causes of climate change by examining these two natural variables or variables that influence them. Assigning a trace gas such as CO_2 as the driving force in climate is simply highly unlikely, and the relatively small proportion of global CO_2 emissions caused by humans are an even more unlikely cause. This is discussed in much more detail in Appendix C.2.

Pangburn projects (see Figure C-25) that global temperatures will decrease until about 2020 based on projections of sunspot numbers in the current solar cycle and then provides a decreasing range until about 2040 depending on assumptions as to the sunspot numbers in the following cycle.

The larger significance of Pangburn's analysis is that it shows that it is possible to reproduce global temperatures since about 1850 using data from two of the most likely sources of multi-decadal-scale climate change without considering changes in CO_2 levels. This does not prove that Pangburn's equations will hold for longer periods of time—only that natural variations can account for temperature observations over this period, contrary to the IPCC conclusions. The IPCC argues that this is not possible. One problem is that he used (and practically had to use) HADCRUT temperature data which may not accurately represent global temperatures.

Section Summary: Natural Climate Hypotheses

This larger subchapter has presented a number of different natural climate-related hypotheses and a little of the data supporting them. More detailed explanations of some of them can be found in Appendix C.2. I believe that taken together they fit together to form a broad understanding as to how climate change may occur naturally and appear to explain many of the observations that have been made concerning climate change. It is quite remarkable that there are hypotheses that explain such a wide variety of observations and suggest the importance of considering climate variability seriously. But there is obviously much more that we do not understand than that we do.

Perhaps the most important problem in understanding global temperature changes is to explain why Earth appears to hit a "floor" during glacial periods and a "ceiling" during interglacial periods. The best available evidence I know of suggests oceanic tropical temperatures are primarily regulated by emergent phenomena but that temperatures in the northern latitudes are determined by a large variety of factors and are therefore much less uniform.

My hypothesis, like that of many others, is that the reason why Earth alternates between glacial and interglacial periods is the Milankovich cycles. The higher northern latitudes are close to the temperature at which water freezes in summer. If anything interrupts the flow of heat from the tropical areas through the oceans and atmosphere, the snow and ice may build up over successive summers. Snow and ice reflect more incoming energy off icy surfaces in the higher latitudes which encourages even more buildup of ice and snow. The result is that the northern polar areas flip back and forth roughly every 100,000 years between comparatively ice free conditions and continuous ice cover as a result of the minor influences of these cycles. The southern polar areas (Antarctica) have permanent glaciers since temperatures do not get high enough to melt the glaciers in Antarctica.

Temperatures in oceanic tropical areas are limited on the upside by emergent phenomena but spread northward during interglacial periods. During glacial periods, on the other hand, oceanic tropical temperatures are limited on the downside because of the "surplus" energy from incoming sunlight at these latitudes compared to the reflected outgoing energy. As long as this "surplus" exists, temperatures cannot go lower. Presumably all that happens is that the cloud "throttle" starts working later in the day in oceanic tropical areas during glacial periods. Since the current ice ages started about 3 million years ago Northern Hemisphere temperatures have been "stuck" between this "floor" and the "ceiling."

The alarmists greatly exaggerate the importance of changes in CO_2; high CO_2 levels in the distant geological past have not resulted in catastrophically higher temperatures. The heat retention on Earth resulting from increases in CO_2 levels may even be offset by decreases in atmospheric water vapor.

Although CO_2 may play a minor role, particularly at non-tropical latitudes, one of the major influences on global temperatures appears to be clouds, which appear to be influenced by solar activity, cosmic rays, and emergent phenomena. The effects of clouds on surface temperatures is great, but is not well modeled by the climate models. That leaves open the important questions of what determines solar activity and ocean oscillations. This Chapter has outlined several ideas concerning this and the ways that these effects may be manifested. The most comprehensive and probably the most controversial such analysis is probably Scafetta's.

Although much remains unknown, there surprisingly appears to be some agreement among the scientists discussed in this Chapter that temperatures are not likely to increase and are more likely to decrease until the 2030s. Eight solar physicists are quoted as agreeing that the sun is becoming quieter and will continue to do so over at least the next few decades and possibly the next century.[288] In the past this has led to lower global temperatures as well. The best thing to do about the proponents of GWD/GEF is to ignore their scientifically invalid CAGW and take a much broader view of the available evidence.

I have summarized strong evidence as to the importance of changes in the incidence of sunlight on the northern latitudes on the ice age cycle. There is substantial evidence that the current Holocene interglacial period will come to an end relatively soon in geological time and may be preceded by more cold periods such as the Little Ice Age. Temperatures on the Greenland ice cap have been irregularly declining for more than 3,000 years.

As mentioned earlier in this Chapter, I am not aware that there is much disagreement about the significance of the Milankovitch Cycles for the ice age cycle except for the claim by

---

[288] Luning and Vahrcholt, 2014.

many alarmists that the effects of these cycles have been overwhelmed by their rediscovery of the greenhouse effects of CO_2. The evidence for future colder global temperatures includes diminishing sunspots compared to earlier cycles, longer sunspot cycles, the Milankovich cycles, the behavior of earlier such cycles, and the pattern of temperatures over the last 3,000 years. All these convince me that the major climate risk we face is much colder temperatures in the next few centuries and millennia in northern latitudes. I am not alone in this view.[289]

The number of sunspots visible during the current solar sunspot cycle is much lower than recent previous cycles, which appears to be crucial to the Pangburn equations discussed earlier in this Chapter. David Archibald, finds evidence beyond his 3,000 year interglacial period analysis in recent changes in the length of recent solar cycles, which he believes indicates that there will be decreases in New Hampshire temperatures by an astounding 4°C by the 2030s. Figure 9-3 shows that the predicted global temperature proxy due to Milankovich Cycles based on 7 different models are all expected to drop precipitously over the next 60,000 years. Figure C-34 shows that only one of the last five interglacial periods lasted longer than the current one, and it came to an end at a point corresponding to about 3,000 years from now. Figure 9-1 shows alternative data suggesting that Northern Hemisphere temperatures were actually colder during the Second Millennium than they were during the First Millennium.

But there is additional and independent evidence provided by a variety of skeptic researchers.[290] As quoted in Appendix C.2, for example, Svensmark cites the rapid recent reductions in the solar

---

[289] For a non-technical summary of such views see Morano, 2014. Obviously, the number of people holding this view does not demonstrate scientific validity, which should be based on the observational evidence for the hypothesis.

[290] For a short, recent, less technical summary of the research underlying the concern about potential global cooling see Dini, 2014.

magnetic field and the influences he believes this will have on cloud formation and temperatures according to his hypothesis. The last time that sunspots were projected to be as weak as they are now was just prior to the Maunder Minimum, the most severe period during the Little Ice Age.[291] Finally, my interpretation of the Scafetta cycles is that he believes that the Earth will experience significantly lower temperatures in coming mid-millennial periods, which presumably means that there are increased risks of new Little Ice Ages during those periods.

The important point, however, is that eight solar physicists expect a downturn in solar activity and most of the skeptics discussed in detail here suggest the same future: **Earth appears to be headed for much lower temperatures, probably a new Little Ice Age, and before the next few millennia are out, the start of a new ice age**. They differ as to why, when and how much, but their general orientation is clear. Scafetta's hypothesis is the least alarming and provides the most sweeping overview of the four. Unfortunately, our knowledge is so limited that one of the other three may well be correct, or even all four since their results are not necessarily mutually incompatible. And to be on the safe side humans should probably assume the worst possible prediction, which is probably Archibald's. Svensmark (quoted in Appendix C.2) appears to be predicting a new Little Ice Age based on recent solar behavior. Archibald is predicting a 5°C drop in New Hampshire temperatures by about 2030 based on recent changes in the length of the solar sunspot cycle.[292] Surely this would pose a grave danger of at least a new LIA and possibly even the start of a new ice age.

I find it somewhat alarming that so many independent hypotheses/skeptics all point in the same direction based on different hypotheses and approaches. The IPCC is clearly using invalid science and continues repeating its global warming meme.

---

[291] Archibald, 2013
[292] *Ibid.*

But leading skeptic hypotheses/skeptics as well as many solar physicists signal cooling. I would put my money on the skeptics and solar physicists, many of whom have at least called CAGW correctly.

The larger perspective is that humans have greatly benefited by the current interglacial period, during which human civilization has developed rapidly during the last few thousand years. Humans, after all, originated in the tropics and have no fur to keep warm. Late in the Holocene, Western Civilization escaped the Malthusian trap of an endless hand-to-mouth existence starting in the late 18th Century and early 19th Century partially as a result of harnessing water and wind power but mainly by using fossil fuels to substitute for human labor. In the late 20th Century these advances have spread to many other parts of the world, particularly Asia. And through further human ingenuity and further increases in human energy use, humans have now reached a level of civilization previously undreamed of during humans' relatively short existence on Earth. Our human civilization is in many ways a creature of the Holocene and an end of the Holocene appears very likely to pose many major problems for us. And as we exit the Holocene over the next few millennia, our problem will not be undue warmth, as the GWD supporters claim, but extreme cold, failing crops, and widespread misery, or worse. Unfortunately, through ignorance or disregard of the science or self-interest the alarmists appear to have gotten their story exactly backwards, and this may have very serious consequences for humans as we face a likely environmental calamity completely unprepared for it.

The consequences of a new ice age or even a new Little Ice Age for humans and all other living things on Earth would be vastly more serious than any likely minor warming due to continued or even expanded CO_2 emissions. In fact, greater emissions might possibly help to delay or even slow a new ice age or LIA by increasing global temperatures slightly. Reductions in CO_2 emissions would have at most minor effects on climate change, which appears to be primarily dependent on emergent and astronomical phenomena.

This Chapter has sketched out not only why the CAGW hypothesis is scientifically invalid, but what an alternative approach that actually explains more of mankind's observations concerning climate change can be explained by the natural workings of emergent phenomena, the solar system, and the cosmos. If we come to understand these natural forces much better than we now do, it may be possible for man to carefully plan whether he wants to intervene to preserve a particular climate and how he is going to do so. This understanding has not improved through the more than $39 billion climate research program of the US Government but has been helped through the dedicated efforts of a few skeptic researchers. This does not speak well for the US climate research program, and makes one wonder how well other US Government research programs are accomplishing their goals. That will be one of the subjects discussed later in this Chapter.

The next ice age will not be kind to northern latitude human civilization or to the natural environment in these areas, and it might be prudent to start thinking about this rather than the minor warming which occurred from the 1970s to 1998, which we actually have some understanding of. In brief, the actions proposed by the IPCC and the Obama Administration are basically wrong in sign because they did not understand or want to understand the real risk—global cooling—and instead claim that catastrophic global warming is imminent.

This failure was in part brought on by the IPCC's charter, which required them only to look at human-caused climate change rather than all causes. And by failing to pursue a SM-based methodology the IPCC and others failed to perceive their mistake, or, if done on purpose, their contribution to a massive hoax or scam. So trillions of dollars have been wasted on mitigation and no government has seriously addressed what is likely to be the real problem. The IPCC and the US Government and the EU appear to have managed to get their analysis wrong both in terms of what the problem is and what caused it. As a result their proposed "solution" is counterproductive and has long been needed to be scrapped so that the world can concentrate on the real problem—global cooling. Until this is done, more trillions will be wasted and little progress will be made in either understanding climate change or addressing our major environmental problem.

About the only thing that the IPCC and other alarmists got right was that minor global warming would occur near the end of the 20th Century. With a little help from data manipulation this enabled them to make a case which all too many accepted. The most likely climate risk lies in the likely onset of a new ice age within the next few millennia. Significant global warming is unlikely to occur until about 90,000 years from now as the Earth emerges from the upcoming ice age and would be desperately welcome at that time by the survivors. The US would be unlikely to escape the ravages of the new ice age but has made no effort to even understand the issue, let alone prepare to do something about it. In fact, it is currently pursuing exactly the wrong policy to avoid it.

It is important to note that the conclusions concerning the likelihood of a new ice age are independent of whether or not the Scafetta or Svensmark hypotheses are correct. The evidence for the end of the current interglacial period stands on its own and is quite compelling. CO_2 impacts on climate appear to be minor. There is no reason to believe that the world will avoid the next ice age without human intervention. We can hope, of course, but hope will not lead to preparations for dealing with what appears to be the inevitable. What the Svensmark and particularly the Scafetta hypothesis do is to propose an explanation for Holocene temperature variations and to suggest when a new ice age is more likely to start assuming no major volcanic eruptions in the wrong places in the meantime. These hypotheses may help us to plan, but do not change the likely outcome. Pursuing the GWD will obviously not help humans cope with the next ice age; it only makes them less prepared to cope with it. Realistic planning needs to include some form of geoengineering, which I will discuss next.

One of the more amusing conclusions is that the more one

believes CAGW "science," the more one should be interested in increasing CO_2 emissions so as to avoid a new ice age. CAGW "science" argues that CO_2 placed into the atmosphere will increase atmospheric CO_2 for as much as thousands of years, just the period that we should be concerned about the start of a new ice age. Some skeptics and I argue that half the increase in CO2 will be gone in 5-6 years.[293] If so, it makes little difference whether more CO_2 is emitted now to avoid a new ice age. The increase will be long gone before the ice age threat becomes an immediate threat. Anyone who really believes CAGW science, however, should also believe that climate can actually be significantly affected by changes in CO_2 emissions and thus that we should already be hard at work emitting as much CO_2 as possible. Consistent with my views concerning CAGW, I doubt that such emissions will have a significant effect in this regard, but believe that it also will not hurt anything to increase emissions anyway. If the GWD adherents should somehow be right about CAGW, increased CO_2 emissions might actually help avoid the next ice age!

This book has included a description of several interesting new hypotheses concerning Earth's climate that were not developed by the US Government climate research program, and have been held back by the invalid science on this subject promoted by the United Nations and more recently by the George W. Bush and particularly the Obama Administrations. It is now possible to understand some of the major influences on Earth's climate, but this is in spite of the continuing efforts of the United Nations and more than $39 billion in US taxpayer funds spent over the last 20 years to promote invalid science, not because of it. Unlike this invalid science, the new hypotheses actually explain how the Earth's climate may be greatly influenced by astronomical variations, ocean oscillations, and thunderstorm activity, all beyond human control. Variations in the Earth's cloud cover, for example, may be influenced by variations in the Sun's magnetic field and cosmic rays from outer space rather

---

[293] Carlin, 2011.

than the puny effects of mankind's efforts to harness fossil-fueled energy to improve its standard of living. Institutions that are responsible for the invalid science and the faulty policy proposals supposedly based on it need to be changed; Chapter 12 describes some of these needed changes. Obviously paying higher prices for energy or placing intolerable constraints on our lives and economic and scientific freedom will do very little to change variations in the Sun. Offering human sacrifices to change the weather did not work very well in ancient times either, but that did not dissuade those who believed they did.

Alternatives for Controlling Global Cooling

Geoengineering involves intentional changes to the planet and its climate system in order to better meet the needs of man. In recent years it has been primarily been studied as a means to decrease or even halt the global warming prophesized by the CIC. Various ideas have been suggested for doing this. The most prominent one is called Solar Radiation Management (SRM) since it endeavors to manage solar radiation so as to decrease (or increase) the amount entering and leaving Earth's atmosphere.[294]

As readers will have realized by now, however, I believe that the future problem is far more likely to be global cooling rather than warming. Fortunately, many of the ideas for geoengineering may be applied to solving this problem too. Attempts to avoid our likely next ice age fate would seem to come down to three basic possibilities. One is to vastly increase the amount of CO_2 in the atmosphere in hopes that the CAGW ideology actually has some validity; the second is to take a geoengineering approach to deliberately engineer a better outcome by warming the planet or at least the northern latitudes most likely to suffer continental glaciation when and if the cooling becomes inevitable, and the third is for humans to attempt to adapt to a new ice age.

---

[294] The technical term is albedo modification. See NRC, 2015.

Obviously if changes in global temperatures are less than a few degrees in the upward direction, as currently appears to be the most that can be expected, the best response is to do nothing, adapt, and be thankful for every added degree in hopes that this will ward off the coming ice age. By moving to urban areas in recent decades, much of the world's population has voluntarily chosen increases of a few degrees due to the Urban Heat Island (UHI) effect with no apparent harm, and can just continue to do so for minor further increases in the unlikely but favorable event that they should occur. Similarly, if there should be only minor cooling, adaptation is the obvious response.

If, however, as appears more likely, a new ice age should start, serious consideration needs to be given to how to reduce its effects or even halt it. Humans originally evolved as tropical animals and are not very well equipped to live in cold climates without considerable infrastructure. If the thunderstorm thermostat hypothesis has validity, presumably the oceanic tropical areas would remain somewhat the same as they are now. Humans and their food sources in the higher latitudes, however, would suffer horrendous damage involving the destruction of much of the infrastructure of Western Civilization and the loss of vital grain-growing areas located in northern latitudes by advancing continental glaciers. In North America, the Laurentide ice sheet would presumably advance down from Northeastern Canada as it has done roughly every 100,000 years over the past million years.

So if and when a new ice age should threaten, I believe humans should seriously consider geoengineering. It is critical that we better understand the world's climate in order to make this possible without undue risk. Attempts to increase atmospheric CO_2 levels would be highly uncertain in their outcome for all the same reasons discussed above in terms of reducing CO_2 in order to reduce global warming, require the cooperation of large sections of the general public in many countries, require a better knowledge of the relation between CO_2 and temperature than we have, and would require decades of advance planning in the face of extreme uncertainty as to exactly when the new ice age might start.

So as in the case of catastrophic warming, geoengineering may be the best alternative for controlling catastrophic cooling as well. SRM is best researched of the geo options,[295] but is not the only one. It would presumably be ineffective in oceanic tropical areas if the cloud regulator/thunderstorm thermostat has validity. But it is the higher latitudes that are probably of greatest concern anyway. SRM involves placing different types of particles in the stratosphere to increase or decrease the solar radiation reaching Earth. The particles can be placed there by a variety of means, of which the simplest is probably the use of cargo planes. The advantage of using the stratosphere for this purpose is that the particles would not be rapidly washed out of the stratosphere by rain as they would in the troposphere. And if science should develop the capability to predict periods of time when global temperatures will be unusually low, SRM might only need to be applied immediately during those years when the risk of an ice buildup is most severe—not continuously. It is inexpensive compared to the monumental cost and uncertainties of increasing CO_2 emissions, could be used by only a few of the most affected countries if necessary (although approval of some multi-country group would be greatly preferable) but may have adverse impacts and needs much further research before any actual implementation.[296]

A second simpler, and probably even less expensive approach, would involve spreading soot on high latitude ice and snow to increase their melting if and when a new Ice Age or perhaps even a new Little Ice Age threatens. We already have the technology in the form of crop duster aircraft, which might even be used in the off season. Depending on where and when the soot was spread, it should increase melting of ice and snow during the summer and might have sufficiently long lasting effects to decrease the build-up of glaciers. But as in the case of SRM considerable research and

---

[295] NRC, 2015; Carlin, 2008.
[296] NRC, 2015; Carlin, 2007.

development would be necessary before actual use.

Many alarmists, of course, are opposed to using geoengineering to decrease temperatures, perhaps because it does not involve their preferred "solution." It would be possible to add the particles/soot primarily in the far northern latitudes since that is where the cooling problem would presumably be most damaging and there is a possibility but far from certainty that they will predominantly stay there. Where the cloud throttle/thunderstorm thermostat hypothesis affects climate, putting particles in the tropical stratosphere would presumably have little effect since the cloud formation/thunderstorms would simply appear a little earlier in the day. But in non-tropical and dry land areas, the particles may have their intended effect of increasing temperatures.

Geoengineering is clearly the ugly duckling of the climate debate, however, since many conservatives often dislike it too in part because it would probably involve increased governmental roles at the national and international levels. If the thunderstorm thermostat should prove to provide an effective ceiling on tropical oceanic temperatures, it would be pointless to try to raise temperatures in these areas. But that might be good because the purpose would presumably be to raise high latitude temperatures to combat an ice age rather than tropical oceanic areas where there is no need to raise temperatures. So the temperature raising particles might be concentrated in the higher latitudes and avoided to the extent possible in the tropical areas. What would be needed is a far better understanding of climate through careful and unbiased research concerning a number of promising geoengineering proposals before actual implementation should be considered and then only if there should be strong evidence that they are urgently needed. It is not needed now, but there needs to be serious research on it. Geoengineering should only be considered when absolutely necessary to avoid imminent climate catastrophe.

There are many other potentially useful geoengineering approaches, including much more expensive but more permanent ones such as placing satellite arrays in space near the inner Lagrange

Point (L1) between the Earth and the sun to regulate solar energy reaching Earth. The whole subject needs additional research, very careful experimentation to resolve important uncertainties, and an international agreement as to how it could best be safely carried out if and when needed.[297] There is probably no immediate urgency but the geoengineering research and international negotiations would probably require many years. Geoengineering is discussed in more detail in Appendix C.3.

Yes, there are dangers to using geoengineering, which needs very careful research and understanding before large scale implementation should be attempted. But there would be even greater and much more certain dangers from a new ice age as well. Attempts to use geoengineering to reduce minor global temperature increases such as may or more likely may not soon occur would obviously be unwise.

Important Question of Science Transparency and Availability

One of the unusual features of Federal funding of academic research is that although the taxpayer funds it the output usually ends up being owned by private journals which charge readers for access to it. Often the charges are enormous so that obtaining individual journal articles is prohibitively expensive unless researchers have access to an unusually large (and wealthy) library. Thus in many cases the taxpayer funds the research but has no practical way to view the results.

EPA's library has been far from ideal in this regard. For example, even in its best days it did not have most of the economics journals, or even the environmental economics journals, or even some of the environmental health journals, and was at one point even abolished along with what journals it had. It was then reestablished with even fewer journals.

---

[297] Carlin, 2007. See also NRC, 2015, which refers to governance issues.

My view is that if the taxpayers pay for research, they should have ready access to it. This is particularly important during periods such as the present when so much of the most useful research on climate change is being done by citizen-scientists with much less access to professional journals than in mainstream research. So although it has proved difficult to change the professional journal culture in economics and many other fields, my grantees were encouraged to publish their findings in the journals. But I insisted as a condition for receiving a grant that they provide EPA a full report on the research they conducted for EPA where I was the Project Officer. In the end every grantee did so (they had to if they wanted payment). Only one grantee resisted until the grants office threatened their funding at my request. With the development of the internet in later decades, I managed to make many of these reports available to the world, which I regarded as the least that could be done to safeguard the taxpayer's interests. This is now easy because the reports are normally submitted in electronic form, so can easily be made available on the EPA website. But for many years prior to that the reports were prepared in typewritten form, which made conversion to an electronic format time consuming and expensive, so that conversion was limited by funding availability. This issue of research report and data availability has only recently become a major public issue, but has long been an issue within EPA.

With the radical environmentalist approach of the Obama Administration, EPA is attempting to impose much stronger regulations for a number of pollutants, primarily air pollutants. In the last few years they have increasingly taken to citing health effects based on research reports done by outside groups for EPA to support the economic and health benefits of their very (and in my view much too) strong proposed regulations such as the new proposed ozone standard and a number of GWD-related proposals. The problem, however, is that the data on which many of these proposed regulations are based have never been made public. So there is no way that anyone outside of EPA can see whether the data actually support the health effects benefits cited by EPA in their proposed regulations.

In 2013 Senator David Vitter, then the Republican Ranking Member of the Environment and Public Works Committee, and Representative Lamar Smith, Chairman of the House Science, Space, and Technology Committee, have made a major issue of this. As of April, 2014, however, little of the most relevant data has been forthcoming. The House Committee even issued a subpoena, with the result that EPA has said that it does not have and in some cases has never had the data. Reportedly, EPA explained that the data was owned by the grantees (Harvard University and the American Cancer Society) and was subject to various restrictions concerning the use of personal medical data. If a more open approach to research findings and data had been adopted much earlier, however, presumably these problems could have been worked out before the grants were closed out, or, much better, before the grants were made. So this suggests the importance of making provision for making available research reports and data at the time the research is being planned.

In my view research and the data on which it is based should be routinely made available and should not be used in regulatory actions if they are not available for review by interested members of the public. If so, there is little point doing the research unless plans are made from the beginning as to when and how it will be made publicly available. The internet provides an inexpensive means for making it available once the bureaucratic issues are resolved. I actually sponsored some of the early research on econometric analysis of effects data, and made the reports available on the internet to the extent that funding could be found to pay for digitizing the earlier reports. To ask the nation to pay hundreds of billions or even trillions of dollars to avoid alleged adverse effects that cannot be demonstrated to any interested citizen is simply wrong. No publicly available data should mean no regulations in my book. Given the hormesis effect to be discussed in Chapter 11, the results of these very low dose exposures may even be positive for human health rather than negative. This needs to be examined and debated as a general issue and in each case.

American Science Has Been Seriously Compromised by Efforts to Promote CAGW

It is evident to me that the response of the American scientific establishment to CAGW has been less than ideal. One observer even argues that climate researchers have reached the point that they attempt to look for the evidence that fits their ideology rather than seeking out scientific truth.[298] It does appear that they have gotten about everything wrong scientifically and will not admit it. They have basically gone for the US Government research money with little regard for scientific validity. The question that needs to be asked is whether this is an unusual case or whether similar outcomes can be expected from other major US Government research efforts. In either case it would seem important for similar outcomes to be avoided in the future.

The list of problems is very long. Based on the Climategate emails, the peer review system at major journals has been subverted by alarmists in order to reduce the number and the degree of skepticism expressed in accepted papers. Skeptic-oriented research is rarely funded by the US Government. There has even been a very visible effort to substitute "consensus" for the scientific method as the criterion for judging science. The alarmists have almost always refused to openly debate the scientific issues raised by skeptics but instead derided them or questioned their motives or source of funding.

To pick just one of these problems, the peer review concept that is the standard method for selection of research grant recipients is widely used in the Federal Government. If reviewers are selected that all hold the same views on the validity of the underlying research effort, it should not be surprising if alternative, more scientifically valid hypotheses, are never considered. This could, of

---

[298] Ridley, 2014a.

335

course, be avoided by having the government officials responsible for selecting the topics to be researched specifically call for research on alternative hypotheses or select reviewers that hold alternative views. Clearly none of these things happened in the case of US Government climate research over the last few decades. In my view, the results have probably been worse than if the research had never been carried out.

This is not to say that additional climate research is not needed. Clearly it is since we must understand Earth's climate well enough to ward off the next ice age. Without such knowledge we may try using a remedy that will not work, with disastrous results. Ignoring any area of knowledge concerning climate that conflicts with the CAGW hypothesis or with the Obama Administration's approved climate science is not the way to learn what we need to know. Given the very slow turnover of Federal scientific personnel, such a change would probably have to come from the top of the Executive Branch.

Clearly, the National Academy of Sciences and the Royal Society and a variety of disciplinary scientific societies have not provided good guidance. Richard Lindzen has provided some thoughts on why this happened.[299]

Importance of Open Science and Scientific Integrity in Government

Although the Obama Administration made an effort to proclaim its commitment to scientific integrity and transparency in its early months, this was rapidly replaced by just the opposite. In my case this shift occurred as soon as I questioned some of their favorite ideological beliefs which they claimed were based on science. Attempts to enforce ideological correctness on science can quickly get out of hand, bring disastrous results, and are inappropriate.

But even if no one dies directly as a result, science and

---

[299] Lindzen, 2012a.

ideological correctness are incompatible. There is a "correct" answer to scientific questions based on the application of the scientific method. If scientists are not allowed to freely express their views on scientific issues, it is highly unlikely that this "correct" answer will be discovered. Instead, all that will be learned is what the ideological "correct" answer is, which is already known. If the ideologically "correct" answer is announced, as it has been during the Obama Administration with regard to CAGW, very few government employees or researchers dependent on government grants or contracts are likely to carry out studies or express ideas which even cast doubt concerning the "correct" answer. The results can still be disastrous even if no one dies. Incorrect policies will be adopted. Billions and trillions of dollars can be wasted on policies that are doomed from the start. The "correct" answer will not be determined because it will never be looked for.

I argue, however, that people have already died as a result of the CIC's invalid scientific pretensions concerning CAGW. These people are primarily poor people in less developed countries. The high price of corn due to its mandatory use for motor vehicle fuel in the US, which the Obama Administration supports, results in higher corn prices in poor countries, where poor people are not able to buy the food they need for lack of income. Energy has or will become more expensive as a result of the CIC's insistence on using high cost "renewables" as a source of energy both in the developed world and on projects in less developed countries funded by certain US and other lending institutions. Since such renewable sources of energy are much more expensive than low cost fossil fuel sources, this retards the use of refrigeration and sewage treatment, both essential for saving lives, and encourages the use of firewood for cooking with resulting hazardous home air pollution and forest destruction. Just because these results of GWD/GEF are largely invisible to Western eyes does not make the loss of life any less real.

This is the reality of the CIC's and the Obama Administration's insistence on their invalid CAGW and EWD science hypotheses. Preventing real debate on these hypothesis both in the press and scientific publications and in discourse both within and outside the government is an important instrument to the CIC's ends, but it carries very high costs. The Obama Administration talks the science integrity talk but suppresses it where it might endanger their radical environmental ideology. I have personal experience in this regard.

It is clear to me that the George W. Bush Administration was much more open to independent science than the Obama Administration, which has been unwilling to consider viewpoints inconsistent with CAGW and EWD science. Again, I am an example of this. No effort was made to restrict my climate inquiries until the Obama Administration came to power. The Bush Administration was unwilling to implement GWD, but that is a policy, not a scientific hypothesis. Science requires a comparison between hypotheses and observational reality. Attempts to impose scientific ideology inevitably results in forcing scientists in and out of government to choose between their scientific beliefs and the official ideology, which affects their future employment. The result is that many leave government service, as I did, and those that remain are too scared to oppose scientific nonsense. As a result major policy errors are more likely.

I see little difference in nature (although certainly in adverse effects) between these repeated efforts by the Obama Administration to dictate scientific truth on CAGW and EWD science and the Soviet Union's promotion of Lysenko's biological theories and the Nazi's "scientific" basis for racial discrimination and mass murder.[300] It is simply not possible for government to dictate scientific truth and have any hope of learning what the science actually says.

---

[300] Agin, 2007; Lindzen, 2013.

10
Radical Environmentalism and Its Shortcomings

ANOTHER MAJOR COMPONENT of the CIC is the environmental organizations, largely in the US, Western Europe, Australia, and other parts of the developed world.

In the years since I initially embraced what is now called the US environmental movement it has changed considerably in several ways. The most obvious change is that it has gone from being primarily concerned about wilderness and other wild lands preservation to primarily restricting fossil fuel energy production and energy use. In the process they have come to view fossil fuel use as inherently evil rather than the huge boon it has been to mankind and, in the longer run, to the environment.

Basically, abundant, cheap, and reliable energy made possible by fossil fuels has allowed the Western world to escape the Malthusian trap by making possible a vastly higher standard of living, longer life, and improved health.[301] The less developed world understandably now wants to do the same thing and will not be tricked into giving up the prospect. At the same time the

---

[301] Goklany, 2012. See also Epstein, 2014.

availability of abundant, low cost, reliable energy has made it possible for humans to greatly decrease their pressure on living resources, which had long provided the energy they needed for their activities, thus greatly decreasing human pressure on the environment.[302] And the improvements in the human condition are what has allowed people the "luxury" of preserving wild lands and the environment in general. Yet radical environmentalists instead dwell on the drawbacks of fossil fuel use (yes, fossil fuel use, like every human activity, has drawbacks) particularly conventional pollution, alleged adverse climate effects, and the fact that fossil fuel resources are not replaced by nature over human life spans. In their telling, the great benefits of abundant, low cost, reliable energy available for human use are made to vanish as if they did not exist. But they do, as the strong market for them has long shown.

Fortunately, none of the drawbacks so often cited by environmentalists to fossil fuel use are very important and/or can be overcome without remaking Western society to do without mpst fossil fuel use, as they insist is necessary. *This means that not only has the energy revolution made possible by fossil fuels allowed humans to escape the Malthusian trap that bound us for much of our history, but that there is no reason to believe that the revolution will come to an end for the foreseeable future unless government imposes energy use limits.* The pollution problems generated by the use of fossil fuels can and have been greatly reduced by government regulations in most developed countries. The climate effects are minor and probably positive, as explained in this book. The lack of fossil fuel renewability can and has been handled by the operation of the market and human ingenuity (and will continue to do so if allowed to operate under the government regulation so favored by radical environmentalists). The environmentalists' wind and solar alternatives cannot provide a useful alternative for reasons discussed later in this chapter and the next one.

A second major change is that environmental protection has

---

[302] *Ibid.*

gone from a bi-partisan effort in the US to a liberal Democratic cause, and most recently to a far left Democratic cause.

An interesting question is why the far left of the Democratic Party (including Obama and Pelosi) have provided so much support for radical climate environmentalism. One possible answer is that they actually believe in the ideology. Another possibility is that they want access to the campaign dollars available from wealthy CIC supporters in order to win elections. One Congressman I discussed this with supported the latter hypothesis. A third possibility is that Obama and some of the Democratic elite believe that GEF is good politics in the long run since they may believe that GEF is inevitable and is supported by young voters they may see as their future.

A third and related major change has been that the movement is now directed by a group of fanatics who are not open to opposing viewpoints on climate science, are convinced that they and only they know the received wisdom to judge what their fellow citizens should believe and do with regard to the generation and use of energy, and have no faith or even interest in or apparent knowledge of how the market can and should be used for reaching policy decisions on their topic. This is somewhat typical of far left (and far right) political movements, but makes reaching reasonable compromises largely impossible.

A fourth and again related major change is that the movement is given to exaggerating its case far beyond what the facts can support yet demands that the government impose sacrifices on the population, often primarily on the less wealthy, for purposes that the rest of the population finds of doubtful or limited value. These proposed sacrifices often involve using higher cost or less reliable energy or reducing energy use.

A fifth major change is that the environmental movement is now very well-funded, which it was not when I was active. At that time the Sierra Club had a small number of paid employees, but the conservation effort depended mainly on efforts by volunteers. Funding the paid staff was a constant problem and resulted in a low pay and low budget operation. The situation today is just the opposite. Most large environmental organizations in the US are well funded by an elaborate hierarchy of foundations and wealthy private funders designed to funnel private "charitable" contributions by these funders into both tax-exempt and non-tax-exempt activities including public relations and political campaigns.[303] The result is that the volunteers have less and less influence and the large funders more and more. In many ways this can even be characterized as a semi-contractual arrangement where the large funders define the goals and means to reach them and the large paid staffs carry out their assigned tasks for pay. This helps to explain the overwhelming leftward drift of the environmental movement— it is being paid for by very wealthy left-wing activists who desire influence and anonymity, but also want their tax deductions. In the process the public interest is often lost and tax laws may be circumvented.

A sixth major change is that the tactics used by environmentalists have changed drastically from polite disagreement and presentation of facts and arguments to aggressive *ad hominem* attacks on their opponents. Attempts to smear opponents rather than debate the issues they raise are now the norm. The first reaction is to attack opponents personally rather than disagree with their arguments.

The seventh and perhaps the most visible major change is that many environmental organizations have adopted an international, formalized ideology and attempted to impose this ideology by trying to influence government decisions. The ideology is based primarily on reducing human impacts on the natural world, and in recent years has emphasized some associated vague but seemingly reasonable goals usually summarized in one or two word buzzwords, including "sustainability," "renewability," "recycling"/"reuse," "biodiversity," and "energy conservation," perhaps in part because they may provide added public relations

---

[303] US Senate, 2014a; Rose, 2014.

support for the movement. Many of these goals were developed by or with assistance of the United Nations, but are vague enough to be molded into support for a number of diverse policies yet provide the appearance of a set of common objectives. I do not know whether their formulation included public opinion testing, but they do have that appearance.

A major part of this section will be devoted to analyzing each of these goals to determine whether they are a reasonable basis for determining natural resource/environmental policy. This will be followed by a discussion of a number of other problems with the environmental organizations' modus operandi which are much less publicized but equally problematic.

How the Environmental Movement Lost Its Way

Between the time of my involvement as a Sierra Club activist and chapter leader in the 1960s and early 1970s the environmental movement has changed enormously. It is now clear that between then and now the movement lost its way and has been taken over by radicals preoccupied with energy generation and use and advocating fantasies that would actually harm the environment as well as the economy largely at the expense of the less wealthy in the US and internationally. As discussed in Chapter 3, I had recognized and opposed this viewpoint in David Brower as early as the mid-1960s.

The American environmental movement was primarily a wilderness preservation/park creation movement prior to the 1970s. Pollution control was discussed and a few campaigns were waged, but the emphasis was not on it. This all changed in the late 1960s and very early 1970s when pollution control became popular among both the public and politicians of both parties. The environmental movement soon followed suit in what started as largely a popular movement. The result was that remarkable progress was made in the 1970s and 1980s towards changing the environmental ethos of the US and many Western nations. It was an issue that had come of age and the environmental world as I had known it as late as the 1960s changed forever. Laws were soon passed in the US and elsewhere greatly improving protections against air and water pollution, toxic substances, and much more. Agencies were created both at the Federal, state, and international levels to implement these laws and encourage further pollution control. Both political parties contributed to this effort in the US. Soon it became both socially and legally unwise to pollute the environment for private gain. This represented enormous progress which I strongly supported and was happy to devote my full time efforts helping to implement.

Significant problems remain after 40 years, but many of the more publicly evident problems, like air pollution, have been brought under substantial control in the US, and as I will argue in Chapter 11, over control. Many of the major remaining problems, like non-point source water pollution, are much less evident to the casual observer and harder to gather public support for control. Construction of wastewater treatment facilities has improved many measures of water pollution but sewer overflows remain a persistent problem. Non-point source pollution has become the largest single obstacle to improving water quality.[304]

So I am not arguing that nothing EPA and other environmental organizations are doing is worthwhile. There will undoubtedly be additional valid environmental problems that need greater attention at various levels of government. But pursuing non-problems or solutions that will not solve real environmental problems or pushing environmental regulations beyond the point of diminishing returns is not something that government agencies should pursue. But that is what they are doing in the US and in Western Europe. Two current examples are CO_2 emission-reduction efforts and in the US excessively strict air pollution regulations such as the new proposed ozone and mercury (MACT) regulations.

An interesting question is how the environmental movement ended up emphasizing CO_2 reduction as its primary goal. One reason may be that the rapid success in controlling the more visible

---

[304] Andreen, 2004.

forms of pollution led to a problem. What were the recently greatly expanded environmental organizations and their now increasingly paid staffs to do next that would pay the increased costs? They could disband or find a new mission. In the end most decided to pursue a much more radical agenda which increasingly involved an anti-energy, anti-capitalist, statist approach. This conversion was undoubtedly encouraged in recent years by the cash available from wealthy environmental activists and foundations. This attracted supporters from the far left of the political spectrum rather than the bi-partisan following that had been active in the 1960s, 70s, and 80s. Left wing supporters may have found it easy to support fighting capitalism because it was allegedly destroying the Earth rather than exploiting the workers (as alleged under Communism), suppressing individual liberties for the benefit of the environment instead of the state, and encouraging the left-of-center media to supply misleading coverage with strong support from privately financed media watchdogs who would immediately denounce anyone who deviated from the environmental "consensus," often in very personal and uncivil terms.

Two other developments may have contributed to this process. One was the demise of the Soviet Union and Soviet Communism in 1991 which had pursued anti-capitalism and suppression of individual liberty for the good of the state (although certainly not with environmental ends in mind) and may have left some far left supporters in Western countries without an intellectual home. The other was the initiative launched by the United Nations starting with the World Climate Conference in 1979, which led to the development of the IPCC, the Rio Conference of 1992, the UN Framework Convention on Climate Change, and the Kyoto Protocol to it. All these developments plus the goal changes enumerated later in this chapter led to the GWD/GEF movement that we know today with its preoccupation with controlling energy production and use.

Whatever the historical process, there is now overwhelming evidence that the radical climate movement draws strong support

Environmentalism Gone Mad

from the anti-capitalist movement, including the Communist Party, since a significant number of the participants in recent climate rallies publicly supported this goal, evidently because they believe that climate change is caused by capitalism and cannot be prevented except by getting rid of capitalism.[305]

I, and I think most other economists, regard energy as a commodity which is a vital input to economic growth and development, but basically a commodity. Radical energy environmentalists too often seem to regard it as the embodiment of evil that must be directly controlled by government rather than left to the market to allocate and develop like most other goods and services.

It is important to note that unlike many of the earlier concerns about conventional pollutants, the new campaign against CO_2 emissions was largely a top-down rather than a bottom-up phenomenon. In other words, the current radical energy environmental campaign against CO_2 emissions did not originate primarily from mass dissatisfaction with CO_2 levels, which are undetectable except by using sophisticated instruments. Rather, it originally came from several conservative politicians (primarily Margaret Thatcher and to a smaller extent George H.W. Bush), but was then pushed by the far left of the Democratic Party in the US, especially Albert Gore, after the Clinton Administration came to power.[306] So it was the political and intellectual elite rather than a grassroots movement that originally pushed for GWD. All the other constituents of the current CIC soon joined in, however.

Just Another "Apocalypse"

If the global warming/climate change "apocalypse" had not been created by the United Nations, the radical environmental movement unfortunately might have created it anyway. But given

---

[305] See Watts, 2014a, and links from it.
[306] Darwall, 2013.

the UN's efforts all it had to do was to seize on it and let the alleged "apocalypse" pay the rapidly increasing bills for a growing professional staff. Simply put, global warming/climate change has paid big short-term dividends to the environmental movement to date. It is ideal for professional staff increasingly employed by most environmental organizations. It gives their efforts a purpose, which many supporters can agree upon. The science is sufficiently obscure that (often justified) claims of bad science have been brushed off by allegations that they come from paid shills funded by evil polluters. The alarmists make no real effort to even understand these allegations let alone to deal with them, which they could not do anyway for lack of much expertise, and appeals to the "authority" of the United Nations IPCC. Other parts of the CIC such as academics supported by Government climate research grants, could always be found to denounce the skeptics' scientific arguments if simply ignoring them does not work. The environmental organizations usually argue that the skeptics do not possess sufficient expertise (despite the evident lack of expertise of most of their leaders) or are in the pay of evil polluters (usually oil and coal companies or their employees) or whatever. In fact, the most common defense by both environmental organizations and the propaganda arms of the CIC is to simply call skeptics nasty names. GWD/GEF has been even more overblown than all the other feared apocalypses from the Club of Rome to Y2K.

Although the Club of Rome's *Limits to Growth* formulation of the feared apocalypses brought on by man did not end up having much influence, their basic idea—that humans are using up their available resources and economic growth cannot continue indefinitely—lives on, particularly in GWD/GEF. The fear of global warming (due to allegedly using up the ability of the environment to absorb CO_2 without causing problems) was one of those advanced by the Club of Rome and lives on by way of the Rio Conference of 1992, the UN Framework Convention on Climate Change, and the IPCC. Like the IPCC, the Club used elaborate computer models to try to "prove" its viewpoint, but with a similar lack of scientific rigor.

The main difference this time is the involvement of the UN, major far left wealthy funders and foundations, many environmental organizations, many Western politicians, particularly in Western Europe and Australia, and more recently the US. These politicians have actually taken global warming seriously and spent money that was badly needed to solve real problems on a non-problem that they could do little or nothing about in the way proposed. Government support has not only resulted in "eco-friendly" policies but also government support for climate research and development. This has gradually changed the intellectual content of relevant disciplines to the extent that alternative approaches are not even being funded. The only good thing is that citizen-scientists have started examining the science and have learned some interesting things which tens of billions of dollars of government-sponsored research has ignored and often actively denigrated. The citizen-scientists may have even managed to advance the science if anyone were willing to listen to what they have to say.

The Absurd Results of What the Environmental Movement Now Advocates

The current Western economy has developed over hundreds of years largely without the direct help of government (beyond supplying a stable, neutral playing field and a legal system to support it). If the public found that new products were useful, they bought them. If the public did not find them useful, they did not. The public found that cars were useful and that most windmills and solar panels were not except for specialized uses. Cars proliferated, windmills largely died out until public subsidies began, and solar panels have largely lived off public subsidies since their introduction.

The essence of radical energy environmentalism is that a small, often wealthy elite with the help of massive mass media publicity should determine what forms of energy are produced and consumed using the power of government coercion to enforce their preferences rather than leaving such decisions to individuals and the market they create through their purchases. It advocates that relevant government policy should not be left to the normal interplay of politics, but must be implemented directly by a less than democratic government executive regardless of the views that political opponents may have. This has happened before, of course, when those advocating other statist ideologies tried to impose their will on society. The Communists claimed that government coercion was required in order to avoid economic exploitation of the masses. Fascists claimed that government coercion was required for the good of the nation. And now environmentalists argue that it is required in order to "save the Earth" from first global warming, then (when there was little warming) climate change, and now (when there has been little more than normal climate change) avoiding extreme weather.

Unfortunately, any time government interferes in the operation of the market, there are unintended consequences. Some of these are so serious that they negate whatever government was trying to achieve in the first place. Such unintended consequences are too numerous to even list in the case of GWD/GEF, but can be found elsewhere,[307] so only a few will be discussed here.

The idea that any small elite knows better than the operation of the market or the operation of normal political processes what people need and want is the basic absurdity of radical energy environmentalism just as it was for the other prominent "isms." The problem is not far right or far left policies, but the basic idea. Government coercion in the economic and political marketplace without careful attention to the need for it has not worked successfully in the past and probably never will. Radical energy environmentalists have joined with the far left to get government to achieve their ill-considered ends: Higher energy prices, particularly difficult for the less wealthy, result in more poverty, decreased economic welfare, slower economic growth, and very little if any

---

[307] Global Warming Foundation, 2015.

change in the climate. Poverty and lack of access to reliable electric power are two of the most serious problems in many less developed countries and radical energy environmentalists are doing everything they can to keep billions of people from leading better, more healthy, and longer lives by getting access to it.

The answer to the world's ills is not a long ago discarded technology with serious adverse environmental effects,[308] windmills, or a new technology that gathers diffuse and unreliable energy from sunlight at very high cost in money and land use.[309] Rather the answer is to make sure that the incentives of the market place fully take into account any environmental "externalities" and let the political and economic marketplaces decide as much as possible. This also avoids the inevitable corruptions of government involvement best exemplified by the US requirement to use ethanol in gasoline. If government was not involved corn ethanol would not be used for fuel and consumers would be far better off for it. But it is involved and the resulting corruption is inevitable and very costly, particularly for the less wealthy in the US and in the rest of the world who must pay more for the corn they need for food.

An interesting question is whether radical energy environmentalists are really motivated by the belief that there will be catastrophic global warming/climate change/extreme weather unless CO_2 emissions are reduced by their proposed 80% or whether their basic motivation is to oppose capitalism/economic growth/material progress in every way they can. Obviously there are mixed motivations. But I believe that they are either being illogical or are simply opposed to economic progress and in favor of increased government control. If they were seriously interested in reducing CO_2 emissions, the best way to do so would be to promote substitution of natural gas for coal by bringing down the price of gas through promoting fracking. This might actually be achievable in countries with large shale resources, just as the US

---

[308] *Ibid.*
[309] *Ibid.*

private sector has shown.

But instead many environmentalists oppose all forms of electric power except those that are prohibitively expensive, unreliable, and destroy vast quantities of land, birds, and bats as well as scarring the landscape for miles around.[310] Many are also opposed to fracking and substituting natural gas for coal despite the greatly reduced CO_2 emissions that would result. They are opposed to building a pipeline from Canada to carry oil to the US Gulf Coast. They are opposed to building liquefied natural gas export terminals which would provide jobs and improve national security in Western Europe. They will not discuss science that is contrary to their ideology and call those who raise such issues nasty names. So instead of pursuing environmental goals that might actually be achievable and justifiable, they appear to be acting irrationally— unless their real motivation is opposition to the continued operation of the capitalist system rather than any reductions in CO_2 emissions. I doubt that they are all irrational.

Besides the total unreality and adverse effects of their energy goals, the environmental movement advocates a number of policies that I find particularly repugnant. In addition, other policies which are equally repugnant have been added by politicians to benefit various special interests on the claim that they will further the same ends. Even when not supported by environmental organizations I hold them responsible for the adverse effects of these policies which would not exist if the environmental organizations had not provided a convenient policy handle for politicians to hang their favorite giveaways to their supporters.

In addition to other adverse effects of GWD/GEF/EWD/AFD, which will be discussed in Chapter 11, I hold the radical environmental movement responsible for the following other particularly repugnant policies:

- *Discouraging/banning the use of DDT to control*

---

[310] *Ibid.*

malaria

- *Increasing the price of food by requiring the use of biofuels made from agricultural products that would otherwise be used as food*
- *Banning the use of genetically modified food that would reduce disease and prolong life*
- *Banning or discouraging the use of GM crops that would increase production of food*
- *Reducing the use or increasing the price of fossil fuels or requiring the use of specified energy production technologies in order to reduce CO_2 emissions*

Absurd Result: Discouraging/Banning the Use of DDT to Control Malaria

As discussed in Chapter 8, the US ban on the use of DDT resulted in similar bans in many less developed countries which effectively ended the effective malaria control programs in some of these countries launched in the 1950s and 1960s. This ban was and to a significant degree still is supported by many environmental organizations. This has resulted in the preventable deaths of tens of millions of people, particularly children, and serious, debilitating, illness for many more.

Absurd Result: Increasing the Price of Food by Requiring the Use of Biofuels Made from Agricultural Products that Would Otherwise Be Used as Food

One result of the radical environmental movement's philosophy (although not generally supported by the movement itself) is that it is used as an excuse for government-required or encouraged use of corn and sugar-based ethanol and vegetable oil-based biodiesel for fuel. This raises the price of corn, sugar, and vegetable oil in many less developed countries dependent on it for food and increases the acreage devoted to their production. This too often leads to soil erosion, wilderness destruction, pesticide use, and most importantly, human hunger as low income people can no longer afford to buy the food they need to survive. And during droughts, such as the US Mid-West suffered in 2012, it further accentuates the huge price increases and human misery that result. The primary winners are the corn and sugar growers and agribusiness, who use the green ideology to urge governments to increase their income at the expense of consumers everywhere and particularly the poorest consumers who can no longer buy the food they need to survive. There can be little doubt that higher prices for food resulting from burning ethanol made from corn and sugar in vehicles, which is justified by the GWD ideology (although not by many environmental organizations) make the poorest in less developed countries hungrier and even poorer. And some of these poorest of the poor end up dying of malnutrition for lack of funds to buy sufficient food. This is unconscionable.

Biofuel use in place of fossil fuels has other adverse effects, sometimes on the environment. One well-known example is the conversion of rainforest to grow palm trees for their palm oil in order to use it as a biofuel.[311]

Absurd Result: Banning the Use of Genetically Modified Food that Would Reduce Disease and Prolong Life

Some environmental organizations, particularly Greenpeace, have strongly resisted the introduction of Golden Rice, which is believed to be the lowest cost way to treat vitamin A deficiency in the less developed world.[312] Vitamin A deficiency causes a quarter to a half million children to go blind every year according to the World Health Organization, of which half die within half a year. A study in the British medical journal *Lancet* estimates that 668,000 children under the age of five die each year from this deficiency. Eye diseases

---

[311] Global Warming Foundation, 2015.
[312] Lomborg, 2013a.

resulting from lack of Vitamin A because of environmentalists' objections to Golden Rice could have been inexpensively prevented by encouraging rather than discouraging the use of Golden Rice, a GM food widely eaten in less developed countries. But I am sure that those responsible for these travesties are happy because they think they are saving the world either directly or through promoting their environmental ideology. I strongly disagree.

Absurd Result: Banning or Discouraging the Use of GM Crops that Would Increase Production of Food

Greenpeace also opposes the use of GM crops to increase global production of food.[313] They claim that there is already enough food for everyone and there is no need to use such food. I argue that the key issue is the price of basic foodstuffs to the poorest people so that they can afford the food they need for good nutrition. Using GM food is one way to lower the costs of food production and therefore food, which I strongly support as long as there is no evidence that it will harm people or the environment, which there is not.

Absurd Result: Preventing the Poorest of the World's Poor from Achieving a Better Life

Energy generated to assist humans is not just another economic commodity. It is the basis for much of the improvement of living and health standards during and since the Industrial Revolution. The availability of low cost, reliable energy appears to be a requirement for further economic growth as well. Even the staff of most environmental organizations are not likely to approve of not being able to recharge their smartphones when the batteries run down. It is an economically justifiable function of government to prevent the adverse economic effects of pollution. But efforts to reduce the positive effects of economic development and energy use and to

---

[313] Greenpeace, 2010.

prevent the normal functioning of economic markets is counterproductive and potentially disastrous in terms of economic growth and welfare and also of environmental protection.

A case in point is that for a decade or more the Sierra Club, which I supported and worked with and in during the 1960s and early 1970s, has advocated halting the generation of electricity from coal, which is in many areas of the world is the lowest cost and one of the most reliable sources. Now they also want to prevent generation from natural gas, currently the next lowest cost fossil fuel source in many areas of the world, too. Most recently, they have advocated civil disobedience to achieve their policy aims, the first time they have ever done so. Their proposed alternative to fossil fuel development is wind and solar generation, despite their unreliability, unsuitability for use in large electric grids, much higher costs, major adverse environmental effects, and their minor or even negative overall effects in reducing CO_2 emissions. If the environmentalists and solar and wind promoters get their way, the turbines and solar arrays will cover millions of acres of farmland and wildlife habitat to provide expensive, unreliable power which must be sent over newly constructed segments of the national grid and backed-up by on-line fossil-fuel or extensive hydroelectric power resources to achieve the reliable power that modern society must have if it is to continue economic growth so vital to the well-being of many citizens and the main hope for an improved environment. Many areas of the world and the US do not have access to extensive hydroelectric power, so are forced to build both "renewable" power sources and fossil-fuel back-up that can utilized on very short notice if they decide to build "renewable" power. The result is much more expensive electrical energy, no significant reduction in CO_2 emissions, and very little or no change in the environment.

The environmental movement often goes beyond this and advocates restricting energy use as a whole despite the fact that energy use has been the primary means of improving the standard of living of humans around the world over the last few hundred years and thereby makes possible much environmental improvement.

Fossil fuels not only save humans from nature's formerly deadly weather whims; they also lower humans' demands on living nature.[314] Plentiful, reliable, low cost energy has made possible the elimination of many diseases still quite common in less developed countries by access to electricity for refrigeration, hospitals, clinics, water purification, and sanitation. Less developed countries without inexpensive, reliable sources of energy pursue some of the most environmentally damaging activities such as gathering and burning firewood desperately needed to keep land from eroding away and endangering the health of those living where the wood is burned.

I have watched the transport of an endless parade of firewood carried by poor people from outlying areas needing it for cooking across the river from the Taj Mahal in Agra, India (see Figure 10-4 at the end of this Chapter), and have seen the resulting air pollution resulting from the burning of such firewood in kitchens. This practice may be "renewable" but is not a good way to safeguard the environment or public health. The availability of low cost, non-wood-based, reliable heat for cooking would be one of the best environmental improvements India and many other less developed countries could make even though it would be directly contrary to current environmental dogma.

Higher prices for fossil fuels needed for warmth and cooking in many parts of the world as a result of attempts to increase prices in accordance with GWD ideology would only make life more miserable for billions of people, particularly poor people around the world, but result in death for many who could no longer afford the food they need to survive.

The broader result of the environmental movement's anti-fossil fuel energy campaign will be to raise the price of energy produced using fossil fuels. This hurts particularly the people who most need low cost energy—the world's poor and economically struggling populations—as well as decreasing the economic competitiveness and economic growth of any country that tries to

---

[314] Goklany, 2012

pursue the movement's proposed policies. In Western countries, increasing energy prices hurt the poor whose expenditures are disproportionately made up of energy costs while being hardly noticed by their more wealthy fellow citizens, the main supporters of raising energy prices. Since many fuel expenditures are for necessities like heat and transportation to work, other necessities will be bound to suffer. I cannot support policies that unnecessarily raise energy prices.

Much of the economic growth over the last few centuries has been made possible using fossil fuel-based energy to substitute for human time and energy and would not have occurred without low cost, reliable electric power since its advent in the late 19th Century. Such energy use initially replaced primarily human muscles starting in the industrial revolution. More recently it has made possible the replacement of human mental effort as a result of the computer revolution. Access to reliable and inexpensive energy is fundamental to the modern economy. Attempts to make it less available, more expensive, or less reliable strike at the very heart of human welfare, economic growth, economic development, and even environmental improvement. Yet whether modern environmentalists understand this or not the effect of what they are trying to do is just these things. I oppose such efforts. The primary losers are those who could achieve a higher level of income and fulfillment if only there were higher economic growth and faster economic development through greater use of inexpensive, reliable energy.

Radical environmentalists essentially advocate that the poor of the world should remain poor by living at subsistence levels in order to avoid emitting a gas that enables plants to grow better and makes it possible for humans to use energy to assist them in their daily tasks and ultimately leave much disease, malnutrition, and other forms of deprivation behind.

I find what radical environmentalists are now trying to do extremely reprehensible and I cannot support such fantasies. This book outlines where I believe the environmental movement has gone astray. Their scientific justification is largely based on speculative computer extrapolations with little or no basis in physical reality and a consistent record of faulty projections, and a number of vague concepts with high public favorability ratings such as "sustainability" which can and are adapted to suit whatever policy outcome the environmental movement currently advocates.

Very recently, in 2012, many environmental organizations, and even President Obama have attempted to substitute alleged extreme weather as the basis for their concerns in place of climate change (which was originally global warming) while leaving the "solutions" unchanged. I note that 2013 and 2014 has brought very little extreme weather to the US. This attempt to redefine the threat is pure political opportunism with even less basis in science. Once again they are often using bad science and computer extrapolations to justify their approach. This is not what is needed to put the environmental movement back on track; it is more of the same. Getting the science right is the first step in successful environmental protection. So the emphasis in this book is on the science and economics, which I believe to be fundamental to environmental protection. Getting the solutions right is critical for environmental protection. This requires great effort and care, which is unlikely to result from mass hysteria or popular imagination or environmental organizations that base their policy prescriptions on such.

A Real Life Example of the Crucial Role Reliable Energy Availability Plays in American Life

As I was writing this book an incident occurred that illustrates the crucial role that reliable energy supply plays in American life. A freak storm, called a derecho, impacted our area on a hot summer evening with 70 mile an hour winds which downed a number of the older, larger trees. Some of these trees took down the overhead power lines that our neighborhood and many others in the US have. Most of the neighborhood's electric power was almost immediately cut off.

This resulted in all the usual coping behavior by us and our neighbors who immediately lost access to many of the necessities of modern life, including the all-important air conditioning, microwave ovens, refrigerators, lights, traffic lights, computers, television, land-line phones, power for recharging cell phones, and the internet. So the next day one neighbor offered to take everyone's hand-held electronic communication devices to a store that was willing to provide power to recharge them. She soon had a plastic bag full of them. Another bought out the stock of ice cubes at a local supermarket and distributed it on the street. My contribution was to talk to a supervisor for the local power company who had apparently been sent out to survey the neighborhood to determine whether to restore power to some or all of it. I pointed out that our street had escaped any damage to its power lines. I noticed that before he left the neighborhood our street and neighboring streets regained power but that the more heavily damaged areas which would require major repair work had not.

For ourselves and our neighbors the loss of power immediately became the preoccupation and immediate concern at the expense of any other concerns or efforts that the residents may have had. Would our food spoil? What damage was the high inside humidity doing? When would the power be restored? How could we tell our families living elsewhere about our welfare with no telephone or internet service? Within a few days, of course, climate alarmists were hard at work claiming that this and other relatively unusual weather events in 2012 were a result of rising CO_2 levels and "proved" the latest version of their climate scare. Along with Superstorm Sandy later in the year, the CIC has attempted to change their objective once again, this time to reducing extreme weather.

So why is this ordinary occasion of American life worth retelling? I do so because it illustrates just how dependent we are on reliable energy supplies. Life as we know it and especially our children know it, comes to a halt without it. Yet environmentalists would have us believe that reliable energy is not a necessity. They oppose every "non-renewable" energy source proposed, and even some of the "renewable" ones. They act as if they were unaware what happens when energy supply does not meet demand even briefly. I have lived in India, however, where this is a routine occurrence and know what happens during periods when your power is cut-off, as it routinely is on an unknown schedule.

Inexpensive, reliable energy supplies are a modern necessity just to maintain our standard of living as well as for further economic growth. We know how to provide such reliable electrical energy except during unusual weather events in areas with above ground electric lines. There are ample fossil fuel energy sources for electrical power for thousands of years in the future, and if we should ever run low, prices will increase and alternative sources will soon be used as long as we continue to use the price mechanism and human ingenuity. The environmental extremists who seek to reduce our capability to provide inexpensive, reliable electric power should be required to show just how they would cope with periods of rolling blackouts that power shortages always lead to and wide area power black-outs they sometimes lead to. Some of the best known figures in the environmental movement, as well as President Obama, are more widely known, however, for energy profligacy than for the energy frugality they so freely urge governments to impose on the rest of us.

Alarmists will claim that their expensive "solution" will avoid global warming, climate change, and even violent storms that sometimes bring loss of access to the energy resources now so essential. They have and will probably continue to generate computer models "proving" that this is the case. But they are selling a "service" that government cannot deliver in the way they propose and will not meet the needs of Americans or any other developed nation that has become dependent on reliable electric power or any less developed country that aspires to an improved standard of living. It is time to find new prophets. The current ones are no better than those who demanded human sacrifices for favorable weather or crops in earlier times but could not deliver either. Paying more for energy or doing without it will have no measureable influence on avoiding unfavorable climate change or extreme weather. The chief losers will be those who do the paying. Unfortunately the Obama Administration and some largely blue states have adopted the radical environmentalist energy ideology/religion. They believe they currently have the power to compel Americans to follow their dictates as long as Congress/state legislatures/courts do not overrule them (which to date they have largely not done), and have formulated plans to impose such a regimen unless they are voted out of power.

Since this extremely serious situation is generally not understood by those who will pay for all this, Chapter 12 will explain how we got into this mess and what we need to do to get out. The radical environmentalists claim that climate change is a problem, that they know how to solve the problem, and that everyone should pay to solve this problem by sacrificing their money and their economic (as well as their intellectual) freedom. I believe that the primary problem is the radical environmentalists and the politicians who are doing their bidding. Unfortunately, the radical environmentalists have managed to get just about everything wrong through lack of attention to the science and the economics or perhaps deliberate distortions of it. Following their bad dictates will only make things worse, not better. This book explains why.

Obviously the US is not in as serious an electric power situation as India, but starving the US system of generating facilities and the coal or natural gas to run them is not a good way to improve the US economy, which needs plentiful, reliable, low cost energy if it is to grow rapidly. And although environmental organizations never mention it, the US has one of the "best" records in the world in terms of reducing CO_2 emissions in recent years even though it fortunately never ratified the Kyoto protocol. Yet the Sierra Club, other environmental organizations, and the Obama Administration are all advocating policies which are likely to result in the same types of problems being experienced in India. I call this madness based on fantasy thinking. And you can be sure that the environmental movement and the Obama Administration will be blamed if non-extreme event weather-related blackouts occur in the US. I cannot think of a better way for radical environmentalists and the liberal fringe of the Democratic Party, which generally support them, to lose power and influence in the longer run.

Ideological Basis for Radical Environmentalism

The general goals of the environmental movement have some common general problems. One is that the one or two word goals are so vague that disputes can and do arise as to their meaning, interactions, and contradictions. Another common general problem is that there is no apparent way to judge how much is too much and how little is too little. These two common general problems will become more apparent as I discuss each of the major goals.

This is where economics could and I believe should fulfill a particularly useful role either through the marketplace or by making it possible to place benefits and costs on policies and activities where that would be useful to humans. But radical environmentalists clearly want economics to play no role in their desired utopia. According to them, all energy use, and particularly new sources, must use their favored list of renewable sources (water power is clearly renewable but usually omitted by environmentalists because they generally oppose new dams), regardless of the costs in terms of wildlife destruction, higher energy costs, and unsightly, often abandoned, windmills on hills and along ridge lines, and generation of electricity when and where it is not needed and lack of generation when and where it is. The economic answer is very simple: Leave these decisions to economic forces fully adjusted for the adverse effects of unwanted side effects such as pollution where justified. Further government meddling in these decisions, which environmental groups consistently favor, almost always leads to bad, often absurd, outcomes. If such decisions are not left to the market, economics provides the crucial answers as to how much energy conservation or renewable use is too much. Where there is no market available, as in areas requiring Government policy setting, economics provides the tools to make informed decisions in the public interest. Without this quantitative use guidance, broad policy prescriptions are not very useful. Somehow economists have too often gone along with these policy prescriptions of environmental groups, perhaps without realizing how their long held principles are being trampled on. The most dramatic examples of the disregard of economics in environmental policymaking are in the energy and the recycling areas, where the environmental magic words are used to justify the economically unjustifiable.

It is evident that the Sierra Club I knew in the late 1960s and early 1970s is no more. An interesting question is why? Like other environmental organizations they chose to adopt the much more ideologically radical goals that I had worked to avoid through my involvement in the 1969 Club election. Instead of emphasizing wilderness preservation, park protection, and park creation, as they did in the first 80 years of their existence with considerable success, during the following decades they and other environmental organizations increasingly adopted the doctrines of what I call radical energy environmentalism, which has a more European rather than an American ancestry, as well as GWD/GEF with its alleged goal of avoiding an alleged apocalyptic global warming, later changed to climate change, and most recently changed to avoiding extreme adverse weather, apparently in response to perceived marketing opportunities.

The shift to more radical ideological goals has been accompanied by the increasing professionalization of the environmental movement. As they became increasingly dependent on paid staff, they also became increasingly dependent on funding from outside their organizations and members. This in turn meant that they constantly had to adjust their goals to meet the interests and desires of these funders, many of whom seem to have wanted primarily to "make a difference." This, I hypothesize, resulted in the Sierra Club becoming increasingly confrontational and less willing and able to understand reality. Much progress has been achieved towards some of the legitimate goals of environmentalism such as air and toxic pollution control in the US some time ago, but by pushing against fossil fuel energy development as a whole and indirectly rejecting modern civilization, and pushing for ever increasing environmental regulation, environmentalists have managed to be confrontational in areas where they will and have gotten strong resistance from many players in the economic world. Perhaps environmental organizations fear that if they are not viewed as confrontational that the donations that pay for the increasing paid staff will dry up.

As guiding principles for resource use, however, the radical environmental ideology is inconsistent and unworkable since as currently interpreted they give rise to a variety of adverse outcomes. The underlying problem is that they do not take into account economics. Ignoring economics leads to a variety of problems. When policies based on their policy-guiding words give rise to conflicts with other policy-guiding words, there is no way to easily discern the "correct" policy. If renewable sources of energy such as wind result in a large reduction of biodiversity by killing endangered raptors and bats, how is this conflict to be resolved? This is more than a theoretical problem since environmentalists sometimes oppose proposed wind energy projects for just this reason.

Ideological Basis: Sustainability/Renewability

If there is a central but very vague concept in radical environmentalism, it is probably "sustainability."[315] Many modern radical environmentalists and their academic supporters believe that environmental decisions should be based on this nebulous concept

---

[315] The term sustainable development is generally credited to the United Nations World Commission on Environment and Development, 1987, often referred to as the Brundtland Report.

rather than on the decisions of a relatively free market corrected for any clear, documentable cases where free market prices do not fully reflect the adverse effects of economic activity on others not directly involved (referred to by economists as negative externalities), whose correction can be economically justified. Environmentalists appear to think that their concept of "sustainability" leads to better decisions than what the operation of predominantly free markets judges to be best.

They often assume that since the total supply of non-renewable resources is fixed in physical terms, human demands must ultimately exceed the supply, usually with allegedly disastrous consequences. In many cases they assume that modern society will soon come to an end because crucial non-renewable resources will run out. This ignores the important role that prices can play in free market economies if allowed to do so in spurring supply from sources that are more difficult to extract and the use of alternatives, and in decreasing demand when demand begins to exceed supply. It also ignores the increasing technical ingenuity and sophistication and decreasing costs of locating and extracting smaller or more difficult to obtain finite resources. Ecologists seem to be particularly prone to believing that modern society will imminently "run out" of critical non-renewable resources.[316] This was the basic idea behind *Limits to Growth*, but persists in the climate debate where ecologists usually claim that the world is fast running out of its ability to use the atmosphere as a repository for the CO_2 generated by our use of fossil fuels. Like all the other Limits to Growth, this ignores the ability of humans to find solutions to raw material problems through innovation and the use of the market mechanism.

The key constraint is not the supply of natural resources. Rather, it is the ability and the opportunity of humans to find and use the technologies and the existence of economic and political systems that allow natural resources to be used to meet human needs. Preserving particular natural resources such as coal or oil by

---

[316] Ridley, 2014.

reducing use or using non-renewable natural resources less intensively (as by employing windmills) on the basis that they will otherwise "run out" assumes that it is the natural resources themselves that are the important constraint on human advancement and ecosystem preservation. This is simply not the case. Ecosystem preservation has and will occur when humans decide to find ways to make it happen. And it will not occur by government regulations forcing humans to use less energy. Ecosystem protection is found either in areas with no humans or with wealthy humans who can afford to worry about it. Human advances occur primarily in areas where wealthy humans live and can afford to invest in better technology to meet their needs.

Sustainability is a concept borrowed from fisheries management, which the environmental movement has attempted to apply to all natural resource use, where it is much less useful. Sustainability as applied to fisheries may be a useful concept because it is dealing with resources which are truly limited in that depletion of fish stocks may lead to making the fishery less productive due to falling reproduction or possibly even to permanently ending it if the fished species go extinct. Non-fish farming fisheries are also generally subject to open access, which means that ownership cannot easily be established and use easily restricted since there is no owner or access cannot easily be controlled; the result is that a purely competitive approach may lead to overuse of the resource in terms of economic efficiency (another externality in economic jargon).

Use of fossil fuel resources, the radical energy environmentalists' major concern in recent decades, as well as other mineral resources, is very different in that such resources will never be totally used up and are not generally subject to overuse through open access since most are owned by someone (although sometimes by various governments). The worst that happens is that higher grade, less expensive to extract, resources will decline so that lower grade, higher cost sources may be the best available. But because of improving human ingenuity, technology, and substitution, the all-important cost of making these resources available for human use often declines over time with continuing or even increasing use. In these cases sustainability offers no useful insights since the normal functioning of the economic marketplace will handle the much feared exhaustion problem as long as it is not interfered with. The sustainability ideology argues for not using any resources that are not naturally renewable. This is foolish and counterproductive. Coal is not a "sustainable" resource according to environmental ideology, but the economic cost of mining it or substitutes for it may well decrease for many years to come. As long as the price of a raw material decreases there is clearly no basis for fearing that the resource will "run out" or that its use needs to be decreased to avoid this.

Even if prices are increasing, as has generally been the case for oil in recent decades (but not late in 2014), the market is signaling that continued use at the current level requires higher prices to find it, extract it, and bring it to market. Normally this leads either to a decrease in use or the use of human ingenuity to find alternative lower cost sources. In the case of oil, use has become more efficient through use of more efficient engines and natural gas has been substituted for some stationary source uses. And the cost of extracting oil from non-reservoir sources such as shale have fallen greatly through the use of fracking and other technological innovations. The assumption that oil use has "peaked" and must necessarily and inevitably fall is simply false if innovation is allowed and if the market is allowed to freely operate. Many environmentalists oppose innovation and free energy markets, which could indeed bring the very results they claim to fear. If fracking and market freedom is curbed by government fiat as they advocate, oil prices will probably increase just as they claim. But if so, it is radical environmentalists and government regulation that will be responsible, not any "shortage" of non-renewable resources such as oil. The substitution of government fiat for the free operation of the market is the major problem and not the solution, as environmentalists usually argue.

Marlo Lewis has expressed the resulting problem well:[317]

> *Because cheap, reliable, scalable alternatives to carbon energy do not yet exist in the vast majority of regions and nations, coercive de-carbonization is bound to be either a 'cure' worse than the alleged disease or an expensive exercise in futility. Cap-and-trade, carbon taxes, renewable energy mandates, and the like will accomplish little except to centralize power and transfer wealth from consumers to special interests.*

The primary concern of the radical environmental movement in recent decades has been fossil fuels. They seem to be fixated on avoiding "overuse" of fossil fuels, particularly coal and oil, or as they call it sustainable use. Their concerns appear to be three fold. The first is that we will somehow use up available supplies; the second is that the carbon dioxide emitted by burning fossil fuels will result in global warming, or later climate change, or most recently extreme adverse weather. Their third claim is that fossil fuels are "dirty" and should be eliminated.

The first concern is not worth serious concern since fossil fuels will be available at affordable prices for many thousands of years in one form or another if the market system is allowed to operate without significant governmental interference (except where justified because of externalities).

A major portion of this book is devoted to showing the lack of a scientific basis for the second concern. In fact, increased emissions of carbon dioxide are a benefit because of the ready worldwide demand for it by plants. Increased plant growth due to increases in the availability of one of their raw materials, CO_2, will help both humans and ecosystems. Any modest increase in global temperatures that may result may help to put off the next Little or full Ice Age with its disastrous consequences for life on Earth.

---

[317] Lewis, 2015.

Radical Environmentalism and Its Shortcomings

The third concern that fossil fuel use is "dirty" and should be eliminated is a gross perversion of reality.. This problem has largely been eliminated by government regulations in developed countries. In less developed countries, anyone who has visited the rural areas knows that the really dirty fuels are wood and animal dung, often used for cooking or heating. Gathering the wood often results in the destruction of trees vital for preventing soil erosion and its use results in serious indoor pollution problems, or even more general outdoor pollution on days with little wind. The substitution of other energy sources such as gas or electricity generated from coal or oil is the best way to reduce pollution from energy use in less developed countries. This will only happen if these energy sources are less expensive than gathering wood in terms of the time and resources of the people who need the heating or cooking. In developed countries, the way to solve pollution problems caused by energy use is not to reduce use but through careful, source unbiased regulation of the individual pollutants produced. When such generation is primarily in large power plants, this is comparatively easy to do.

At its heart, the sustainability movement is based on the alleged need to avoid future scarcity of natural resources through maximal conservation of existing resources used by humans and other life by limiting current usage by humans. It seeks to achieve this through the use of government regulations. In doing so, this basic concept of radical energy environmentalism disregards and attempts to replace the workings of the competitive market system. Somehow this result has rarely been directly attacked by economists even though it is contrary to most of the tenets of their discipline.

Substitution is an important aspect of the market mechanism. Petroleum from large underground reservoirs was substituted for whale oil in the mid-19th Century as the whale population fell precipitously and prices of whale oil soared. As petroleum from large underground reservoirs became more expensive in the late 20th Century, human ingenuity made it possible to obtain oil from some shale formations using hydraulic fracturing (fracking) and horizontal drilling. Particularly if and when this source becomes more expensive, another alternative is to create oil from coal or even natural gas, both of which are proven technologies. In addition, natural gas can and will be substituted for some uses of petroleum; natural gas is now also being obtained in large quantities from very widely located shale deposits in the US in addition to the large reservoirs that have been the primary source in the 20th Century. All this has happened because of the operation of the market mechanism and until recently very little involvement by government. There appears to be little reason for concern that human innovation will come to an end if allowed to operate free of unwarranted governmental interference.[318]

Just consider what improved technology accomplished in the 50 years prior to 2000 with regard to the availability of fossil fuels:[319]

> *Proved oil reserves today are estimated to be fifteen times greater than the original 1948 estimates despite interim production of eleven times this amount. World natural gas reserves in the last thirty years have increased five-fold despite interim production that has been 80 percent above the 1967 estimate. World coal reserves today are estimated to be over four times the amount calculated nearly a half-century ago.*

And since this was written in 2000, new technology has achieved even more remarkable expansions of oil and natural gas reserves.

Innovation is not limited to the supply side of energy use. Innovation is also occurring on the demand side. Fossil fuel energy use is gradually becoming more efficient. The internal combustion engine running on gasoline is pretty hard to beat and is improving

---

[318] An expansion of this argument can be found in Bradley, 2000.
[319] *Ibid*, p. 31.

all the time,[320] contrary to the assumptions of environmentalists and the Obama Administration. Whether we jump to electric cars should not be decided by government regulations but rather by the operation of the market. Government regulations may encourage increases in combustion efficiency, but are highly unlikely to do so in the most cost-effective way.

In their preoccupation with energy and sustainability, the radical environmental movement has attempted to drive up the cost of energy by supporting the use of cap and trade laws, carbon taxes, and regulations to shut down or at least reduce the use of cheaper energy sources such as coal, natural gas, and hydropower, and by outlawing the use of fracking. One motivation for this may be that they believe that this will result in making their ideologically chosen sources, wind, solar, and biomass, more economically attractive. In fact, radical environmentalism is basically about changing the mix of energy sources through government intervention so as to favor their preferred "renewable" sources so that the market will no longer encourage use of the most efficient sources, as it will if not interfered with.

Ironically, current efforts by many governments to promote "green" energy sources have coincided with the fracking and horizontal drilling oil and gas revolutions. The "renewables" advocated by environmentalists are a splendid example of governments' propensity for picking losers that cannot survive in a market-based economy.

The arrival of new technologies for obtaining raw materials and energy and using them more efficiently is not always a smooth phenomenon since new technology does not arrive according to some bureaucratic schedule. Often it is stimulated by higher prices for an increasingly scarce commodity. Thus higher prices for raw materials do not always indicate that the market has failed or that the raw material will become unobtainable. They rather show that the market is working and that alternative, initially often more

---

[320] Levine, 2015.

expensive, technology will be needed to compete against older, better established technology. In the case of oil, prices started rising in the 1970s as demand for oil began to come closer to supplies from large underground reservoirs. Some observers questioned whether oil supply had "peaked" and would inevitably decline with correspondingly ever higher prices. But technology came to the rescue by combining "fracking" with horizontal drilling. This is a more expensive way to produce oil, but if allowed to be used may supply oil for many years to come. Still more expensive technology involving the conversion of coal and natural gas to oil have been proven and will provide an even larger source of oil when justified by prices.

Sustainability is an abstract concept irrelevant in terms of deciding how to meet human needs outside the fisheries area where it originated. There is no current basis for arguing that human needs cannot be met now and for the indefinite future assuming that human ingenuity and markets are allowed to work with minimal government interference. The real danger is that markets will not be allowed to work and that governments will impose the environmental movement's ideologically-based "solutions."

In the case of energy use, radical environmentalists have usually argued that energy must be obtained only from wind, solar, and biomass sources since fossil fuels sources are not "sustainable," nuclear is too dangerous, and hydropower is too ecologically harmful. Lack of "sustainability" is often an excuse for interfering with the operation of a free market, which usually results in supply-demand imbalances rather than being a solution to them. One amusing aspect is that environmentalists seem to think that it is best to save our inventory of fossil fuels since the current higher grade deposits currently being used cannot be replaced. While it is true that they cannot be replaced except by continued exploration, the lowest cost alternative sources will be automatically used if the choices are made by the market. It should also be borne in mind that in high northern latitudes fossil fuel deposits not accessed before the next ice age will be unavailable for roughly 90,000 years.

It would seem better to access them now rather than save them for use 2,700 generations from now, when far better technology is likely to be available.

The market mechanism is a far better way to select resources for use than "sustainability." The free market takes into account the usefulness of the output produced, not just whether the resources are "renewable." This problem is of considerable importance and leads to many of the problems with "renewable" sources. One of the major problems with government favoring "sustainable" sources of energy is that this drives up the cost of such sources. This is a particular problem in the case of food used for fuel. If governments create incentives for food to be used as a source of energy, this makes food more expensive. And more expensive food leads to less use, malnutrition, increased risk of starvation, and, in the more poverty-stricken parts of the world, death.

There is a reason why windmills were until recently largely abandoned for generating energy. The reason is that wind is unreliable since the wind often does not blow strongly enough to meet the minimum requirement for energy generation. Even worse, wind tends to blow when it is least needed for electricity generation.[321] The result is that there must be a spinning reserve source to provide the reliable, continuous power required by electrical networks and human applications such as manufacturing and computer use. The major exception to this is areas where there is abundant hydropower available that can quickly and cheaply be substituted for wind energy when not available. One of the major problems with substitution of wind for fossil fuel sources is that it assumes that substituting wind energy sources for fossil fuel sources will result in a significant decrease in emissions of CO_2. This has never been proved in any rigorous way,[322] but is really quite basic to any sort of rational determination of whether quite large investments, often at the expense of taxpayers and ratepayers, are

---

[321] See Lomborg, 2013.
[322] Boone, 2010.

worthwhile even under the assumption that CO_2 emissions make much difference.

The situation for solar power is very similar because of nighttime darkness and daytime cloudiness. The result is that except in a few areas with abundant hydropower it is necessary to build and pay for continuous operation of spinning backup fossil fuel plants to provide the essential continuous electrical energy modern society requires to operate efficiently. The costs of providing this backup supply makes wind and solar energy extremely expensive and drives up the cost of electrical energy as well as the enormous cost of transporting the energy from the generally less inhabited areas with good wind and sun to urban areas where most of it is used, is far beyond the cost of fossil fuel electricity.[323] All of this is taken into account by the price mechanism if it is allowed to operate, but not by the doctrine of sustainability. Hence the continuing request for government subsidies to subsidize wind and solar despite the lack of any or at best minor savings in CO_2 emissions taking the system as a whole. It is very evident that wind and solar development will largely stop just as soon as the government subsidies disappear even after decades of subsidies. Many of the wind and solar sources still operating then are likely to be abandoned in these cases (as illustrated in Figures 10-1 through 10-3 at the end of this Chapter), creating huge long-lasting eyesores, as has happened in the past with earlier wind energy developments.

The cost of the subsidies paid for wind and solar power are enormous and are paid by taxpayers and electricity ratepayers. One estimate puts it at $70 billion per year worldwide but others put it much higher,[324] which could find much more useful uses by governments and ratepayers. These subsidies achieve no purpose except to support an ideology that has appealed to modern environmentalists and at best make some people "feel good" even though the commodity, reliable electricity, is exactly the same from

---

[323] Hughes, 2012.
[324] Nova, 2012.

renewable (after suitable back-up sources have been added to insure reliable supplies) and non-renewable sources. If alarmists actually believe in CAGW and should ever see the need to avoid an environmentally disastrous new ice age, they should actually be arguing for maximizing fossil fuel use so as to decrease the chances of a new ice age and minimizing the construction of renewables. Or better yet, they should be arguing to just leave the allocation of energy sources for electrical generation to the market with huge savings to both taxpayers and ratepayers.

In order to determine the sustainability of a policy or activity it is necessary to consider whether the policy or activity can be continued in the future. Although it is unclear how this analysis is to be carried out, it is generally done by using a static analysis. If the resources needed are finite or in any way limited, the policy or activity is often declared by environmental groups to be unsustainable and rejected. This approach ignores the dynamic nature of the competitive market economy. Economics is based on how the economy responds to scarcities. It responds by changing prices, innovation, and substitution, not by changing the policy or halting the activity. So far Western nations have rarely had to change a policy or halt an activity because of resource scarcity. Rather, nearly every scarcity critical to economic development has been resolved by price changes, innovation, and substation, almost always through the automatic operation of the competitive market. There is no reason to believe that this history will suddenly change in the future. Radical environmentalism, on the other hand, argues that this time it will be different despite the failure of the Club of Rome's "Limits to Growth" to attract a significant following.

Renewability is probably the most expensive of all these environmental slogans. It is claimed that a product or service that can be provided using only renewable raw materials is better than those that cannot make such claims, regardless of its cost or other characteristics. In other words, a product or service that provides the same user services is inferior if more non-renewable raw materials are used to make it and make it available for use. The best known example is electricity made by burning coal or natural gas rather than using wind or solar sources. But large public subsidies have been and are being provided to encourage generation of power from renewable sources even at the expense of other urgent uses for these funds. Although never discussed by proponents, the solar and wind sources are much less useful and valuable per unit of theoretical generating capacity. So consumers get less reliable, less useful energy at far higher prices. And for what purpose? Western Europeans have become particularly enamored of these approaches, but much money has been squandered in the US as well. In most cases the resulting facilities have been/will be abandoned before long because it costs too much to maintain them and there is little market for their high cost, unreliable energy. Once abandoned, they often become eyesores although at least their bird-and bat-killing days will have ended in the case of wind.

Ideological Basis: Energy Efficiency/Conservation

There are related problems with many of the other environmental "buzz words." Energy efficiency/conservation has been perhaps the most successful of these environmental, possibly PR-based, goals. In fact it has long since become accepted policy by many politicians in both US parties. Who could be opposed? More energy using products and services for each unit of energy expended. Thrifty people see merit in this; environmentalists see it as the easiest way to reduce fossil fuel use. Hence we have government regulations telling us what kinds of electric light sources we can buy, what the energy characteristics of electrical appliances must be, and the average miles per gallon new cars must provide in order to be sold in the US.

Yet almost no one appears to have analyzed why these decisions should not be left to the marketplace, even though such analyses are required by Federal regulatory guidelines[325] There is

---

[325] Gayer and Viscusi, 2013

substantial reason to believe that the results are far from optimum. It is harder for top loading washing machines to meet Federal efficiency standards than front loading ones. Consumers must now buy expensive, hazardous fluorescent lights even for locations where they are used only briefly and infrequently where they are almost certain to be far more expensive because of their high initial cost. Soon consumers will be forced to buy either very small, dangerous, and less family-useful tiny vehicles or to pay much more for hybrid vehicles that get over 50 miles per gallon on average. And for what purpose? Environmentalists claim it is to prevent climate change, which has been discussed at length in this book; politicians have found a market for such regulations as a "service" of government. Vehicle manufacturers will be able to sell many more, more expensive hybrids. As is all too frequently the case, the only losers are consumers. These are good examples of decisions that should be left to consumers and the marketplace whenever possible. I support providing efficiency information to consumers, but that has generally been the case for many years in the US. More recently the exact trade-off between energy use and efficiency has been set by governmental edict without regard to the many other characteristics that these energy using products have. Yet few economists have pointed this out. They need to.

But what is optimum energy efficiency? Why not push mileage standards even further to the point that everyone except the ultra rich drives small motorized golf carts offering no protection in a crash, little acceleration, no room for children or groceries, and few of the other characteristics desired by owners? The answer, once again, is economics and the preferences of consumers. If car buyers preferred golf carts they would buy them for general use without government dictates. But they do not; they evidently much prefer mini-vans that can accommodate the whole family and its luggage comfortably and provide additional safety in the event of a crash. So why not leave energy efficiency to the marketplace? It may be (although this has not been demonstrated) that Americans do not give sufficient attention to the national security aspects of gasoline use (although this argument is rapidly diminishing as a result of the surge in shale oil production); it is probably more likely that they do not give sufficient attention to the safety implications of increasing mileage standards since lighter vehicles are more dangerous than heavier ones. Everyone is in favor of reducing injuries and deaths from motor vehicle accidents but seem not to understand that there is a strong connection between accident injuries and deaths and motor vehicle weight. And this is never discussed by environmentalists when they strongly advocate greater vehicle fuel efficiency. No evidence has been presented of a market failure, so why is the government involved at all?

Such unbalanced discussion may help environmentalists pursue their ideology but does not lead to good public policy or wise consumer choices. There is nothing wrong with allowing the market and private decisions to dictate resource use. In fact, it has proved remarkably successful in the US and other countries where it has been tried for several centuries. Why do we suddenly assume that Washington can make better decisions concerning energy efficiency? How does Washington know whether I will use a particular electric light once a month (in which case incandescent bulbs are usually much cheaper and almost as energy efficient) or continuously (where fluorescent lights are sometimes cheaper and more energy efficient despite their higher initial cost and dangerous mercury) and mandate that I can no longer buy incandescent bulbs? How does Washington know whether front-loading or top-loading washing machines better meet my needs? Why not let me decide these questions for myself? Why should energy efficiency decisions be made in Washington or by any other level of government? I strongly object to governments making energy efficiency decisions for individuals. Unfortunately, Federal law and many state laws do exactly that. The environmentalists have been entirely too successful in bringing this about, and it is time to get government out of mandatory energy efficiency requirements. Environmental ideology says energy efficiency is more important than economics; I say that economic decisions by individuals will consistently produce the most economically efficient and democratic decisions provided that externalities are reflected in the prices paid. Our energies should be put into correcting energy externalities where they exist, not mandating energy use decisions, which can be better done by individuals responding to accurate prices.

Ideological Basis: Recycling/Reuse

Another environmental PR slogan goal is recycling. This juggernaut has become the law of the land almost everywhere in the US and some Western European countries at the local level and is expected to be soon in the European Union as a whole as well.[326] People who would rather avoid the effort of separating waste into various ill-defined and uncertain categories now violate the law if they fail to do so in many areas. This requires added effort, of course, with no profit and little satisfaction in return. All this is done in the name of the environment but at the expense of consumers. Does it really make economic sense to separate trash and have it picked up separately at much greater expense in fuel and vehicle use and human labor? Surely this should be left to the consumer, presumably by offering him/her a higher priced trash service with an added recycling feature supplied by the service. This would seem likely to fail commercially, but that just makes the point—there is little economic value in recycling. But the widespread response of local governments in recent years has been to make it mandatory and to set one price for trash collection. The consumers lose again by being required to pay in both time and money for services they would not be willing to pay for in an open market.

As in so many other goals of environmentalism, recycling has been imposed by governmental fiat, often based on advocacy by the environmental movement. But how much recycling is optimum? The environmental movement generally argues that the more recycling the better, regardless of cost. But is this really so? Are

---

[326] Delingpole, 2014. See also Tierney, 1996.

more hours spent by homeowners, more miles driven by trash trucks, and more fuel and labor used to operate them always worthwhile? My answer is that the optimum is the amount that would exist in a competitive economy adjusted for any externalities such as the possible adverse external effects of expanding landfills. In my view recycling has generally gone far beyond what would take place under competitive market forces. The result is that trash trucks make multiple trips to each home to pick up a variety of specified types of household wastes, burning much more fossil fuels than they would if there were no recycling regulations. Here again there is a trade-off between two radical environmental goals, energy conservation and recycling. In this case, energy conservation is the loser. And households spend much more time and effort to separate the household trash into similar components. Is some recycling desirable? Of course. But how much? The recycling/reuse mantra does not provide a clue. Competitive markets or economic analysis would if allowed to operate freely without intrusive, unjustified government regulations.

Why Radical Environmentalism Exposes Humans to Greater Future Dangers and Destruction of Wilderness

Chapter 9 argued that a new ice age is less than a few centuries or at best a few millennia away, and that the best chance to avoid this, one of the worst possible environmental disasters, is the use of geoengineering techniques such as SRM. A major problem that is likely to occur in implementing this solution to this potentially devastating environmental problem is likely to be that the environmentalists of the future may well object to it and possibly even make an all-out effort to avoid using such geoengineering techniques. Although it is possible that these techniques would not be sufficient to solve the problem (we need a far better understanding of what influences climate), the more likely scenario is that environmentalists will try to prevent them from even being tried even though there may be no other realistic alternatives other than hoping and praying that greatly increased GHG emissions might help. But environmentalists of the future might well object to that too. Many environmentalists already oppose use of geoengineering for solving their alleged global warming/climate change problem and an elaborate rationale for this policy position has already been developed; Albert Gore strongly agrees.[327] So presumably the same arguments will be put forward against future use against the next ice age. This would be extremely unfortunate in my view given the likely disastrous results for both humans and the environment. There is some precedence for this, however, since an environmentalist lawsuit halted the construction of a hurricane barrier in New Orleans prior to Hurricane Katrina.[328]

The underlying rationale for environmentalist objections to the use of geo presumably will be that humans should make no changes in nature but preserve it the way it is. This may work during interglacial periods but will not do so during the onset of the next ice age. This thinking could have immensely negative consequences for both humans and the natural world if we sit idly by as the northern latitude countries are wiped off the face of the Earth by the unstoppable advance of continental glaciers. It also reveals a fundamental problem with one of the basic components of modern radical environmental ideology. Maintaining that nature should rule could result in inaction when strong action may actually be needed to save a significant portion of mankind and plant and animal habitat from one of the most horrible fates imaginable—being forcibly moved and/or frozen to death.

From a larger perspective, the best option available to humans concerning climate change is to prepare for the worst likely outcome in a situation of extreme uncertainty. If we plan for warming, as we are now doing, and it cools, adaptation will be much more difficult and the effects of the cooling more destructive. If we plan for cooling and it warms, we will be less prepared for it, but the results will be less destructive. The alarmists strongly

---

[327] Goldenberg, 2014.
[328] Vartabedian and Pae, 2003.

advocate the former but have thus far proved consistently wrong in their predictions. Their approach defies common sense and risks making the more likely cooling even more destructive than it otherwise would have been.

In the shorter run, the radical environmentalists' favored energy and food production policies are also very environmentally damaging. Substituting bio for fossil fuels results in a huge expansion of cultivated land, often using more marginal, easily damaged land. Prohibitions on the use of genetic and agricultural chemicals and substitution of organic farming has the same result. Wind and solar energy production facilities are unsightly and kill vast numbers of bats and birds, including many rare and endangered ones. They also result in the destruction of wilderness in the construction phase with their need for access roads, their huge land expanse, and their new and longer transmission lines. If all these policies were reversed very large land areas would revert to wilderness and conservation reserves.

Environmentalism as a Religion/Ideology

Environmentalism was largely unknown in terminology before the 1970s although a related concept, conservation, had been used for many years prior to that. It has now taken on the characteristics of a religion and ideology.[329] while attempting to portray itself as being based on science. Michael Crichton[330] has characterized the resulting problems well, and strongly supported reducing the role of environmentalism. He strongly supported a scientific approach to environmental decision making. I add to this an economic approach as well to insure that the country as a whole is made better off economically by each regulation or other government action actually adopted. I also favor a look-back approach to learn from prior regulatory mistakes that may have been made and correct

---

[329] Beatty, 2012.
[330] Crichton, 2003

errors. The look-back mechanism has very effectively been used in the air transport industry and it should help here too. It is only by understanding and acknowledging past mistakes that future ones can be avoided and future improvements made. Unfortunately there is no institution that currently has the responsibility and resources to perform this look-back approach.

A Broader Approach Towards the Problems Posed by Radical Environmentalism

There is a definite pattern to many of the major campaigns conducted by the environmental movement, which often follow the following series of claims:

1. *Humans have sinned and caused damage to the natural environment.*

2. *Capitalism is evil and has resulted in/contributed to this damage.*

3. *Another vital aspect of Western civilization needs to be banned/regulated by the government.*

4. *The government can do a much better job of handling this aspect than individuals can working through the market mechanism.*

5. *Human needs and freedom are much less important than "saving" this aspect of the environment.*

6. *The scientific/economic case for government regulation of this aspect is overwhelming and does not need to be debated.*

7. *Anyone who opposes regulating this aspect is evil and in the pay of polluters or unqualified to express their views and must be called disparaging names.*

8. *Major polluters provide overwhelming financial support to opponents on this issue.*

9. *Because environmental problems are so serious, action is needed before we really understand the science and economics (the precautionary principle).*

Environmental organizations have increasingly used part or all of this approach for some time. It started with the claim that raptors were declining as a result of using DDT (use allegedly resulted in egg shell thinning), but has since included many other claims such as the decline of the snail darter (fish allegedly imperiled by dam building in Tennessee), decline of the spotted owl (allegedly imperiled by cutting old growth forests), increase in the ozone hole over Antarctica and skin cancer caused by human-caused emissions of chlorofluorocarbons, global warming/climate change/extreme weather caused by human-caused emissions of burning fossil fuels (evidence is lacking, particularly since 1998 for warming), decline of the polar bear (allegedly imperiled by global warming but actually increasing in numbers), and drinking water contamination from fracking (but has never been shown to have occurred in the US). In each case the evidence was flimsy but immediate governmental action was called for which adversely affected the rest of the population, particularly the less wealthy. Although the science is far from settled in each of these cases, it appears that the environmentalists were either wrong or had overstated the danger. In two cases, DDT (which was discussed in more detail in Chapter 8) and fossil fuel use, immense economic and environmental/health damages have occurred/will occur as a result of governmental actions resulting from taking their views seriously.

If as seems almost certain radical environmentalists are going to continue following this script and making irresponsible policy recommendations on the basis of hasty or faulty scientific claims it is vital that their proposals and science be carefully evaluated before government action is actually taken. And there needs to be a strong bias against government intervention except when there is a demonstrable externality that can be compensated for efficiently by adjusting regulations, or better prices. This careful balancing cannot be done by agencies that have been "captured" by either environmentalists or by polluters. Only truly independent agencies have any hope of fairly evaluating environmental claims and counter claims. And they need to have vigorous internal checks and balances to make sure that the balancing is carefully done. This has clearly not been the case at EPA and Interior during the Obama Administration; EPA had such checks and balances prior to the Clinton Administration, but they were then dismantled at the insistence of the environmental organizations. When biased groups control such agencies the agencies are at risk of doing more harm than good, and in some cases enormous harm. That is what has been happening, particularly during the Obama Administration, and will likely continue until more neutral and objective groups control these agencies.

One obvious question is whether the radical environmentalists are providing a useful public service, or whether we might be better off without the efforts of the more radical among them? It is now a very large enterprise with multiple organizations and offices in most Western countries. Some of their efforts, often the ones least covered by the news media, have provided useful results, of course. But a number of the efforts by radical environmentalists cannot withstand closer examination.

They have a perfect right to express their opinions, of course, but hopefully in the future public and public officials will examine their pronouncements much more carefully and critically before taking action based on them. Particular attention needs to be given to whether each new campaign has a firm foundation in good science and economics. Each of these efforts provide useful lessons as to what went wrong and how to avoid similar problems in the future but retrospective studies of environmental control activities are few and far between.

One of the many ironies concerning this whole situation is that in their zeal to "save the planet" from the very recent and somewhat predictable global warming over the warming phase of the last 60 year cycle, the Obama Administration is putting into place all the regulatory pieces they think will be needed to build an all-encompassing environmental bureaucracy to decrease carbon dioxide emissions. This will primarily have huge adverse consequences for the American economy and economic freedom in America—but end up making the environment worse! It might be described as all pain for negative gain. There is no environmental catastrophe worse than a new ice age other than a super volcanic eruption, but that is what we may be headed for in coming centuries or at most a few millennia. Many of the economic freedoms that Americans enjoy to use energy when and where they believe it will help them will be sacrificed supposedly to improve the environment, mainly in the rest of the world, but in the longer run the real losers are likely to be countries located in the northern latitudes, especially the United States, Canada, Scandinavia, and Russia, because they are likely to bear the brunt of a new ice age, which would be made more, not less likely by curbing CO_2 emissions. So not only have the environmentalists gotten the science wrong, they have even gotten the proposed solution wrong too even given their definition of the problem. All because they were so anxious to be the saviors of the world that they could not be bothered to learn and understand the science and/or did not want to. Unfortunately, the same goes for the US Environmental Protection Agency since they simply used the IPCC and US reports based on it without taking the time to question and understand the science, as I pointed out at the time.

All this regulatory activity is happening through regulations published in the *Federal Register* which few people read and fewer still understand their significance. If the Obama Administration understood the science, however, I believe that they should be doing exactly the opposite of what the Administration and the radical environmentalists are trying to achieve—encouraging the burning of fossil fuels, which requires no expanded bureaucracy, just low prices for fuel. This, in turn, would mean encouraging fossil fuel production on both private and Federal lands. The Obama Administration has attempted to and succeeded in reducing availability on Federal lands and cannot claim (as they have tried to) any credit for increasing it on private lands, which has responded to high oil prices and new technology. Unfortunately, although radical environmentalism attempts to cloak its views in the trappings of science, it has shown itself to be remarkably resistant to any scientific ideas that do not conform to its pre-conceived ideas concerning what the science should say, not what the observations and the application of the scientific method actually say.

Energy use is one of the defining characteristics of modern civilization. The history of energy conservation is not supportive of the idea that humans will willingly substantially reduce its use. The economic value of increased energy use is simply too high to forego it. Although radical environmentalists have emphasized how easy they believe it would be to reduce energy use through energy conservation and use of renewable energy sources, neither of these approaches has proved either economic or very effective in reducing energy use. Among other effects, increases in energy efficiency results in lower fuel costs, which results in increased use, partially offsetting the original reductions in energy use.[331] In the past, most of the reductions in motor vehicle miles per gallon have been effectively used to increase horsepower rather than save fuel.[332] So reducing energy use is very hard, not easy. And increased energy use generally results in increased human productivity, which benefits everyone. Given the benefits of increased energy use, efforts to drastically reduce use which receive wide publicity and public attention (such as the proposed cap and trade bill in 2009-10) appear likely to be overturned by either popular decision (as in the case of the Australian election of 2013) or widespread defiance of the law (as during prohibition). In early 2015 Switzerland held a referendum on substituting a carbon tax for their current value added tax system. It lost by 92 to 8%. Perhaps the Obama Administration believes it can sneak their GWD/GEF campaigns through EPA without anyone noticing, but I doubt it.

The radical environmental movement and particularly their academic supporters are not just guilty of promoting bad science

---

[331] Bailey, 2012.
[332] *Ibid.*

and unachievable solutions to exaggerated problems. They are also attempting to change a number of the ideas and principles (to be detailed below) that have allowed humans to achieve such dramatic progress in living standards in recent centuries, and as a result, in environmental protection in recent decades. The results are two-fold. The first is that it makes it easier for them to try to justify past, present, and potential future doubtful policy goals economically and scientifically. The second is that it corrupts both economics and science in general with unknown but potentially widespread effects in many other areas where the same principles and ideas may need to be applied.

As explained in Chapter 9, the climate issue may not be the first or last time something like this has occurred as a result of the environmental movement's efforts. A continuing series of environmental issues based on similar changes in ideas and principles is very likely to prove even more damaging than the huge and growing wasted investments in soon to be abandoned solar and wind farms it has managed to get others to build using other people's money. Whether the environmentalists and their academic supporters intended to change these ideas and principles is unclear. It is possible that they just got in their way and had to be sacrificed on an ad hoc basis in their attempts to "save the world." It is clear that in many cases the principles made it hard for them to claim scientific validity and economic merit for their proposals. In any case, it is these pretensions that appear to have given rise to many of their unfortunate attempts to corrupt these principles and ideas.

But regardless of their motivation, the time has come to come to the defense of these tried and true principles and ideas before still more damage is done to what may be the very heart of the success of Western Civilization. It is important to note that the damage will initially occur whenever and wherever modern environmental ideology is applied without conformance with the general principles and ideas, which is currently largely in the Western developed countries. But there is a significant risk that they will spread more widely to other unrelated issues and areas of the world due to the influence of and the example set by the Western world. It is hard to quantify the effects of corrupting science and economics, but it is likely to be enormous over the longer term.

There are a number of these fundamental principles and ideas, but the two most fundamental ones are reliance on relatively free, competitive markets to allocate resources and the use of the scientific method (SM) to determine what the nature of the world is. Both are fundamental to much of neo-classical economics and modern science. And there are interactions between them in that I believe that economists should not count effects that cannot be verified as scientifically valid as economic benefits of environmental controls. I should explain that by free, competitive markets I refer to markets that are predominantly free of governmental interference without economic justification and competitive in their structure since there are varying degrees of freedom and competition in each market and few are totally free of government interference or influence of some sort or purely competitive. Further, relatively free markets can exist even when there is government intervention to correct clear externalities or because government resources are involved.

Unfortunately, freedom from government interference and competition in energy distribution are more the exception than the rule in many countries in part because of the long noted added cost of duplicating distribution facilities (a legitimate reason for government intervention). Energy production has increasingly had more governmental interference than is likely to result in efficient decisions perhaps because of the difficult time politicians have in keeping their hands off such decisions. Energy use decisions have usually been relatively free in developed Western countries, but environmentalists are trying desperately to change that.

The SM has made possible much of civilization's scientific progress in recent centuries by making it possible to distinguish between valid and invalid science. Without it, it would be very difficult for civilization to make much scientific progress since it would be unclear which hypotheses were incorrect and should be abandoned. Unless it is allowed to fully operate, there is a very real threat that objective scientific inquiry will give way to scientific fads and political ideologies. It is already happening in the case of climate change/global warming in the Western World and has happened on other issues in totalitarian countries. Various promising approaches to understanding climate change such as the effects of solar variability on cosmic rays have been purposely ignored since this is inconsistent with the anthropogenic explanation favored by environmentalists. This not only holds back the development of the climate science, but leads to proposed "solutions" that will probably solve nothing.

Relatively free, competitive markets have resulted in consumer sovereignty and freedom to select products that buyers believe will most meet their needs. Such markets resulted in the rapid increase in the use of energy in recent centuries (as well as the laptop computers, cell phones, and other electronic gadgets that seem sure to further increase energy use). Together, the operation of relatively free, competitive markets and increasing energy use have resulted in an explosion of technological progress and living standards never before experienced by mankind. No person or government planned or mandated the increased use of energy (at least prior to the development of communism); it rather evolved because humans found it useful and were relatively free to exercise their choices through their purchase decisions in the free market. But the result is that the economy serves the varied needs of consumers in a timely and efficient manner by responding to what buyers want at minimum cost, not what someone else thinks they should want based on ideology.

In the Twentieth Century various less democratic approaches to managing economies were tried involving much greater degrees of government control, particularly in the former Soviet Block and in countries sympathetic to their centralized governmental approach to managing economies. The Soviet approach also had some problems with using objective, SM-based science as well. These centralized governmental approaches have been found wanting, however, usually because living standards did not advance very rapidly or because the economies involved were uncompetitive with economies that pursued a more free market-oriented approach. I do not claim that it is impossible for a centrally-planned economy to achieve rapid growth or considerable consumer satisfaction—just that this is unlikely to happen, difficult to find historically, and is quite unlikely to result in optimally meeting consumer demand. The most important reason is that central planners (or even those applying other governmental interventions) would be highly unlikely to manage to achieve exactly the outcome that would be achieved under a freer, more competitive market since they do not know what that outcome would be and probably have other goals in mind besides consumer satisfaction. I also do not claim that capitalist economies have no faults. One of the most obvious is economic cycles that result in periodic unemployment. Attempts to create "green jobs" have thus far proved less than successful in reducing overall unemployment and appear to offer little chance of future success, however. It is worth noting that most of the Soviet-style economies have also generally shown worse environmental records than relatively free market economies.

The major counter-example that is sometimes given to these generalities is probably Communist China. Since substantial economic reforms starting in 1979, China has experienced some of the most rapid economic growth the world has ever known but is clearly still far from a free market economy. The best explanation for this appears to be that since the reforms started China has experienced unusually strong productivity growth.[333] Presumably this results from a rapid movement of farm labor into manufacturing made possible by abundant export demand. In addition, the central Government has concentrated on those aspects of the economy where government intervention is least likely to be counter-productive and most likely to be helpful such as infrastructure development, which is much needed under

---

[333] Hu and Khan, 1997.

conditions of rapid growth. Some observers have estimated that about 70% of the economy is in the hands of smaller private enterprises which are largely unregulated by the central Government and account for much of the economic success of the economy. The Government does own a number of larger enterprises which are not always as productive, however. What this suggests to me is that partially freeing the economy has paid rich dividends. Completely freeing it would do even more.

China's approach to environmental industries seems to have had some of the same problems as that of the Obama Administration's. In its 12th Five Year Plan, the Chinese Government identified solar, wind, and electric automobiles as "strategic emerging industries" that would receive particular state support.[334] By 2012 all three industries were in serious difficulty, with falling prices, large losses, overcapacity and the need for bailouts.[335] China's high speed rail investments have also encountered major problems.[336] Although a number of China's state investments have proved useful and productive, it appears that these green investments have not been. A wiser approach in both the US and China would have been to stick to justifiable environmental regulations and allow the market to sort out how to respond to them and to consumer demand.

There are a variety of intermediate approaches between centralized governmental approaches to managing economies and relatively free, competitive markets. One of these is sometimes referred to as (1) industrial policy. In this approach the government does not directly control the economy—it rather makes what it believes to be strategic investments using taxpayer funds or uses other non-regulatory incentives to favor particular sectors and firms in hopes that this will lead to whatever objectives the government favors but the free market would not. Another approach (2) is "well

---

[334] Chovanec, 2012.
[335] Ibid.
[336] Ibid.

intentioned" but non-justifiable governmental regulations that do not involve formal central planning but may end up strangling economic growth and living standards just as effectively. A third (3) intermediate approach is direct government manipulation of prices with the intention of favoring certain activities and discouraging others. Most Western economies influenced by modern environmentalism have thought about or actually attempted one or more of these intermediate approaches. These do not involve totalitarian central planning and direct manipulation of science, but the results may not be very different in terms of economic results and scientific freedom of inquiry. Under the US Obama Administration, all three of these "intermediate" approaches have been seriously proposed or tried with regard to environmental issues. Europe has generally gone even further in these directions.

US environmentalism started largely as an attempt to preserve the natural environment through restrictions on land use, usually through proposing government action to prevent development of areas with considerable natural values. Some free-market economists have argued that these land use decisions are best left to the free market too. The most prominent of these actions— creation of national parks and similar preserves—appears to have been widely accepted and have been expanding to larger areas and more countries. These and other environmentally-inspired land use restrictions appear unlikely to be reversed.

In recent years, however, radical environmentalists largely in developed countries have increasingly taken the view that they want to impose major restrictions on the prevailing (free) market economic system more generally, particularly on issues involving energy production and use. The main similarity to the earlier efforts is that they continue to try to make these changes through government action, that they are advocating further restrictions on the free market, and that they attempt to use scientific arguments to advance their cause. Their argument is presumably that these changes in the basis for much of Western Civilization are necessary to avoid alleged alarming adverse effects of man's activities on the environment, particularly through global warming or climate change. The economic and scientific basis for this view has not been made.

Science and economics have evolved over a number of centuries a set of guidelines for deciding how to determine what rules nature actually follows and what exceptions should be made to a free market approach to resource allocation based on market failures. Not surprisingly, many environmentalists and their academic supporters want to disregard these guidelines while often pretending that what they are proposing represents valid science and sound economics. In the next Chapter I will suggest a number of ways in which modern environmental proposals for governmental action violate these guidelines and other economic findings.

I argue that alteration of these well founded guidelines is likely to have much more adverse effects on human society than any of the alleged adverse climatic effects that will actually occur as a result of ignoring the actions proposed by modern environmentalists. As will be discussed in Chapter 11, if modern environmentalists truly want to improve the natural environment, they are actually pursuing exactly the wrong basic approach towards economic development and energy use.

Some Specific Areas Where Radical Environmentalism Ignores the Guidelines/Tenets of Science and Economics

This Section raises the question of the compatibility of sound economics and the scientific method with modern radical environmentalism. I propose to base this discussion on what most radical environmentalists have actually advocated in recent years. Here are a number of their economic/scientific guidelines/assumptions that are inconsistent with the some of the basic tenets of science and neo-classical economics:

Scientific issues:

- *Use of physical "science" which is not supported by*

application of the scientific method and the most relevant observational data

- *Use of precautionary principle*

Economic issues:

- *Attempt to achieve "sustainability" and other PR slogans rather than allowing relatively free, competitive markets to function*

- *Assume people's behavior should be changed by government regulation / incentives when inconsistent with environmental ideology*

- *Use of unreasonably low discount rates in determining economic feasibility*

- *Support of government intervention on resource issues of interest when there is no economic basis for it.*

I have discussed these inconsistencies elsewhere in this book and in a journal article.[337] Modern radical environmentalists and their academic supporters are very anxious to cloak their policy proposals as if they were dictated by science. But they do not appear to want to have the underlying science judged on the basis of the scientific method, the traditional and only reasonable arbiter of scientific validity. They largely act as if the SM does not exist. When discrepancies are pointed out they accuse the authors of being paid by Big Oil or other supposedly environmental "evil doers" or of being unqualified to judge rather than addressing the specific discrepancies raised. They attempt to justify their views on the basis that there is an alleged "consensus" among qualified experts, even though they are surely aware that the relevant criterion is the SM. Then they just continue to make the same "scientific" arguments which have just been shown to be invalid using the SM.

Although it is true that some conservative groups have funded

---

[337] Carlin, 2013.

some of the organizations opposing GWD/GEF, this funding has been anything but robust and is totally dwarfed by the expenditures by the alarmist side, which has long enjoyed strong support from the US Government and the taxpayers that support it. Two of the major organizations opposed to GWD/GEF, the Heartland Institute and the Competitive Enterprise Institute, also work on many other conservative issues and have only a few people devoted to this one topic. If funding were so plentiful, their efforts in this area would be many times their current size. Apparently the alarmists find it convenient to perpetuate this big-polluter funding myth—perhaps to explain why their efforts have been so ineffective—but have never managed to document it. The truth is that most of the skeptic cause is self-funded by those working in it—often retired scientists or meteorologists. In other words, these are people such as myself who are pursuing this effort out of strong conviction, not immediate or even future personal gain. The alarmist myth in this regard is not conducive to understanding their opposition or making rational decisions as to how to pursue their cause, and is probably self-defeating.

The heart of the SM is to make inferences from the hypothesis being examined and then to determine whether these inferences are supported by observations. To my knowledge, no alarmist study has attempted to make their case using the SM. Surprisingly few "skeptic" studies have explicitly done this, but the results are not supportive of some of the basic hypotheses concerning the environmentalists' GWD "solution."[338] As discussed in Chapter 9, the GWD supporters have rather largely relied on elaborate but unvalidated computer models of the atmosphere, as if they proved scientific validity rather than their architects' views of the system being modeled.

Excluding effects not supported by the SM has an

---

[338] I am the author of perhaps the most systematic such article in the refereed literature (Carlin, 2011). The published literature contains a variety of relevant input analyses, of course, some of which I have used.

overwhelming effect on the economic benefits of reducing CO_2 emissions and therefore on the economic feasibility of GWD "solutions." A recent paper of mine[339] concluded that excluding these effects results in drastically reduced economic benefits of control, with the result that net benefits are highly unlikely to be positive.

Problem

It is all too evident that major American and international institutions have failed to provide sound advice on whether and how to control the alleged global warming/climate change/extreme weather problem. Although there are important environmental problems to keep environmentally-concerned people usefully occupied for decades if not centuries, major parts of the environmental movement have instead chosen to follow a self-destructive and ultimately doomed strategy of trying to undermine the primary basis—substitution of fossil fuel energy for human labor—that has made possible many of the advances in human civilization beginning with the Industrial Revolution. Their rationale for this is their one and two-word phrases that constitute their ideology plus the GWD/GEF. There are still billions of people on Earth who would do anything to obtain the advantages of using fossil fuel energy which the more developed world enjoys. They will not be deceived into following the course advocated by the environmental movement. The only ones that will are some people living in the more developed world that have failed to understand the increasing role that reliable energy supplies play in their lives rather than retrogressing to some imagined pre-industrial era nirvana that they themselves would not actually want to live in.

The radical environmental movement claims that modern human needs for energy can be satisfied by using "renewable" sources, especially solar, wind, and biomass, and that the use of

---

[339] Carlin, 2011.

fossil fuels must be drastically reduced. Unfortunately many people and governments in developed countries, particularly in Western Europe, have subscribed to this propaganda. Their principal arguments appear to be three-fold: that "sustainability" and reduced emissions of GHGs are important attributes of a fuel source and that some fossil fuels are "dirty." In addition they often argue that humans will soon run out of fossil fuels and that renewables will soon price fossil fuels out of the market. Taking these arguments in reverse order, these additional two arguments are simply false.[340] Much of the rest of this book argues against the alleged importance of CO_2 and other GHG emissions. Conventional pollution from burning fossil fuels can be a major problem, but it has been controlled in much of the developed world, and even over-controlled in the US.

With regard to "sustainability," as argued elsewhere in this book, we need, if anything, more GHG emissions from a purely environmental standpoint as well as an economic viewpoint. Sustainability is a nebulous concept that should play no role in selection of a fuel source, as argued earlier in this Chapter. "Free" energy sources are not necessarily better or more economic. It is necessary to take account of all the benefits and costs of different energy sources and this is normally best done by leaving the choice to the energy market rather than for government to dictate the answers.

In making the case for "renewable" energy the CIC often misleads the public by making several assumptions. One is that renewable energy can simply be substituted for energy generated using fossil fuels and that if a wind or solar generator is rated at the same capacity as a fossil fuel source the two are equivalent in terms of their usefulness to users. Such is not the case, however. Although fossil-fueled electricity generators have occasional outages, they are few and far between and can be greatly reduced by periodic maintenance when demand is predicted to be low. Wind and solar,

---

[340] Ridley, 2015.

on the other hand, have partial and even full outages often on an unpredictable basis because of low or no wind or clouds, so their effective capacity is much lower than their nameplate capacity much of the time. The result is that achieving a given guaranteed generating capacity requires building much higher capacity wind/solar and providing fossil fuel or hydro backup which is constantly available. And when the wind falls below the minimum levels for operation of the turbines or the sun is completely obscured by clouds or fog or nighttime, which often happens when demand is highest, the output of wind/solar is almost zero. Solar/wind capacity also degrades much more rapidly with age than fossil fuel capacity. All these problems increase the equivalent costs of wind/solar compared to fossil-fuel energy, but are rarely taken into account in comparisons, particularly by those made by the CIC.

Wind and solar energy sources are not serious answers to obtaining the reliable energy sources needed by modern society. Their output is undependable because sunlight and wind are unreliable. In addition, solar has high maintenance costs if it is to achieve its rated output[341] (as illustrated in Figures 10-2 and 10-3 at the end of this Chapter). Wind also has high maintenance costs, which result in abandonment of many wind farms (as illustrated in Figure 10-1), even in high energy cost states such as California and Hawaii. Unreliable energy sources must have reliable backup sources since supply and demand for electricity must be balanced at all times. In areas without abundant hydro power this requires less efficient simple cycle natural gas or other fossil fuel plants in spinning reserve—which means substantial additional fossil-fuel backup online in addition to the "renewable" source.

One experienced system operator estimates that because of the unreliability of wind, it is necessary to have 60% of the rated capacity for wind backup available at all times.[342] But even this may

---

[341] Gunderson, 2012.
[342] E.ON Netz, 2004.

be too little given German experience with simultaneous calm wind and cloudy days. Even without this 60% redundancy, the costs of wind and solar are much higher than fossil fuel sources in most areas of the US, so the result is much higher energy bills for consumers and/or taxpayer subsidies. This situation is becoming even more evident as a result of recent reductions in the price of natural gas in the US, which provides an even lower cost alternative to wind and solar and sometimes even coal. One advantage of natural gas compared to coal is that it is more flexible in its use, but only if adequate gas pipeline capacity is available. Unlike coal, simple cycle plants can be brought on line much more rapidly when needed, which makes them more suitable for use in coping with the fluctuating levels of wind and solar. Such plants are much less efficient compared to combined cycle or steam generation gas plants, however. And from a longer term viewpoint, standby plants must be kept in ready to operate status if they are to provide backup power even on an interim basis. This costs money.

The fundamental problem with wind and solar is that they are only available when the wind blows and the sun shines. To be useful, electric power is needed when humans need it. The resulting mismatch means that fossil fuel or hydro power must be available at all times in case the wind or solar is not available. This means paying for both wind or solar and fossil fuel/hydro backup. This is an important reason why wind and solar is so expensive, and appears likely to continue so. They are both inherently more expensive and require expensive backup since there is currently no known way to economically store large quantities of electricity until it is needed. Wind and solar are just not a practical source of reliable electric power despite the CIC's many claims that they are.

The example of German use of renewable generation of electricity is instructive. During the cold winter months, renewable energy often fails to generate any appreciable amount of electricity for weeks or even months. This occurred, for example during December, 2013 after the passage of storm Xavier, which was followed by doldrums and fog.[343] What this means is that full fossil fuel backup or hydro must be available during these periods if blackouts are to be avoided. Since Germany has little hydro, either Germany must maintain availability of full backup for these periods or rely on neighboring countries to do it for them. Maintaining full backup is extremely expensive and makes investments in already expensive renewables even more uneconomic.

A particular problem with wind is that its peak availability rarely coincides with peak demand for electricity both on daily and seasonal basis. During heat and cold waves, which are often associated with high pressure systems, for example, wind turbines make almost no contribution.[344]

Another problem is that the best wind and solar sites are usually not found near major load centers. This means that expensive additions must be made to electric grids to bring the power to the load centers.[345] Unfortunately, these costs are often not included in cost comparisons. Fossil fuel plants, on the other hand, can be located near major load centers without incurring such high costs for grid expansions.

Finally, solar and wind have high environmental costs. Wind turbine blades kill many bats and birds, including raptors and endangered species, and creates huge tracts of unsightly towers before they are abandoned and derelict towers afterwards. Windmills are also very noisy for nearby humans, can throw off ice shards that form on their blades during cold weather, are frequently abandoned, creating a huge mess when the subsidies run out, and ruin the view for anyone viewing them. Unlike wind power, solar is at least quiet but also kills passing wildlife. Photovoltaic panels pose the same hazard to them as windows and buildings and are mistaken for water. Solar thermal facilities literally incinerate large numbers

---

[343] Wetzel, 2013.

[344] *Ibid.*

[345] *Ibid.*

of passing birds,[346] which are attracted by the bright lights and insects. Solar power stations are not very beautiful to look at but at least are not usually built on ridgelines for all to see. Finally, when solar panels are manufactured, discarded or abandoned, they result in the creation or discharge of hazardous substances. Like wind turbines, however, they are often built in areas that are not easily connected to the existing grid, which means that ugly high tension power lines must often be built to transport the power to load centers.

It is not generally known that the wind industry appears to have been systematically trying to reduce the bird/bat death count by promoting guidelines that result in undercounts of bird/bat deaths by a factor of 50 or more[347] and that wildlife agencies have often tried to ignore the slaughter.

One interesting question is whether the use of wind power, the main non-hydro "renewable" source for electric power, actually reduces CO_2 emissions. The wind industry likes to pretend that every kilowatt hour (kwh) generated by wind turbines represents emissions saved at a fossil fuel plant. This may be the greatest misconception concerning wind energy. This is simply not the case.

Two recent papers suggest that wind turbines do not save much, if any, CO_2 emissions, particularly as they age. One of the major problems with wind turbines is that they not only operate at only a fraction of their rated capacity due to wind variability and other problems but that even this load factor declines rapidly over time even for those built within recent years. Seddon[348] finds that relative output (actual output as a percentage of rated capacity) in Australia is from 22 to 42%, with an average of 33%. *He also finds that where relative output is less than 32% there are no net savings of emissions because of the greater inefficiency of operating gas-fired plants (the most efficient fossil fueled plants for backup purposes) on an*

---

[346] Kagan *et al.*, 2014.
[347] Wiegand, 2013.
[348] Seddon, 2013.

intermittent basis. Since the average relative output in Australia is 33%, there is tiny net gain, but probably within the margin of error.

The decline in relative output over time is particularly a problem for larger turbines. A study of British and Danish wind turbines[349] concluded that the decline is so rapid that it is rarely economic to operate wind farms for more than 12 to 15 years rather than the 20 to 25 years often projected by the wind industry and government agencies.[350] UK onshore wind farms had an average load factor per megawatt of about 25% of their rated capacity at age 1; this dropped to about 7% at age 10 and just 3.5% at 15 years.[351] It is important to note that costs per megawatt hour are often quoted on the basis of rated capacity, not relative output, which greatly understates actual costs and is a major misconception concerning wind energy.

Wind and solar are not serious answers to the continuing need for reliable, inexpensive energy. Money used to build and maintain them could be much better spent elsewhere. Very little large scale solar and wind would be built without government subsidies and requirements as a result of the fad created by modern environmentalism. A relatively free market for electrical energy would provide the most efficient and environmentally sound approach.

I believe that environmental organizations, national governments, and state legislatures should not get into decisions as to which fuels should be used by power plants or whether no plants should be built at all on private land. Environmental organizations should ideally concentrate on improving the siting of power plants provided that all resulting pollutants are carefully regulated to the extent justified by sound science and economics so that decision makers are faced with correct incentives. Each situation is different

---

[349] Hughes, 2012a.
[350] Renewable Energy Foundation, 2012.
[351] Hughes, 2012a, Figure 10, page 33.

and the factors influencing the optimum choice of fuels are always changing as the economics of each fuel respond to supply and demand factors for individual fuels as well as to new understandings of the risks involved in using each particular fuel. The idea that environmental organizations or Washington bureaucrats can make better choices for the entire country or even some far off areas that they do not understand in detail strikes me as presumptuous at best. But that is what the environmental organizations and now EPA are doing when they oppose all plants using a particular fuel or even all new plants. Attempts to specify fuel use allocation, which a number of states do, are counterproductive in my view. I do not see exactly what these states and countries have gained by these fuel allocation dictates other than to increase the costs of reliable energy. Reliable energy is an input to most modern economies and the price needs to be kept as low as possible for the sake of everyone involved.

There are overwhelming reasons why the world has largely changed to using fossil fuels rather than wind, solar, and biomass. Wind and biomass supplied energy for many generations prior to the use of fossil fuels. The world has largely abandoned these earlier energy sources because they do not provide nearly as reliable or inexpensive energy as fossil fuels, nuclear, and water power do. Solar, wind, and biomass energy require comparatively large amounts of land to yield for the energy produced. This land might otherwise be used for agriculture or ecological preservation, which are both important. There is no reason to revise the decisions of the market economy in this respect, which take into account a wide variety of characteristics of each proposed energy source unless there are uncompensated adverse effects involved. These adverse effects appear to be fairly small and are already the subject of many regulations in the more developed countries.

Government regulations to reduce fossil fuel use so necessary for modern civilization and economic growth may seem to some to be one way to achieve broader anti-capitalist, anti-growth, and anti-urbanization goals, but I regard it as a monumental waste of money which will never achieve its utopian goals. Instead it will cause endless problems in achieving economic and environmental progress. One of the major casualties has been abandoning good science and economics.

In following the UN-sponsored movement radical environmentalists either explicitly or implicitly adopted the full GWD with all its contradictions and junk science. The GWD/GEF movement is fundamentally a "solution" looking for a rationale. The solution is always the same—governmental action to drastically reduce CO_2 emissions. This "solution" is very expensive, requires decades of advance planning in order to have any chance of achieving a specified CO_2 level even if all their assumptions are correct, and may be practically unattainable for modern civilization if there is to be continued economic growth. They have put forth a succession of rationales. The first was a reduction of global warming; the next was reduction of climate change; the most recent is reduction of extreme weather. But defying all logic, the solution is always the same. I argue that this whole approach is backwards. The first step should be to determine what the problem, if any, is, and only then try to find solutions to it if any are possible, worthwhile pursuing, and would leave the world a better place.

It is likely that the succession of alleged fears promoted by the CIC has resulted from pragmatic recognition of public interest. The reality appears to be that the very minor global warming that occurred from the bottom of the 60 year cycle in the 1970s until 1998 has come to an end, and will be replaced by decreasing or flat temperatures into the 2030s. It is becoming increasingly difficult for GWD supporters to claim that global warming has occurred since 1998 let alone that it is something to worry about. So they switched their concerns to climate change. But although this is a more flexible, more all-purpose, concern, it suffers from the same reality that there has been very little climate change visible to non-GWD supporters since 1998. The current emphasis on extreme weather in the wake of Hurricane Sandy, the Midwestern drought of 2012, the derecho in the Eastern US of 2012, and the 2013 Philippine Typhoon Haiyan responds much better to current public concerns but suffers from a total lack of any credible research to support a connection with these weather events and CO_2 emissions. Supporters will no doubt seize on any other relatively unusual and destructive weather events as they inevitably occur in the years to come. This represents a new low for the movement in terms of research and intellectual rationale. It suggests an opportunistic willingness to adapt the message to fit current sales opportunities. But it also highlights how the "solution" remains the same despite the continuing shifts by GWD supporters as to what the problem is. All this shows the intellectual bankruptcy and increasingly desperate attempts by the CIC to justify their "solution" and the environmental organizations' more general slogan-based goals.

Al Gore attempted to support the GWD cause as Vice President during the Clinton Administration, but reportedly did not have much support from Clinton because of its political unreality. The Bush Administration gave verbal support to the CAGW hypothesis but refused to implement proposed CO_2 control. But the Obama Administration has been willing to give GWD/GEF its full budgetary and political support. The Western European support for GWD appears to have weakened as their energy supply situation has worsened as a result of increasing fear of nuclear power due to Fukishima in 2011 and the failure of large wind investments to meet demand. The Obama Administration appears to have deliberately held off implementing the more draconian GWD-inspired regulations until after the 2012 election. Preferably sooner but possibly later government support for the GWD will falter to the point that most of the current GWD supporting organizations will find it advisable to abandon the GWD ship. Before long after that happens, the environmental organizations that joined the now sinking ship will have to abandon it even if they do not acknowledge error. The gross discrepancy between the GWD predictions and the actual temeratures will become ever more obvious and difficult to explain.

What has changed is not me but rather the environmental movement, which went from a minor special interest with a somewhat sensible agenda to preserve those parts of the natural environment that could reasonably be preserved without endangering the economic progress that has made the United States and the OECD countries successful to a major special interest that has increasingly adopted an anti-growth, anti-economic progress, big world government agenda which is quite infeasible and would, if adopted, have a negative impact on the environment and a huge impact on economic freedom, economic welfare, the standard of living, and the less wealthy. Opposing all fossil fuel-based energy development and economic development and higher living standards is not a responsible or rational viewpoint. Their excuse that wind and solar-based energy can satisfy needs may have public relations value since it seems to provide an answer to the question of how they propose to supply the energy needed by modern society, but is quite impractical for all the reasons just discussed. It is also pointless since it will have no measurable effect on any climatic variable.

In many ways the adoption of CO_2 emission controls as one of their principal objectives has laid themselves open to exactly the same concerns that environmental organizations had long expressed about major projects by development interests. Environmental organizations had pointed out that these projects ignored the larger environmental and ecological effects of these projects, which needed to be assessed on a broad basis rather than narrowly by individual specialties and special interests. Now the environmental organizations are endorsing a particular set of broad social engineering policies proposed by the United Nations with little or no careful analysis of the feasibility or broader impact of the proposals. This has opened them up to the same objections concerning the lack of careful analyses or problems with them that the environmental organizations have long made with respect to major development projects. And this is exactly what has transpired.

At the same time many environmental groups have abandoned or at least greatly decreased many of their earlier efforts to decrease needless environmental damages consistent with continued economic growth. Almost as bad, the willingness of the GWD activists, many professional scientific organizations, and even national science academies to condone, if not encourage, the abandonment of the scientific method, one of the cornerstones of modern civilization, to accomplish their GWD ends is simply unforgivable. I do not believe that any such abandonment is worth any other goal. Further, I believe that their misuse of the EPA (and many other parts of the Federal Government) in pursuit of their GWD ideology may destroy any faith that the public may have had in it and thereby make EPA's important job of finding scientifically sound and economically feasible ways to balance environmental and economic aims even more difficult than it already was in years to come.

Finally, I believe that implementation of CO_2 emission controls would further greatly decrease the standards of living and economic growth for those countries that adopt it. I have seen firsthand what government bureaucracy with largely good intentions can do to a country like India, and I do not see why a green bureaucracy in America or other developed countries world would do any better. Various Indian state agencies own and manage much of their electric generation industry. The result has long been frequent interruptions, theft of service, and a history of inadequate capacity. Is that what we want in an America increasingly dependent on reliable electric power for almost everything? Britain's Chief of the National Grid, Steve Holiday, recently stated[352] that the public should realize that reliable service was no longer possible. And for what purpose? None that I can see other than the growth and welfare of the CIC. This is madness and needs to be ended before it does even more serious damage to economic welfare, human freedom, and the scientific method. If it succeeds, just imagine what the next effort to misuse science and economics would do with this as an example for future bureaucratic and

---

[352] Eschenbach, 2011.

political entrepreneurs!

As I had immediately recognized when presented with the anti-energy use viewpoint by David Brower in the 1960s, and as I had advocated in the Sierra Club Brower-slate election of 1969, environmental groups would be much more successful and the environment would be more rapidly improved by generally taking a selective pro-energy, pro-economic development viewpoint, not the opposite. Somehow during my long absence from the Club at EPA the approach that I and many of the older generation of Sierra Club leaders has been dropped. I do not recall any Club election over this specific issue comparable to the 1969 one, but somehow the anti-energy, anti-growth approach took over not only the Sierra Club but most other national environmental organizations as well as in other Western nations and the United Nations. The Sierra Club is now one of the more radical environmental organizations in this regard with their opposition to coal and natural gas use and the development of reasonable standards for fracking, and their support of civil disobedience (which results in decreasing the rule of law) to achieve their ends.

So I believe I have been consistent: Energy and economic development are generally good for the economy and therefore for the longer-term impact on the environment and need to be selectively supported by environmentalists, governments, and the public when suitable environmental safeguards are observed. The current GWD/GEF/EWD, alarmist approach taken by environmentalists, many Western governments, and the UN is counterproductive for everyone involved and needs to be soundly defeated. The task at hand is not to stop energy and economic development. It is to minimize the adverse environmental effects of this development while encouraging the development because of the overwhelmingly positive effects it has on the quality of human life and the ability it provides for society to undertake environmental improvements.

The answer as to what to do about the GWD/GEF/EWD is very simple. Simply ignore these activists and their political supporters and do nothing except question their invalid "science," their bad economics and law, and particularly their economically and environmentally damaging and wildly improbable "solutions." With slight variations big government solutions have been tried before by various East Eastern European countries during the Soviet era and their sympathizers such as Nehru's India, with disastrous results. Unfortunately, when GWD/GEF/EWD advocates head major Western nations and use this ideology in making their decisions it is not possible to ignore them. They need to be fought on this issue.

Perverse Approach by Radical Environmental Organizations to GWD/GEF

One of the many ironies concerning radical energy environmentalists is that they appear to be blind to the obvious fact that their solution (government-imposed restrictions on energy production and use) for alleged global warming/climate change/extreme weather problems has not worked as well in achieving their goals as the simple and much lower cost alternative of leaving energy policy to market forces. The US did not join the Kyoto Protocol and has not imposed "cap and trade." The Obama Administration is working to propose EPA limitations on CO_2 emissions; these regulations will not take effect before 2016. Yet the US has actually met the goals of the Protocol, apparently unlike any other developed country.[353]

Part of this is due to the Great Recession, of course, but a significant portion is due to allowing market forces to encourage oil and natural gas development on private land, which has resulted in a major reduction in oil and natural gas prices and the substitution of natural gas for coal in electrical generation where it is economical to do so—largely to meet variable load. Part of this demand used to be supplied by older coal plants, but as natural gas has come down

---

[353] Watts, 2013.

in price it is more economical to meet this demand by using natural gas plants that can quickly be brought on line, which coal plants cannot do if not already operating. Because of its chemical composition, burning natural gas in the place of coal greatly reduces CO_2 emissions for a similar energy output. Between 2007 and 2012 US emissions of CO_2 decreased at the highest percentage of any country or region in one recent study.[354] Despite the efforts of most environmental organizations who have fought the use of fracking, this new technology has made possible a huge expansion of natural gas production, the reduction in gas prices, the rapid substitution of natural gas for coal in electricity generation, and quite substantial reductions in CO_2 emissions.[355]

One would think that the environmental organizations would rejoice that the reductions in CO_2 emissions in the US have come about and support the new technology that has made it possible, but I have seen no mention of this by any environmental organization. In fact, just the opposite is the case. Apparently, it is not the end that matters but the method by which it is achieved that is the important thing to them. Unless the solution involves government-imposed mandates on private market decisions apparently no "progress" counts. I suspect that the last thing they would ever want to admit is that leaving energy decisions to market forces is generally good public policy. After all, such an admission would suggest that their basic "solution," government-imposed constraints on emissions, has been wrong all along and that there really is no useful role for them in broader energy policy. The US reductions in CO_2 emissions are almost entirely the result of market forces, not government intervention. The fact that the Sierra Club and some other environmental organizations have attacked both fracking and the use of natural gas for electricity production makes no possible sense if they are adamant about reducing CO_2 emissions. It is perverse. Although the Obama Administration has supported

---

[354] Based on data from British Petroleum, 2013.
[355] de Gouw et al., 2014.

fracking rhetorically, I have not seen a concerted effort by them to strongly support its use such as by encouraging its use on Federal lands.

The US has actually shown the way towards reducing CO_2 emissions (assuming that this is a desirable goal, which it is not): Encourage fracking for natural gas, which results in greater production, lower prices, and substitution for coal. But instead of trumpeting this to the world, the Obama Administration continues to believe that it should pursue a regulatory approach towards energy production, which it believes would show world leadership in reducing CO_2 which together with large payments to less developed countries would, they believe, would result in similar regulatory efforts in other countries. More likely is that even if a new climate protocol is adopted by the UN that included mandated, significant, verifiable emissions reductions by less developed countries as a result of US "leadership" and possibly payments, the required emissions reductions by less developed countries would simply not implement their commitments. By pushing the advantages of fracking for natural gas they could instead present the less developed countries a profit opportunity which would have the indirect effect of reducing CO_2 emissions. This might actually work, which the current Administration effort never will. Better yet, of course, would be to get the US Government out of all energy policy where it is not required for economic reasons.

Decreasing Public Support for Environmental Movement

Public support in the US for the environmental movement in terms of active participation and sympathy have decreased from about 70% in 2000 to about 60% in 2013.[356] The fraction of Americans who said the movement has definitely or probably done more good than harm decreased from 76 percent in 1992 to 62 percent in 2010. Conversely, the percentage of Americans saying the

---

[356] Tobin, 2013.

movement has definitely or probably done more harm than good rose from 14 percent in 1992 to 36 percent in 2010. These changes in public opinion suggest that support for the environmental movement has decreased significantly but that it still enjoys substantial support.

Problems with the Environmentalist Party Line

In the last few years a number of large oil discoveries have been made in the Gulf of Mexico. With careful development they promise a significant increase in US oil production and a significant decrease in US petroleum imports. The response of one environmental organization to this development was as follows:[357]

> "Recent trends in US energy consumption and production suggest we don't need to find more oil offshore," Cindy Zipf, director of Clean Ocean Action, Inc., in Sandy Hook, N.J., wrote in the Wall Street Journal in 2013. "Our investment dollars and energies are better spent on renewable energy, conservation and efficiencies such as improved mass transit, smart grids and clean-emission vehicles—an approach that creates jobs, doesn't damage the environment and addresses fossil-fuel-driven climate change."

I quote this because it may be typical of current radical environmentalist rhetoric. I trust the market to allocate resources much more than either the government or the environmental organizations. Assuming that suitable safeguards have been built into Federal leasing of these new sites, which hopefully has happened after the leasing changes instituted after the Macondo blowout, the risk of serious damage to the Gulf of Mexico has hopefully been reduced to a low level. Federal efforts to allocate investment to "clean" innovation as part of the American Recovery

---

[357] Jonsson, 2013.

and Reinvestment Act has proved to be an unmitigated disaster with many bankruptcies and nothing to show for many of the investments. Green jobs have proved to be a phantom that never seems to show up. And fossil-fueled global warming is minor and good, not bad. Both government and environmental organizations need to leave such decisions to the market and stop trying to use government for these purposes. There is no reason to think that the market will misallocate investment dollars in this case, as Ms. Zipf appears to believe.

One of the many ironies of green energy ideology is that while they claim that CO_2 emissions pose a catastrophic threat to human civilization, they oppose most of the real policies that could actually reduce such emissions. Burning natural gas rather than coal will reduce CO_2 emissions relative to coal. But they are generally opposed to using natural gas too. Hydropower is clearly CO_2 free, but they are opposed to that. Geoengineering may be able to reduce (or increase) global or regional temperatures, but they are opposed to that. When posed with the question as to how CO_2 should be reduced, their claim usually is that energy conservation, and wind and solar are the answer. But they are not. The US has spent quite large amounts on energy conservation and the result has been that energy use has remained roughly stable in recent years. Wind and solar are very expensive and unreliable, particularly as their use reaches higher levels. So essentially the environmentalists want to abandon technology that works for technology that could never work or at best only unreliably at a very high cost. This is a recipe for a long term disaster both for the environmentalists and for any country that adopts their policies. The results of following such policies can be found in Great Britain and Germany, which have rising energy prices, dangerously low electricity reserves, and are now building diesel and coal backup generating plants.

General Conclusions Concerning Radical Environmentalism

As discussed in Chapter 11 and contrary to the current tenets of environmentalism, the key to improving the environment is not decreasing the emissions of a highly beneficial trace gas (carbon dioxide) fundamental to life on Earth, but rather continued improvement of human standards of living which have been repeatedly shown to be the best way to bring about environmental improvements. Increasing productive energy use is the surest way to improve living standards, and thus the surest way to an improved environment. The continuing failure of the environmental movement and the Obama Administration to understand this well established finding of environmental economics is leading us towards an increase in dictatorial intervention by governments in the private lives of their citizens and the private sector, a huge transfer of wealth from the less well off to the wealthy, and continuing economic problems for the United States and other developed nations that adopt the GWD/GEF. And all this will ultimately result in a worse, not a better environment compared to what it would otherwise have been. Some individual environmental standards may be made more stringent, but the overall standard of living will not improve as much as it otherwise would, with a resulting loss of the overall environmental quality in terms of what it would otherwise have been.

The radical environmental policies of the Obama Administration are likely to have another very unfortunate effect. The US Governmental agencies charged with improving the environment appear likely to lose much of their credibility with the public and may well find themselves penalized as a result. This is not a desirable outcome in my view but may in the end prove to be the only solution to bringing sanity into the environmental improvement effort. These agencies have been "captured" by the environmental movement and are no longer able to provide a balanced approach towards environmental regulation and improvement reflecting the interests of the nation as a whole. This reflects a fundamental flaw in their structure which needs to be remedied.

The Obama Administration appears to regard almost any tightening of environmental standards, particularly if they result in fewer coal-fired power plants, as favorable. A case in point is their attempts to tighten the ambient standard for ozone. This is not the first time that this has happened, but their attempt is ridiculous in its severity. Unfortunately, ambient air standards are not supposed to consider the cost of regulations, only the health effects. There is obviously a trade-off between the two; previous administrations have understood this and largely refrained from pursuing standards that have astronomical costs for very little health gains. The best way to solve the problem would appear to be to rewrite the Act to allow for consideration of cost in setting ambient standards. Given that this is unlikely to happen, it must rest with each Administration to exercise restraint and judgment, something which the Obama Administration appears to be unable to do when it comes to environmental regulation. Imposing huge added environmental regulatory costs with doubtful benefits on an economy finally beginning to overcome the effects of the most serious recession since World War II is clearly not an optimal policy.

The basic question is whether the new energy-related activism of the radical environmental movement has and will accomplish more or less than the older, more moderate approach typified by the environmental organizations of fifty years ago. I maintain that it has not and will not. By pursuing broad anti-growth policies typified by the GWD/GEF/EWD/AFD ideology the environmental movement has mainly insured a strong backlash such as that the strongly adverse Republican Party views on climate change "control." Despite all the lavish funding by liberal foundations and the Federal Government on their GWD-inspired programs the radical environmental movement has long since gone so far beyond rationality that it is counter-productive in achieving its own ends. This was my fear and the basic reason that I opposed

Radical Environmentalism and Its Shortcomings

David Brower in the key Sierra Club Board election that resulted in his departure. There is a natural tendency in any sort of radical movement for the movement to become more radical as time goes on since the adherents of a more radical approach can always claim that the less radical factions are not true believers. Ultimately, however, they are likely to become so radical that their views are rejected by other important segments of society, with a resulting reduction of influence rather than the increase they so fervently desire. The early goals of the environmental movement were largely accomplished because of general public support for them. The newer, more radical proposals opposing energy development and use and thus economic growth do not and will not ever win broad public support. This is what I believe we are seeing in today's environmental movement.

During my long absence from the movement while working at EPA, the radicals took over the Sierra Club as the older, much more pragmatic generation that I favored retired or died. In more recent years environmental organizations have never even been exposed to a more moderate, broader world view. Brower was certainly sympathetic with the more radical views now so common in environmental organizations, but unfortunately his departure from the Sierra Club's paid staff only served to temporarily postpone the Club's espousal of a much more extreme environmental ideology and created two other organizations better representing his more radical views (Friends of the Earth and later the Earth Island Institute). The radical environmentalists have abandoned sound science and economics as a basis for their goals and instead pursued goals that are not politically achievable and will actually slow progress towards introducing an environmental perspective on societal choices. If adopted, these policies would at the very least slow economic growth and thereby longer-term environmental improvements.

The major current goal of the movement—reducing emissions of CO_2 and other GHGs—would prove very expensive for those countries that actually adopt it, would depress economic activity,

417

would retard the growth of so-called C4 plants (which are better able to use higher levels of CO_2), and might make the Earth somewhat more prone to disastrous cooling in the future. Since a new ice age would almost certainly have more adverse economic effects than minor warming that is unlikely to occur, it is cooling that most needs to be avoided, particularly in northern latitude areas such as the United States. Historically over the last 3 million years the Earth has been much colder most of the time than it now is. Human civilization has understandably developed primarily during the current interglacial period, the Holocene, rather than during the previous ice age period. Higher levels of atmospheric CO_2 may provide some protection against future colder periods and that is what humans really need protection against, not allegedly slightly higher temperatures in the next few years. There is no evidence that higher levels of CO_2 will result in anything more than minor increases in temperatures. These should be welcomed rather than feared based on current knowledge. An improved knowledge of how climate change occurs is important, but is highly unlikely to happen when only one avenue of research is pursued (climate model building) which as so little promise. Given all this, attempts to implement measures to reduce CO_2 emissions were premature at best on the basis of the science and now appear ill-advised given the larger context developed in this book. Even if it should be necessary to reduce world temperatures this is probably much more effectively and efficiently done through geoengineering than through reductions in CO_2 emissions. More likely is that humans will find it advisable to increase world temperatures rather than decreasing them, which is also best done through geoengineering, and will be harmed, not helped, by CO_2 emissions reductions if the CAGW believers should be at all correct.

So it is hard to see what the environmental movement has gained by largely abandoning its previous important efforts to decrease the adverse effects of human activities on sensitive parts of the environment and instead expending its resources on a hopeless enterprise which in all probability would have adverse rather than positive effects on humans and the natural world. And it may well have lost the support of many thoughtful people. It certainly has lost my support.

It was obvious to me that the pre-1970 environmental conditions in the US were not adequate and needed improvement. I devoted most of my career towards supporting this end. Attempts to dream up imaginary environmental problems and "solutions" to them in order to maintain interest in the goals of environmental organizations and income for climate researchers is at best counterproductive and more likely will result in a weaker environmental movement because of its inevitable loss of credibility. Important traditional pollution problems remain to be solved, but are receiving little attention, but should. It has now been over 17 years since global temperatures increased; sooner or later people will notice. The peak oil hypothesis is similarly being disproved by increasing oil production from shale rock. The environmental movement has backed the wrong horses and is likely to pay for their mistakes as this becomes more evident to the public. To the extent that state and Federal Governments continue to pursue these wrong horses, the American public will pay dearly.

General Conclusions Concerning the Impact of GWD/GEF/EWD on the Environment

Implementation of GWD/GEF would make the world worse off environmentally, not better off as the radical environmentalists claim. Unfortunately, the UN IPCC, the environmental organizations, and EPA all got the science wrong—backwards, in fact—in two respects. First, as explained in Chapter 9 and Appendix C, the major environmental risk Earth faces appears to be not global warming but rather global cooling. If we attempt to implement the CIC's "solution" of reducing CO_2 emissions by 80%, the result would be to make our major environmental problem, global cooling, worse, assuming, of course, that it had any measurable effect at all. Second, the IPCC assumes that global temperatures are determined by CO_2 levels when it now appears, as explained in Appendix C, that CO_2 levels may be determined by temperatures and soil humidity levels. Finally, the effect of human-caused emissions is fairly minor because of the very small greenhouse effect of changes in CO_2 levels at current elevated atmospheric CO_2 levels. The IPCC "solution," in the unlikely event it could be implemented, would move temperatures in the wrong direction to the extent that it would have any effect at all.

In other words, the CIC "solution" is based on faulty science and would have very little effect if implemented, and even that would be in precisely the wrong direction. Failure to carry out valid science can have major effects on environmental protection as in many other areas.

The adverse environmental effects of wind turbines, the leading CIC-favored technology for generating electricity, have been evident for many years but unfortunately not recognized by either radical environmentalists or until very recently by wildlife regulatory agencies. I do not see why a rare bird killed by oil leaking from a pipeline should result in a serious penalty while the same bird killed by a wind turbine blade is almost always excused or ignored. In the first case the oil leak was accidental whereas in the second it is the inevitable result of using the technology, but the result is exactly the same—a dead rare bird. Of the two, I find the windmill death the more reprehensible since it was bound to happen whereas the oil leak death was at least not supposed to happen.

The CIC Chain of Command

The CIC defined in Chapter 1 is not just a theoretical abstraction. Rather, it is now clear that major parts of it actually exist as functioning units that are very carefully coordinated by a large and complex group of organizations whose purpose is to promote a far left radical environmental agenda. According to a 2014 minority US

Senate report,[358] a major portion of the CIC is funded and indirectly and anonymously commanded by a group of very wealthy individuals and foundations who espouse these views, manage to make use of the charitable tax exemption laws, fund coordinated political action campaigns, and substantially influence if not control the actions and the strategy of the Environmental Protection Agency. This would explain how it is that the CIC public relations campaigns are so large, so duplicative, and on theme, as discussed in Chapter 7. It would also explain how the CIC as a whole is so well coordinated and on theme. Although much of the CIC is funded by government and thus the taxpayers, much of the rest of it is funded by the private groups. Not surprisingly, deviation is apparently not tolerated. In my view all this can be reasonably characterized as an organized conspiracy.

Unfortunately, the CIC is not looking after the interests of the US public or even the Democratic Party, but rather those of very wealthy far left radical environmentalists. This is why lower income groups, like coal miners, are being thrown out of work and electricity rates are increasing, hitting mainly lower income groups. The wealthy liberals apparently do not care about lower income groups. It would also explain how it is that the professional staffs of environmental organizations have undergone such a dramatic increase in recent years. In my day, staff was very hard to pay for and were a very scarce resource. But now it is apparently often paid for by wealthy far-left liberals. Finally, it helps to explain why the environmental organizations have drifted so far left. It is not just due to a leftward drift among the membership of such organizations (in the case of those few environmental organizations that actually have members that exercise any real influence), but rather the influx of huge quantities of cash from wealthy funders—all or mostly carefully laundered so that the wealthy liberals can claim tax

---

[358] US Senate, 2014a. See also Arnold and Gottlieb, 1994 for a much earlier exploration of some of the same parts of the CIC as they existed then.

exemptions for "charity" even though what they are really doing is trying to influence legislation and elections. The environmental organizations, who were hardly wealthy when I worked with them, appear to have been widely bought. Perhaps that is why they are so certain that skeptics have been bought by carbon "polluters." After all, that is what they may have been—bought—and they cannot imagine that the opposition might be primarily motivated by simple civic virtue and a shared common vision of what needs to be done.

This excerpt[359] from the Senate Report may be particularly significant:

> *EPA Administrator Gina McCarthy recently told Congress that the Agency's proposed Existing Source Performance Standards for coal fired power plants, which is widely believed to be the death knell for coal as an industry, was, in fact, an opportunity for economic growth: "The great thing about this proposal is that it really is an investment opportunity. This is not about pollution control. It's about increased efficiency at our plants, no matter where you want to invest. It's about investment in renewables and clean energy."[360] In fact, multiple sources, including the New York Times, have attributed the authorship of the proposal in large part to the "NRDC mafia," including David Doniger, Daniel Lashof, and David Hawkins.[361]*

As this report reveals, NRDC obtains a significant amount of donations from the Energy Foundation, which is heavily funded by Sea Change Foundation, whose major donors are reported to be heavily invested in renewable technologies. This report offers a new perspective on the "opportunities" McCarthy was referring to which

---

[359] *Ibid.*, p.67.
[360] Testimony before the Senate Committee on Environment and Public Works, July 23, 2014
[361] Davenport, 2014.

are the economic opportunities of millionaires and billionaires who are part of the far-left environmental machine heavily invested in helping EPA advance such regulations. It is surely not an opportunity for Americans living in Appalachia or the Powder River Basin who depend on coal for their energy supply and livelihood, nor is it an economic opportunity for Americans already struggling to pay their energy bills.

Figure 10-1: Abandoned Wind Turbines, Tehachipi Pass, California[362]

---

[362] Walden, 2010.

Figure 10-2: Solar Panels Near Markranstaedt, Germany, 18 Months after Construction[363]

---

[363] Gosselin, 2011. Photo taken by Prof. Knut Löschke, June 2011.

**Figure 10-3: Solar Modules at Solar Sewage Treatment
Plan in Fehmarn, Germany, after 20 Years** [364]

---

[364] Klimakatastrophe.wordpress.com, 2009. Photograph by webmaster.

Figure 10-4: Women Carrying Firewood Along the Yamuna River Across from Taj Mahal, Agra, India[365]

---

[365] Photo by Alan Carlin, 2009.

11

An Economic and Security Perspective

UP TO THIS point the reader may say so what; if a few climate modelers want to promote their ridiculous CO_2 hypothesis and try to take credit for "saving the world" while taking public funds to pursue more unproductive research and some politicians spend public money to subsidize wind and solar power, what difference does that make to me? This Chapter will explain why it makes a significant economic difference to ratepayers and taxpayers in the US and abroad and to the national security of the US.

The fundamental economic issue posed by GEF is that there is currently no basis for governmental intervention in the market for production and use of energy beyond programs to reduce real conventional air pollution, and even these have gone much too far in the case of US air pollution. As discussed in Chapter 9 and Appendix C there is no significant market failure and certainly none that would justify the cost of trying to intervene in the markets involved. The scientific hypotheses offered for CAGW do not withstand scrutiny using the scientific method. Accordingly the economics of proposed such interventions will be unfavorable and the interventions should not be attempted from an economic viewpoint either.

GWD/GEF has had major effects in a number of Western European countries and will increasingly have important economic effects in the US if the Obama Administration is successful in its attempts to impose its climate agenda through regulations from the Environmental Protection Agency and other Federal agencies.

This Chapter will start by examining the underlying problem of why the EPA air office is experiencing rapidly diminishing returns from their ever more stringent and dubious regulations. Following this, I will present an overview of how EPA's air regulations have lost touch with reality, including a very brief summary of an economic review of the proposed climate regulations.

This will include a general discussion of their economic efficiency. This is important because it provides an economic yardstick as to whether the proposed regulations represent a good investment from a societal viewpoint. If society as a whole receives fewer benefits than the policy costs, society as whole will be worse off economically. The second aspect is the distributional effects, or who loses and who gains. This is important because even if society gains from an investment it may disproportionally impact the poorest or the richest members of society, which might change people's evaluation of it. An economically efficient policy that benefits entirely the very rich may not be favorably received by the many losers. A third perspective concerns whether a regulation will stimulate or retard economic growth, or what the larger effects of the regulation are likely to be. The Chapter will end by outlining the importance of avoiding abandonment of sound economics in pollution control.

Diminishing Returns from Most New Air and Energy-related Environmental Regulations

It seems odd that despite more than 40 years of effort by major environmental and governmental organizations at all levels, the demands of the environmental organizations and the Federal Government for still more air pollution control are more insistent and shrill than they were before the first Earth Day. How could it be that the more that is accomplished, the more needs to be done from the viewpoint of these organizations? I believe that this arises because the environmental organizations and the US EPA air office have become steadily more radical, not that the problems have become worse. In fact, pollution data suggests that air pollution has decreased greatly since 1970. One question is why this leftward shift of the movement's views has occurred. As the former very real problems have been partially or wholly solved, the organizations must either scale back their efforts (and therefore their reason for being) or find new reasons for concern. They have clearly chosen the latter approach.

The commonsense observation that the usefulness of further pollution control decreases as control increases is what would be expected from economic analysis. As environmental regulations become more strict, the costs of additional regulations become higher and higher and the benefits less and less. Yet particularly under the Obama Administration, no regulation is strict enough. More should always be done and be based on environmental ideology, not careful scientific and economic analysis.

In my view many, if not most, of the more recent EPA air pollution control efforts have gone well beyond the point of rapidly diminishing returns. The political reason for this is that EPA air regulations were allowed to exceed reasonable bounds by the Clinton and particularly the Obama Administrations, presumably in exchange for political support from the environmental movement. The bureaucratic reason is that the economic efficiency analysis requirements for new regulations imposed at the beginning of the Reagan Administration have been increasingly circumvented in recent years and that the Clean Air Act is overly strict by requiring in many cases that pollutants be controlled without regard to the costs. An additional reason is that EPA has continued to use an overly generous assumption concerning what the effects of pollutants are at very low levels.

Although the story of how the EPA air regulations got out of control would not seem to have much to do with climate, it does because EPA's climate efforts have been built on the foundation provided by the Air Office's earlier and continuing efforts to use fine particulates to justify its non-climate regulations so that many of the problems in climate reflect the problems in their non-climate regulations. The story of how the requirements for scientific and economic analyses have been circumvented and ignored needs to be told so as to understand what has happened with the climate regulations.

Much of this story can be found in a Senate Minority report,[366] which details the EPA air "playbook" developed in part by two Air Office employees, one of whom retired because of an ethics problem and the other who was sent to prison for not showing up for work or performing many of the duties for which he was paid. First of all they participated in the development of a new-to-EPA pollutant called $PM_{2.5}$, which is particles smaller than 2.5 microns in size. They are so small that they are not visible to the unaided eye. They then asserted that there were adverse health effects which they claimed were life threatening and widespread so that many lives could be saved if it were reduced. Finally, they used two epidemiological studies for which the data have never been made public as the basis for these effects even though important confounding variables were not controlled for and at least six other studies reached different conclusions.[367] These studies have now been dubbed by Republican congressional critics as "EPA secret science."[368] Clinical studies of key $PM_{2.5}$ components emitted by power plants have failed to find any adverse effects, even at higher levels than found in US air.[369] If there are no clinical effects from exposure to a substance, there are no benefits from reducing it. They then helped to push through regulations to control this

---

[366] US Senate, 2014.
[367] Goodman, 2012.
[368] US Senate, 2014.
[369] Schwartz, 2008; Green, 2003.

pollutant with an accelerated schedule and less than thorough scientific review.[370]

By law these regulations are to be set at a level such that the pollutant will not pose a health risk "with an adequate margin of safety." But, importantly for our climate discussion, they used coincidental further reductions of this pollutant beyond that called for in the $PM_{2.5}$ regulation to argue for health "co-benefits" from reducing 30 other conventional pollutants.

Although they are no longer employed by EPA, EPA is still using the "playbook" they developed to push through many of its proposed new air regulations,[371] including those for its proposed climate regulations. Without their "playbook" "co-benefits" many if not most of the air regulations proposed in recent years, both climate and non-climate, could not be made to even appear to be justified economically. A number of studies[372] have pointed out what the problems are, but the Obama Administration would be the last one to do what needs to be done to bring back verifiable science and justifiable economics to air pollution control. The Clinton and Bush Administrations were a little better than the Obama Administration in this regard, although the Clinton Administration originally approved $PM_{2.5}$ regulations despite a lack of careful scientific verification and the Bush Administration allowed the problem to fester.

One of the persistent aspects of EPA's attempts to reduce conventional air pollution in general and now CO_2 emissions from existing power plants is their reliance on these so-called health "co-benefits" based on coincidental reductions in other pollutants to justify their proposals. What they claim is that if you cease to operate coal plants there will also be reduced emissions of real conventional pollutants. This is undoubtedly the case. Fewer coal

---

[370] US Senate, 2014.

[371] A list of regulations using $PM_{2.5}$ co-benefits since 1999 can be found in Appendix A of Senate, 2014.

[372] Smith, 2011, 2011a, 2012, and 2012a; see also Senate, 2014.

plants operating mean slightly less sulfur dioxide, particulate matter, and all the rest, despite existing standards that are some of the strictest in the world.

EPA's air "playbook" describes the various strategies used by the EPA Air Office to subvert the various checks and balances built over many years to prevent uneconomic and scientifically unjustified regulations from being promulgated by EPA.[373] What is needed is not continued use of the EPA's $PM_{2.5}$ "co-benefits" in new and recent proposed air regulations but a very careful review of the health basis for the $PM_{2.5}$ regulations.

With regard to the "secret science," EPA has been unwilling or unable to provide the public with the original data and analyses that led EPA to believe that these "co-benefits" exist and their magnitude despite persistent inquiries and even a subpoena issued by the House of Representatives. I maintain that "secret science" has no place in environmental protection and any regulations justified using it should be rescinded.

I am suspicious of using co-benefits to justify regulations that cannot be justified without them. Such use runs the risk that regulations that are primarily for the purpose of say reducing CO_2 emissions are justified by using benefits from unrelated pollution effects reduction for pollutants already strictly regulated by EPA (in this case $PM_{2.5}$). The result, as in this case, is that an economically valueless regulation like reducing CO_2 from power plants is propped up in terms of economic analysis by claiming that it has health "co-benefits" which *EPA has chosen* not to achieve directly by tighter regulation (of $PM_{2.5}$) even though it is mandated to reduce air pollution levels to a safe level with an adequate margin of safety.

Given the Clean Air Act requirement, EPA has already said that the current mandated level is safe with an adequate margin of safety. It cannot be both safe and unsafe at the same time. EPA is trying to have it both ways and appears to be seeking to game the system. So a "co-benefit" is used to justify a regulation that has no

---

economic value except for other pollution-reducing effects it happens to have by accident. In defending the proposed EPA CO_2 power plant regulation, both Obama and EPA strongly emphasized the alleged health co-benefits, not the CO_2 reductions, their real aim. Instead Obama visited a hospital with sick children. Perhaps Obama realized that fewer sick children, even though unproven, might sell better to the public than ephemeral extreme weather improvements, abandoned power plants, and unemployed workers. Electricity users would have to pay for these proposed regulations with substantially higher electric bills, but this was not really explained.

There is great scientific dispute as to whether further $PM_{2.5}$ health co-benefits actually exist and should be used in benefit-cost analysis if they do.[374] EPA assumes what are called linear no threshold effects, which are based on assumption, not experimental evidence; this is a general problem of many EPA regulations which will be discussed further shortly. Critics have raised the issue of how very low levels of $PM_{2.5}$ could possibly result in 22% of all deaths in many parts of the US[375] even though EPA cannot say what causes of death actually result from low level $PM_{2.5}$.

In brief, $PM_{2.5}$ "co-benefits" appear likely to be a scam perpetrated by EPA to justify expensive and often inefficient air pollution control regulations which have not been shown to have real value in themselves. I believe that "co-benefits" that have not been rigorously analyzed and found to have scientifically verifiable adverse effects should be counted. And I believe that scientifically dubious "co-benefits" which are based entirely on assumptions as to effects at such low levels that they are unmeasurable should not be counted in "co-benefit" calculations.

The proposed GWD/GEF-inspired regulations are bad ideas because they are based on invalid science, so should not be undertaken in the first place. But a number of the proposed air

---

[374] Smith, 2011a.
[375] *Ibid.*

pollution regulations which do have some grounding in valid science are just too expensive for what they are likely to achieve and are justified using dose-response functions which are grossly exaggerated and often $PM_{2.5}$ "co-benefits." The way to catch such regulations, in my view, is through the use of careful and independent economic analysis. Such analyses were introduced during the Reagan Administration and were the subject of much of my work at EPA. The EPA effort was greatly weakened in the early days by a decision that the same ("program") offices that prepare the regulations should evaluate their economics. This was bound to lead to problems and it has. Even if responsibility for regulatory economic analysis were moved to another office, such as the policy office, I believe that there still exists a significant danger that the analyses would be compromised. So it may be better to move economic analysis of proposed regulations completely out of EPA to a more independent agency other than the Office of Management and Budget, which currently reviews all regulations but is too easily influenced by the President.

Three concrete examples of all this are the proposed EPA rules/proposed rules on ozone, mercury, and CO_2 reductions from existing power plants. In the case of mercury, it is called the Utility MACT rule and establishes maximum achievable control technology (MACT) standards for hazardous air pollutant emissions from power plants. The rule is based on protecting an assumed population of pregnant subsistence fisherwomen who consume hundreds of pounds of self-caught fish yearly exclusively from the most polluted inland bodies of fresh water.[376] Whether any such people exist has not been established. The EPA does not estimate the benefits for other hazardous air pollutants, but finds that for mercury the estimated costs exceed the estimated benefits by a ratio of 1600 to 1 or even 19,200 to 1. Instead, the EPA justifies the rule economically on the basis of the health "co-benefits" primarily from reductions in $PM_{2.5}$, which would occur accidentally from

---

[376] Lewis, Yeatman, and Bier, 2012.

implementation of controls on the hazardous air pollutants.[377] There is no way to know whether the proposed controls are the most efficient way to achieve reductions in $PM_{2.5}$ since EPA did not directly consider this objective. Very expensive regulations which in this case have led the operators to consider shutting down about one-fourth of existing plant capacity of US coal power plants[378] on the basis of a dubious justification and research that no one outside EPA can check is an example of unjustified overkill that EPA is increasingly engaged in for air pollution and energy-related control. It may also be dangerous for US citizens who may not be able to access electric energy they may need in the next few years during adverse weather events, particularly prolonged hot or cold weather such as was experienced in the winter of 2013-14 in the Northeast. Most of the plants considered for shut down were called into service during the winter of 2013-14, but this would no longer be possible once they are closed.

The proposed ozone rule would be the most expensive in EPA history, would lower the ozone limits close to background levels, and would be another revision of the standard despite the absence of significant new research suggesting the need for tighter standards and doubtful benefits.[379] Since the standard cannot take into account economic costs as per the Clean Air Act, economic considerations cannot be considered. I have always regarded this as a major weakness of the Clean Air Act, which the Obama Administration clearly plans to exploit. EPA's calculations, however, once again use similar co-benefits to those used in its analysis of the Utility MACT rule to make the economic case for the proposed rule. I believe this proposal is an example of EPA's refusal to recognize that its proposals would take regulations far beyond the point of

---

[377] Smith, 2011a. See also her testimony before various House Committees, namely, Smith, 2011, 2012, and 2012a. See also Lewis, Yeatman, and Bier, 2012.

[378] US Energy Information Administration, 2014.

[379] Milloy, 2011.

diminishing returns. Past administrations understood the adverse effects that ozone overregulation can have, but not the Obama Administration.

The third (proposed) regulation requiring reductions in CO_2 from existing power plants is analyzed in Appendix C.4, but suffers from many of the same problems as the other two in addition to clearly wrong climate benefits.

In addition to the simple diminishing returns to increasingly strict environmental regulations there is an important scientific issue concerning whether EPA has been routinely using an overly conservative model to estimate the effects of pollutants at very low levels. The problem arises because it is very hard if not impossible to measure the effects of very low levels of pollutants simply because the effects are so small, so regulators must make assumptions as to what occurs at these levels. But the actual human exposures to many pollutants regulated by EPA are believed to occur at just these very low levels, particularly as exposures at higher levels are reduced.

In order to estimate pollutant effects at very low levels, EPA has long used a linear no threshold (LNT) assumption. This model assumes, for example, that if one hundred aspirins is a fatal dose, then out of 100 people taking just one aspirin, one will die. This is absurd, of course. LNT is the basis for arguing that minute levels of pollution will cause health effects, as EPA argues in the case of $PM_{2.5}$. An argument for using it, which EPA has long advanced, is that this model is the most "conservative" of three common models in the sense that it is more likely to overestimate the effects more than the other two. Thus regulations based on it would be the least likely to underestimate the problem being examined. But this advantage comes at a high price in that very low levels of most pollutants are beneficial to human health (as illustrated in the following charts by the area below the line) and decreasing them will actually harm health This can be seen in the following figure, where the harmful area above the line is largest for LNT and lowest for the hermetic model. The smallest such area is for the hormetic model, where there is a significant beneficial area below the line for low pollution levels. This means that at these low concentrations the pollutants benefit health, rather than there being a need for control. Use of LNT also leads to much higher cost regulations, which is becoming an increasing problem as regulations have become ever more strict.

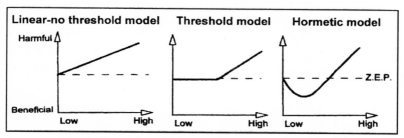

Three Prominent Dose-Response Models[380]

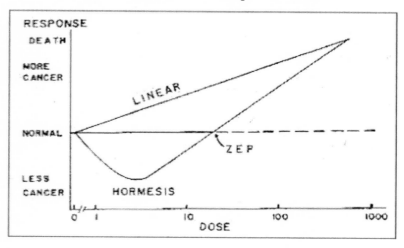

Illustration of Superimposed Linear No Threshold and Hormetic Dose Response Models[381]

---

[380] Brook, 2011.
[381] Oakley, 2006. Adapted by Oakley from Luckey, 2001. ZEP refers to

The companion figure illustrates two of these models with the LNT and hormetic models superimposed. It is assumed that the lowest concentration data point is at the right in the figure. It makes a substantial difference as to which model is used, but EPA has never seriously considered moving to one of the other models.

There is now increasing evidence that one of the other models is more accurate than the LNT model in the sense that biological responses appear to fit this model in an increasing number of cases studied.[382] That model is the hormetic model. Although this may seem a rather obscure, academic issue, it is actually a major game changer in the world of regulatory pollution control, and deserves urgent attention even if it is necessary to delay relevant EPA and other regulations to do so. The reason it is so important is two-fold. The first is using the LNT model results in much more expensive pollution control than using either of the other prominent models—the hermetic and the threshold models—since it effectively assumes that the adverse effects of pollutants continue even down to the lowest concentrations. The second reason is that available evidence indicates that the results are much less protective of public health. Thus pollutants have adverse effects at any concentration and must be completely removed if all adverse effects are to be avoided. Removing the last small quantity of anything becomes astronomically expensive. In many cases very low concentrations cannot even be measured. There are a number of other related issues such as radiation limits and pollution levels when evacuations are necessary which also need urgent attention.[383] Obviously the environmental organizations will be strongly opposed, but that is another reason why regulatory decisions need greater insulation from politics and greater input from objective

---

zero equivalent point or the level at which the pollutant does neither harm nor good.

[382] Mattson and Calabrese, 2009.

[383] Conca, 2013.

science.

One illustration of the hormetic dose response relationship is provided by sunlight. In small quantities it is beneficial to humans by stimulating the production of vitamin D. In much higher concentrations it results in sunburn, and at extremely high levels can even result in death. The hormetic response pattern has been found even for a number of types of nuclear radiation.

In other words LNT is an assumption EPA (and other regulators) has made that has not been shown to be scientifically valid, and increasingly appears to be invalid based on scientific research on a large variety of pollutants. I believe that the time has come to question its validity and consider changing to more accurate assumptions in order to avoid making health worse and wasting very large resources on what appears to be an invalid scientific assumption. As EPA continues to tighten its pollution standards this becomes ever more urgent because it is increasingly likely to be attempting to control pollution in areas where the beneficial hormetic effects predominate and further tightening would be harmful, not beneficial to health. Like the CAGW question, much of the scientific and bureaucratic establishment is resisting such a change, but it is becoming increasingly important to examine the issue before ever more expensive and possibly even harmful pollution regulations are approved.

I need to emphasize that I do not believe that EPA is systematically exaggerating all the risks of the pollutants that it has jurisdiction over. Although there needs to be a reassessment of the basic methodology used to estimate low level effects, that does not mean that every pollutant EPA has regulated has been over-controlled. Each pollutant needs to be examined on the basis of the best information available for that pollutant. And they need to be periodically reexamined in light of new research and experience.

One example is radon, a naturally occurring radioactive gas found in many homes in the US and abroad. There has been some scientific uncertainty as to whether the low levels often found in US homes constitute a significant risk for lung cancer. New research suggests that current EPA standards may not be adequate.[384] I would argue that radon deserves more, not less attention by EPA, particularly compared to mercury emissions from coal-fired power plants. Although the Administration has directed EPA to address a non-problem, CO_2, there are other risks that do deserve greater EPA attention in my view.

Economic Efficiency

One of the things that economists often look for is whether a proposed policy change will make people better or worse off as a whole. In economic jargon this is called economic efficiency. Obviously a given policy change is likely to make some people better off and some worse off. People who make windmills or solar panels will be better off if coal-fired power plants are torn down and replaced with their "renewable" power sources, as EPA has now proposed. On the other hand, those who are forced to pay for the windmills and solar panels will usually be worse off since their electric rates will increase and they will have to pay more for the same amount of electric power.

Economists measure economic efficiency by comparing the changes in total benefits of the policy change with the total costs, regardless of who receives them. This type of analysis is called benefit-cost analysis, and has a very extensive literature and history. Since the benefits are often received far in the future and the costs often start immediately, it is often necessary to use a discount rate so that the benefits and costs are discounted back to a common basis. This is because humans usually have a strong preference for benefits received now over those received sometime in the future. In the market place this is done through charging an interest rate, which is a type of discount rate. Loans to buy houses or cars always carry specified interest rates. Economic analyses use discount rates in a similar way and for a similar reason. Failure to discount back to

---

[384] Wang *et al*, 2013.

a common basis would result in huge overinvestments in projects or policy changes with benefits far in the future compared to the costs. This is a constant problem in the natural resource and environmental areas because of the high likelihood that costs will overwhelm the benefits in the early years. This has led to a longstanding issue as to what discount rate to use. Fortunately, in the US the Office of Management and Budget has set standards, although as we will see in the case of reducing CO2 emissions from power plants, they are sometimes flouted by agencies intent on justifying projects and policies which they believe support their particular "missions."

If there were a real basis for believing that increasing CO_2 emissions would bring about CAGW, this should certainly be taken into account. But as argued in Chapter 9, there is no such basis. Some have argued that even a small risk of CAGW makes it worthwhile to take precautions by limiting CO_2 emissions in some way. I argue that there is actually a higher risk—even a modest probability—that the world is in a gradual process of plunging into a new Little Ice Age, or even worse, a new ice age, in which case more CO_2 in the atmosphere might help if there is any validity to AGW science, not hurt. CAGW is not only unproven but quite unlikely. Hence there is no need to adjust benefit-cost analyses for the highly speculative and unlikely CAGW risk. And even if there is a significant CAGW risk the proposed GWD/GEF "solution" would not be effective even if it could be implemented.

Prior to 2014 there have been two major studies[385] of the economic efficiency of GWD, one in Australia and one in England. Since both were written by alarmists, it may not be surprising that both reached favorable conclusions for investments in GWD. I once asked one of the authors why he had ignored all the scientific evidence against CAGW and he responded that he had simply assumed that the IPCC and related studies were correct without checking. What is telling, however, is that in order to make a

---

[385] Stern, 2006 and Garnaut, 2008.

positive economic case for GWD both studies had to use extraordinarily low discount or interest rates, in fact, almost zero.[386] The reason this helped their alarmist cause is that the alleged benefits are in the distant future while the costs (largely increased costs for the use of fossil fuels) start immediately. By using an ultra-low discount rate they minimized this very real problem. In practical (but not academic) terms that non-economists can relate to, this is as if you could borrow money for buying a house for about 0.25% interest. Even during the current post-great recession era of unusually low interest rates, such rates are not available in the housing market. And they should not be used in public policy analyses either despite various sophisticated efforts to try to justify their use.

The two studies assumed that the IPCC models accurately represented the future of climate and the effect of reducing CO_2 emissions on global temperatures. Obviously if temperatures are going to fall rather than increase anything spent to make them fall would at best only make them fall even faster and serves no GWD purpose. What the use of low discount rates means is that even under these favorable assumptions concerning future global warming, the investment is not worthwhile economically. The authors of these two studies almost certainly realized that they would be criticized for using such ridiculously low discount rates, but presumably could find no other way to reach a conclusion consistent with their alarmist view of the need to "control" climate change/global warming. And with more realistic assumptions about either future natural temperature changes or the effects of reducing CO_2 emissions, the benefit-cost conclusion would be negative, in fact, strongly negative.[387] This means that GWD is a bad investment from a societal viewpoint even if the solution could actually be implemented. That has been and continues to be my conclusion. So despite these two major alarmist studies, I believe that it would not

[386] Carlin, 2011.
[387] Carlin, 2011.

be economically efficient to pursue GWD.

But the GWD economic case is much worse than I just portrayed. The problems with these two major studies are not just their unrealistic scientific assumptions. We can now state with certainty that global warming can take at least 17 years longer to occur than assumed in most studies of global warming mitigation as a result of the pause in global warming starting in 1998. That means that there have been no benefits of climate mitigation for an additional 17 years later than hypothesized in these mitigation studies. At a more justifiable discount rate this is fatal and even at the ridiculous 0.25% assumed by the two major studies this takes a substantial toll. The costs are incurred starting immediately and we have a documented instance (1998 to 2014) where there are no benefits for 17 years since there was no warming during that period based on any of the major global temperature series.

I wonder if the pause (or more accurately shelf or plateau since we do not know whether the mild upward trend in temperatures will resume later) may explain the switch from GWD/GEF to EWD and the increasing elevation of the CIC *ad hominem* attacks and name calling, which in 2013 reached the presidential level? The risks of making this change were large for the alarmists since there is little or no literature to support their new EWD cause, and much opposed to it. But sticking with GWD/GEF had its own risks, which increase with each year that there is no significant increase in global temperatures (readers of Chapter 9 will realize that there are probably another 15 to 20 years of no significant change or even decreases ahead of us). GWD/GEF-inspired warming mitigation makes no possible economic sense even assuming that the IPCC were completely correct in all their absurd assumptions/model results. Mitigation makes no sense even under these circumstances. Adaptation is the only reasonable possibility, and will likely not be necessary in terms of global warming for the reasons discussed in Chapter 9. And adaptation does not require hugely expensive, politically impossible, and ineffective CO_2 emissions reductions, as the CIC continues to advocate. The emperor has no clothes but refuses to recognize reality.

On June 2, 2014 EPA released its Regulatory Impact Assessment (RIA) for controlling CO_2 emissions from existing power plants.[388] It unsurprisingly reaches the conclusion that the economic benefits are strongly positive (just imagine what would have happened to EPA if the conclusion had been the opposite!). This will be discussed in more detail at the end of the next three subsections on economic benefits and economic costs in general; a discussion of EPA's benefit-cost analysis of reducing CO_2 emissions from existing electric power plants is contained in Appendix C.4.

Efficiency: Economic Benefits of GEF

An important point which I have not seen in an alarmist study is that minor warming appears to be economically beneficial. A survey of climate control benefits by a prominent researcher active in the IPCC concludes that there are net economic benefits, not damages, until global warming exceeds $1.1°C$,[389] which has not occurred and appears quite unlikely to occur in the next 20 years, if ever. So to date the minor warming since the 1970s has resulted only in net benefits according to this result. Richard Tol describes the reasons for this as follows:[390]

> *The initial benefits arise partly because more carbon dioxide in the atmosphere reduces "water stress" in plants and may make them grow faster....In addition, the output of the global economy is concentrated in the temperate zone, where warming reduces heating costs and cold-related health problems. Although the world population is concentrated in the tropics, where the initial effects of climate change are probably negative, the relatively smaller size of the economy*

---

[388] US EPA, 2014b,
[389] Tol, 2009.
[390] *Ibid.*

in these areas means that—at least over the interval of small increases in global temperatures—gains for the high-income areas of the world exceed losses in the low-income areas.

I would argue that one of the major adverse effects of CAGW would be from the possibility that it would lead to a significant rise in sea levels. The reason is that rising sea levels may result in economic damages because humans have built expensive developments near sea level which may be adversely impacted if sea level should rise significantly. In the early stages of an interglacial period this is exactly what happens, but near the end it becomes increasingly unlikely. Obviously if temperatures jumped to significantly higher levels and this resulted in rapid sea level rise, this rise would be something to worry about. But we are a long way from the Holocene highs. So there are no net benefits and only net costs from decreasing global temperature increases less than about 1.1°C. Since the chances of increases greater than that before the 2030s is very small (as discussed in Chapter 9 and Appendix C), there is very little chance of any benefits from mitigation (assuming it could be achieved)—only costs before the 2030s at the earliest. So rather than being fearful of possible minor increases in global temperatures we should happily accept them and hope for still more modest increases. Even if the CIC could mitigate temperature increases, all that the CIC is left with are foregone benefits and very large costs from mitigation efforts to date and probably until the 2030s.

Since humans evolved in tropical Africa and do not have fur, we are not likely to suffer from minor increases in global temperatures. But we are not so well equipped to handle cold. It should be kept in mind that the alleged damages from increased global temperatures are far harder to document than damages from decreased global temperatures. Higher levels of CO_2 appear to have

445

many benefits for plants[391] since CO_2 acts like fertilizer for them. It is only at much higher temperature increases that the economic costs exceed the economic benefits, especially if rising sea levels should ever be large enough to make any significant difference.

The social cost of carbon (SCC) is a concept used by some environmental economists to determine the societal costs of using fossil fuels containing carbon. They usually assume to begin with that the market for the energy produced by using fossil fuels does not fully reflect the adverse effects of using these fossil fuels and the carbon contained in them. In other words, they assume that there are negative externalities not fully captured by market prices, even though at least inside the US very stringent regulations are already in place to reduce many of these externalities. These regulations effectively increase the prices paid by utilities to generate electricity and thus the prices paid by consumers. The analysts make assumptions as to what these externalities are and try to calculate what these added societal costs are for a given unit of carbon in CO_2 emissions. The answers that they come up are often based on sophisticated economic models that, like all models, depend crucially on the assumptions made (such as the discount rate, the effects of changes in CO_2 emissions on global temperatures, and the effects of changes in global temperatures on humans). Thus the economic models are in turn dependent on the climate models and the many and too often invalid assumptions made in building them. So a simple set of assumptions often influence the SCC in complicated ways, with the result that the original assumptions are often obscured. So rather than highlighting each of these assumptions, which are the real issue, the user is often left to take all these assumptions on faith since the assumptions are only briefly discussed or not at all. And if they are discussed it is usually in terms that only economists can understand.

As discussed in Chapter 4, I sponsored some of the early research to determine the social costs of carbon from the most

---

[391] Idso *et al.*, 2014.

common fuel sources by taking into account the non-climate adverse effects of their use. That does not mean that I support the use of the current implementation of SCC by the Federal Government.

On the contrary, the attempts by the US Government to derive the SCC have had various problems and have not been very useful in my view. Claims that there are net benefits from decreasing the growth of CO_2 emissions are unlikely to be realized. Two problems are that the climate models do not take into account the 60 year temperature cycles discussed in Chapter 9 or changes in the estimated climate sensitivity factor as we learn more about it. If global temperatures continue to hold steady or even fall at least until the 2030s, as I expect, there will be no "benefits" from reduced temperature increases before then. And those that might occur after that would be reduced greatly if reasonable discount rates were used. It is also evident (as discussed in Chapter 12) that the climate models are much too sensitive to changes in CO_2; downward adjustments will also decrease SCC estimates. An additional problem is that two of the models used for this purpose have proved very unstable[392] and the third is proprietary so that there is no way to verify what has been done. Still another problem is that EPA often tries to model the benefits of GEF/EWD by counting alleged benefits outside the US even though the costs of the proposed regulations are incurred entirely inside the US,[393] thus making the benefits and costs non-comparable.

Appendix C.4 outlines an alternative and very different set of benefit-cost calculations for the EPA proposed regulations for existing plants using the ideas discussed above.

---

[392] Dayaratna and Kreutzer. 2013 and 2014.
[393] Gayer and Viscusi, 2014

Efficiency: Economic Costs of GEF

If the benefits of GEF/EWD are negative, it is not economically efficient to incur any costs, but it may still be of interest what they may be so as to see the magnitude of the waste. Unfortunately, the definitions of various cost studies differ, do not exactly coincide with GEF costs, and are far from comprehensive.

One study estimates that world capital flows to implement GWD/GEF were about $1 billion per day in 2012[394] or about $365 billion per year. It appears that this total does not include expenditures for research as opposed to adaptation and mitigation, which is the basis for the $1 billion per day.

A second study tracks world expenditures for renewables. Although there are other GEF costs, the major one is likely to be renewables. This shows new investment in "clean energy" as with a high of $318 billion in 2011, $286 billion in 2012, and $254 billion in 2013.[395]

A third interesting calculation for one country, Australia, is that in the first year of their carbon tax ending in September, 2013, Australians paid 7 billion Australian dollars[396] (roughly $6.3 billion US in early 2014). That comes to about 310 AU$/ per person or 1,250 per household of 4. Emissions fell from 543.9 to 542.1 million tons of CO_2, or 0.3%. The cost per ton of reduction was almost AU$4,000. The change in global atmospheric CO_2 levels is unmeasureably small.

A fourth study puts US Government expenditures on climate at $185 billion between 1993 and 2013, of which $133 billion were for direct Government expenditures and $52 billion for special tax

---

[394] Buchner *et al.*, 2013.

[395] Bloomberg New Energy Finance, 2014. BNEF describes their estimates as including corporate and government R&D and spending for digital energy and energy storage projects.

[396] Nova, 2014b,

incentives to selected industries and activities.[397] $39 billion of the $133 billion were for climate science. $43 million is believed to have been given to the IPCC since its founding.[398] Haapala reports 2013 expenditures as follows:[399]

> *Of the total of $22.2 billion in expenditures; 11% ($2.5 billion) went to the US Global Change Research Program (the US science); 26% ($5.8 billion) to Clean Energy Technology (developing alternatives to fossil fuels, or zero CO2 emissions); 4% ($0.9 billion) to International Assistance, including funding of IPCC activities; 23% ($5 billion) to Energy Tax Provisions That May Reduce Greenhouse Gases (tax breaks for special interests and activities); and 36% ($8.1 billion) to Energy Payments in Lieu of Tax Provisions (cash payments to special groups such as developers of wind power). The last two categories account for the $13.1 billion in soft expenditures and are more than the $11 billion the National Institutes of Health spent on clinical research of known diseases and other human health issues.*

The 22.2 billion is roughly $70 per person in the US in 2013 and is roughly double expenditures on border security.[400]

The US Government has spent $7.45 billion in Fiscal Years 2010 through 2012 on assistance to climate change mitigation activities in less developed countries.[401] The Administration believes that the US has thus met the US's "fair share" of a collective pledge in 2009 by developed nations of nearly $30 billion in "fast start

---

[397] Haapala, 2013.

[398] Arnold, 2014.

[399] Haapala, 2013.

[400] Bastasch, 2013.

[401] Goodenough, 2014. See also US Government Accountability Office, 2013.

finance." It has promised to work towards a collective pledge of $100 billion per year by 2020 to address the climate change needs of developing nations. It is far from clear exactly what the US taxpayers have accomplished with the $7.45 billion expenditure or would achieve with even larger expenditures by 2020. If the purpose is to pay less developed countries to support an enforceable international agreement for CO_2 emission cuts including less developed countries, it is not likely to achieve its goal. If the purpose is to reduce global temperatures the results will not be measurable and would probably be counterproductive anyway as argued in this book.

Clearly the costs of actually achieving the CIC-proposed 80% reductions in human-caused CO_2 emissions would be very much higher than the estimated costs presented here since world emissions are increasing, not decreasing, as the CIC desires, currently. So the world has been estimated to have spent more than $286 to $365 billion per year in 2012 or $2.86 to $3.65 trillion per decade; much more and perhaps infinite amounts would be required to "solve" the "problem" in the way proposed by the CIC.

The world is reported to be spending between $0.78 to $1 billion per day on a scientifically unjustified fear; these expenditures probably have little effect on global temperatures. If they have any effect, the "benefits" of these expenditures would be negative since according to Tol global warming would provide benefits up to about a 1.1°C increase. Documented warming from 1979 to 2012 was about 0.3°C[402] and 0°C from 1998 to 2014, which is clearly less than 1.1°C.

However unreasonable as it may be, the CIC appears to be reluctant to give up the more than $22 billion annual payments

---

[402] See Figure C-12. This includes only global temperatures measured by satellites; because of repeated and often unknown adjustments and the inherent difficulty of measuring surface temperatures using non-randomly spaced thermometers, I regard these measurements as unreliable and best ignored when satellite data is available.

from the US Treasury alone. Substantial additional payments come from ratepayers through higher electricity prices both in the US and abroad as well as from foreign governments.

There can be no doubt that legal requirements to use "renewable" sources of energy have and will result in higher energy prices and lower electric power reliability. These added costs should be added to the yearly costs of GWD/GEF/EWD but usually are not.

A study for the energy industry found that annual electricity and gas costs to all sectors would increase by 60% (or 37% in real terms) from 2012 to 2020 with CO_2 reduction regulations in place as a result of a number of new or proposed EPA regulations, with household costs rising by 35% (15%)[403]

One of the important potential costs of the proposed EPA CO_2 reductions from power plants is the risk that the closure of reliable coal plants and their replacement by less reliable types of plants may destabilize the electric grid or result in forced load shedding. The cost of one large outage in which a regional grid goes down is so high in our society that it would probably exceed the alleged benefits, assuming there were any. There are reasons to believe that this is a very real danger from earlier EPA regulations[404] and especially the new proposed regulation on existing power plants. Yet this risk has not been factored into the benefit-cost analysis. It needs to be. Deciding on an arbitrary 30% reduction suggests that this risk has not been given the importance that it so badly needs, since this is a real limiting factor that requires careful adjustments to meet, not arbitrarily set goals. The EPA Administrator stated in 2014 that she was tired of hearing about the problem. But just imagine what her successor will hear if there actually should be a grid collapse or even persistent load shedding as a result of EPA's new proposed regulation.

The increased risk to the stability of the grid and the necessity

---

[403] Energy Ventures Analysis, 2014.
[404] Yeatman, 2011. For a more recent analysis see Yeatman, 2014a.

for load shedding during particularly hot or cold weather is more than idle speculation. The Obama Administration's attempts to reduce coal use for power generation may have substantial adverse effects on the reliability of the US electric grid and its ability to respond to strains put on it during periods of particularly hot or cold weather and other unexpected events. Forcing the closure of coal power plants for whatever reason puts more strain on natural gas plants. This situation was already marginal at times during the colder-than usual 2013-14 winter when there was competition between the need for gas for home heating and power generation in several areas, particularly the Northeast.[405] Unlike coal plants, natural gas plants have no stockpiles of fuel that they can call on in emergencies when the demand for home heating is high (as during the winter of 2013-14) and are limited by the capacity of existing gas lines that also need to supply homeowners and business users. In the end the situation was saved by the use of additional coal plants,[406] *90% of which are scheduled for closure to meet EPA's Utility MACT (mercury) standard,*[407] which suffers from many of the health "co-benefits" justification problems just discussed. This was prior to EPA's new June 2014 proposed regulations which would have the effect of closing still more existing coal plants. If those previously scheduled for closure had already been closed, the Northeast would have suffered blackouts, resulting in load shedding and rotating electric blackouts.[408] This would have had huge adverse economic effects on the areas affected, and possibly even human misery or the death of vulnerable older citizens as well during adverse weather. The EPA proposed regulation on existing plants and Utility MACT do not include any provision I know of to avoid the closure of critical plants in terms of grid reliability; rather the CO_2 reduction

---

[405] For a summary see American Association for Clean Coal Electricity, 2014.

[406] Bezdek and Clemente, 2014.

[407] Yeatman, 2014.

[408] American Association for Clean Coal Electricity, 2014.

rules simply prescribe percentage reductions for each state designed to reach an arbitrary overall CO_2 reduction. Such arbitrary bureaucratic goals appear likely to produce grid unreliability with substantial potential economic costs that have not been factored into EPA's analysis. If, as appears likely, US temperatures fall over the next two decades, such grid reliability problems will become more likely; forcing the closure of the most reliable plants under these circumstances appears foolhardy and irresponsible.

Thus during the winter of 2013-14, many of the Northeastern coal plants that are planned to be shut down as a result of the Utility MACT (mercury) regulation were the ones that saved the day. These plants had coal stockpiles onsite that they could call on for added generating capacity. Natural gas plants, on the other hand, could not obtain added supplies because they were needed to meet homeowner heating needs and because of pipeline capacity problems. Shutting down such coal plants, as EPA envisions under its Utility MACT regulations as well as under its new rules for existing power plants could well lead to an energy disaster and even the death of vulnerable older people who could not heat their homes adequately as a result of load shedding or the increased risks of a grid collapse during unusually hot or cold weather. Surely avoiding such tragedies is more important than arbitrary reductions in power plant emissions intended primarily to impress other countries to possibly follow EPA's misguided efforts.

Efficiency: Benefit-Cost Analysis for Proposed EPA Reduction of CO_2 Emissions from Power Plants

The EPA claims that the benefit-cost analysis of their CO_2 reduction regulation for existing power plants is strongly positive. Despite all the pages written by EPA analyzing the question, this is simply not the case, as explained in Appendix C.4. For all the reasons just discussed under the benefits and costs of CO_2 emissions reductions, the EPA CO_2 regulations should be abandoned since they do not meet the cost-benefit requirements established by the relevant Executive Orders. But instead of admitting this, the EPA

453

Regulatory Impact Analysis presents a flimsy and misleading attempt at an economic justification for the proposed CO_2 power plant regulations. The detailed analysis in Appendix C.4 explains why the net benefits are actually strongly negative. The proposed EPA power plant emission reductions would result in large increases in the cost of electricity for Americans. So no climate benefits and minor if any health benefits so that the economy as a whole would be worse off with the regulations than without them.

Distributional Effects of GWD

One of the other things economists look for in analyzing a proposed policy change is whether different income groups are made better or worse off or whether all income groups are treated equally. The reason for this is that even if a policy change is economically efficient as a whole, some segments of the population may be treated much better than others. Economists do not prejudge whether such differential effects are good or bad, but have found that knowledge of this may well influence how people judge whether they want a particular policy change or not. Thus if high income people are helped by a policy change while low income ones are hurt, many people may conclude that it is not worthwhile to make the change since they would prefer that income levels be more equal rather than less equal. For this reason economists often try to determine the differential income effects of a policy change. This involves analyzing who gains and who loses from a policy change.

It is very clear that the economic effects of implementing GWD/GEF are strongly disadvantageous to lower income groups. It is they that end up with most of the costs, which are a much larger share of their income (and of course receive the same, in this case, negative benefits). GWD/GEF can in fact end up killing people who can no longer afford higher energy prices. Typically there are a substantial number of excess deaths during the winter due to the lower temperatures. Many of these can be prevented by the use of space heating. But the poor often cannot afford the higher price of fuels resulting from GWD/GEF for this purpose, so must choose between using less fuel for heating or less food to eat or less medical care.[409] So some of these excess deaths can be attributed to GWD/GEF.

Not only has GWD/GEF led to repeated government financial disasters in attempts to stimulate green energy, it is also disproportionally economically damaging to less wealthy groups compared to more wealthy groups since energy costs are a much higher proportion of the expenditures of the former than the latter. In the case of the US the lowest 20% of US households paid 24.4% of their incomes for energy costs in 2011; the highest 20% paid only 4.3%.[410]

Yet it is the left of center political parties in most Western countries, including the US, that support the GWD/GEF cause while at the same time claiming that they are more generally representing the interests of the lower income groups than the right of center parties. This is particularly evident in the case of carbon taxes, but is also the case for most other GWD/GEF policy proposals as well. Carbon taxes are very regressive because although energy usage is higher among higher income households, they represent a much lower percentage of expenditures by the rich. The wealthiest households use more air travel and heat and cool larger houses, but their lower income fellow citizens also have to pay a much larger proportion of their income for transportation to and from work and for heating and cooling. There are ways to shield lower income groups from some of these adverse effects by rebating part or all of any carbon tax but that does not help those who want to use such taxes to raise revenue and creates added distortions by rebating the same amount to people with vastly different energy usage.

The use of environmental regulations (such as those now proposed by USEPA) instead of carbon taxes to reduce carbon

---

[409] Global Warming Foundation, 2015.
[410] Furchtgott-Roth, 2013, Table 1.

usage has the same regressive distributional effects. Requiring that new cars meet stringent mileage requirements drives up the cost of cars, which makes cars less affordable to lower income individuals, and increases injuries and deaths from traffic accidents because of the reduced weight of more efficient vehicles; these injuries and deaths are more likely to occur to those that drive the most—lower income groups. It is even harder to devise compensation mechanisms for lower income groups in the case of CO_2 control by regulation, particularly since there is no added government revenue to rebate.

The mystery is why parties who claim they represent the interests of lower income groups would even consider such carbon reduction schemes. Presumably they are trying to represent the interests of other groups in their coalitions such as the CIC or wealthy environmentalists, so try to avoid mentioning the adverse effects on other groups in their own coalitions. The practical reality is that radical environmentalists attempted to form a blue-green alliance including blue-collar union members in 2008-9, but the environmentalists have continued to actually offer policies that are very detrimental to such workers. They oppose fracking, which would create many well paid blue collar jobs and lower natural gas prices (important to lower income households). They aim to reduce if not end the use of coal to produce US electricity, which would reduce all the coal mining jobs that go with it and the low electricity prices charged for power produced using it. So why would blue-collar union members support the Democratic Party that advocates such policies? Some coal states have in fact reduced their Democratic Party representation in Congress as a result of the Party's support of GWD/GEF.

Lower income for those who already have lower incomes appears to have adverse effects on morbidity and mortality by reducing expenditures of these groups for life saving investments. One study found that although there are some regulations with large life-saving benefits, the majority of 24 Federal regulations studied resulted in an unintended increase in risk.[411] Where the proposed EPA regulation on existing power plant CO_2 emissions falls in this spectrum is uncertain but given the dubious nature of the health benefits discussed above, it probably falls into the less favorable category. This cost of the proposed regulation is not accounted for in the EPA RIA.

At the international level, the adverse impact is similar—and much more deadly. Over the period 2009-13 under the Obama Administration the US Overseas Private Investment Corporation has invested in more than 40 new energy projects, and all but two were renewables. Excluding fossil fuel projects greatly increases the costs of energy to less developed countries, which means that less energy will be generated for the funds provided. That means that more people will have to continue burning wood and dung for cooking and heating and forego the electricity they so badly need to lead a healthful and productive life. One estimate is that indoor air pollution kills 3.5 million people each year, thus making it the world's deadliest environmental issue.[412] Obviously additional power plants would only save a portion of this 3.5 million, but there is a trade-off between CO_2 emissions reductions (to the extent that there are any using renewables) and indoor air pollution deaths. I find the Obama Administration's and the rest of the CIC decision to back GEF/EWD to be unconscionable and immoral.

Effects on Economic Growth and the Economy

There are at least six major economic issues posed by GWD/GEF. The first is that the economics of what the alarmists propose is hopeless since most of what they propose includes efforts that will lose money for the indefinite future. As a result, they will continue only as long as those paying the subsidies to support them are willing to incur these costs, whether they be taxpayers or

---

[411] Hahn et al., 2000.
[412] Lomborg, 2014.

ratepayers, but they are likely to sooner or later to pull the plug. The second is that GWD/GEF has retarded the recovery of the economy from the Great Recession. The third is that almost certainly because of the supposed need to justify GWD-inspired regulations the Federal Government has increasingly resorted to economically incorrect procedures to carry out the benefit-cost analyses required for major Federal regulations. This last threatens to destroy decades of efforts to insure that valid benefit-cost procedures are used for analyzing such regulations and is of particular concern to me because of my many years of effort to improve and support the use of such analyses in regulatory analyses. Lastly, GWD/GEF "solutions" would put the US at a competitive disadvantage economically compared to other countries that do not adopt such "solutions."

From a larger viewpoint if there is any basis for CAGW (which is very unlikely as explained in Chapter 9) there could be a trade-off between the unproven risk of such warming and the ability of the less and more developed world to make it possible for their less-well-off citizens to enjoy an improved standard of living through the substitution of reliable energy use for human labor. Recent research,[413] although primarily on Canada, concludes that:

> *Energy use in Canada is not a mere by-product of prosperity, but a limiting factor in growth: real per-capita income is constrained by policies that restrict energy availability and/or increase energy costs, and growth in energy abundance leads to growth in GDP per capita.*

I believe that similar research in other countries would be likely to reach similar conclusions. This is a very important conclusion. Driving up the cost of energy or decreasing its reliability will have adverse effects on economic growth. Thus the prescriptions of the CIC, if followed, will have adverse economic effects on countries

---

[413] McKitrick and Aliakbari, 2014.

where they are tried. This is directly opposed to much of the CIC propaganda that claims that using wind, solar, and biomass will not have significant effects on economic growth and thus the welfare of economies as a whole.

More energy use results in more CO_2 unless only renewable sources are used. But renewable sources are much more expensive and much less reliable and must be supplemented with fossil fuel or nuclear sources except where abundant water power is available. The resulting more expensive and less reliable energy will necessarily result in lower use and hence lower economic growth.

So why would anyone want to reduce economic growth? Only people who have no understanding of the real problems faced by many average people even in the US who live from paycheck to paycheck and borrow at exorbitant rates when they do not quite make it to the next paycheck or the old car breaks down. People who live in less affluent societies face the very real tradeoffs between necessities and early deaths.

This book has argued that CAGW is not only unproven but highly unlikely, particularly at the current stage of the ice age cycle. Yet the adverse effects to humans of lack of access to inexpensive, reliable electric power is proven and persuasive. I find it unconscionable that so many CIC supporters would prefer to reduce an unproven, unlikely risk (CAGW) at the expense of condemning so many people in the world to a lower standard of living, which among other things results in reduced health relative to what it could be with access to inexpensive, reliable electricity. Yet this is exactly what Western radical energy environmentalists advocate, although they never mention its adverse effects.

The less developed countries particularly well understand this since they have lived with lack of reliable energy for many years and will never agree to GWD/GEF as a result, regardless of any sacrifices/payments that the developed world might possibly offer. So the Obama Administration is wildly unrealistic to think that their proposed radical reductions in CO_2 emissions using EPA regulations will result in any significant reductions in the less developed world.

I do not even believe that it is worthwhile to make such a trade-off in the developed world either since there are many low income workers who will have to do without other important necessities in order to pay for expensive, unreliable renewable power (excluding hydro, which the environmentalists oppose). Fortunately, there is no evidence that such a trade-off between preventing CAGW and development is a real possibility. So the answer is what it always has been—promote responsible development rather than worry about unproven and unlikely (and at worst minor) climate risks.

Government actions to achieve GWD/GEF decrease economic growth by forcing the economy to spend more on energy infrastructure and production, which leaves less for other purposes such as producing consumer goods and exports. The energy produced provides exactly the same benefits as before, but costs users much more. Further, users cannot tell how the electricity they use was generated unless all electricity in a region is generated in the same way. Higher energy costs discourage industrial investment, which is what is happening in Western Europe, which has undertaken major investments in "renewables," which has resulted in very large increases in electricity prices and reductions in reliability. Discouraging fossil fuel development leaves countries and regions rich in such resources poorer than they otherwise would be. Western Europe does not have vast petroleum resources but may have very large natural gas resources in its shale formations.

Prominent GWD supporters have claimed that it would provide plentiful "green jobs" that would make GWD worthwhile. Skeptic studies have correctly argued just the opposite, of course, and seem to have carried the day by arguing that while some employees would be needed to build and maintain the solar and wind plants, that many more workers would be lost in other parts of the economy as a result of much higher prices for energy and decreases in fossil fuel plant use.[414] I strongly believe the skeptic

---

[414] Green, 2011.

studies.

Growth: Hopeless Efforts to Stimulate Our Way to GEF

The best known results of the attempt by some western governments to impose GWD/GEF "solutions" on their economies is the repeated financial failures that have resulted. In the US the best known of these disasters are the loans/loan guarantees made by the Department of Energy to attempt to promote "green" enterprises such as solar generation and electric automobiles under the Stimulus legislation. Those responsible for the implementation of this Act apparently thought that selected "green"-favored enterprises could be brought into existence by government fiat and money whether the market was willing to buy the products or not. Funding production of products for which there is no or little demand always results in the same outcome: waste and abandonment.

But probably even worse has been the repeated efforts over many years and at both the Federal and state levels to promote "green" sources of electrical energy, particularly wind, solar, and biomass. The result for all or almost all of the older efforts has always been the same regardless of where it has been undertaken—failure leading to abandonment of the effort at huge financial cost. Government cannot stimulate favored green energy sources and expect them to survive after the subsidies are discontinued because the "green" sources are higher cost and much less reliable than fossil fuel sources and are very likely to remain so for many years and probably forever. When and if these renewable sources become competitive, the market will build them without government subsidies. The windmills pictured in Figure 10-1 were not built during the Obama Administration, but the US Government obviously learned nothing from the message they convey. It does not matter whether the "renewable" sources are directly subsidized by governments or by ratepayers; sooner or later the money is likely to run out because the people paying the bills get tired of paying more than others for the same product—electrical power.

There may be a "feel good" benefit from using electrical energy from "renewable" sources, but that wears thin before long. And for the reasons discussed in this book there are no actual net benefits—just costs. To make matters worse, abandoned windmills and presumably before long solar plants/panels (see Figures 10-1 to 10-3) will litter the landscape for generations to come as they are slowly abandoned—too often with no provision for their disposal. Abandoned solar panels are likely to leave even more lasting effects as toxic substances.

Growth: Competitive Disadvantage Resulting from GWD/GEF "Solutions"

Most, if not all GWD/GEF "solutions" will end up making any country that unilaterally adopts them economically less competitive with those countries that do not adopt them. That may be one of the reasons alarmists have long advocated worldwide government-enforced restrictions on CO_2 emissions. It is now evident, however, that the less developed world will not accept such restrictions unless possibly if they are paid very large sums to do so. Since such outlays would largely have to come from Western democracies, this is highly unlikely to happen because of disapproval by their taxpayers. This was the message of the failed Copenhagen conference in 2009. Hence there appears to be no way that developing countries will approve such restrictions on CO_2 and therefore their own growth and any Western nation attempting to carry out GWD/GEF unilaterally will not only harm its own economy but have no significant impact on CO_2 emissions.

Making energy more expensive, as Obama proposes to do, is generally a very bad idea from an economic viewpoint. It was a particularly bad idea when the economy was weak, unemployment was high, and the US was having great difficulty fully recovering from the worst recession since World War II. It matters, of course, how his proposed climate policy would increase energy costs, but regardless of how it is done it will reduce economic growth by raising energy prices, raising the prices of any product utilizing energy in its production, and thus make the US less competitive in the market places where the items made are sold. Not only would it reduce employment in coal mining, oil and gas extraction, and petroleum refining,[415] but it would also particularly discourage energy intensive manufacturing where energy prices are crucial.

If a Western nation raises its price of electricity, the manufacturers of goods that are very sensitive to this price are likely to move elsewhere where the price is lower. Particularly because of the pursuit of higher energy prices in Western Europe for similar GWD/GEF reasons, a number of companies are considering or making major energy intensive investments in the US in the belief that energy costs will be lower here. As of 2013 these included[416] France's Vallourec Star, Luxembourg's Tenaris, China's Tanjin Pipe, Germany's BASF, Austria's Voestalpine, and South Africa's Sasol. The surest way to discourage this very welcome development is to do anything that will change the energy cost expectations of these and other companies considering such investments. But that is what Obama plans to do by reducing the use of the often lowest cost energy source, coal, and by implementing other environmental controls that will also raise energy prices. As argued here and elsewhere in this book, we need to encourage CO_2 emissions, not reduce them, for both economic and environmental reasons.

If production is moved to another country, And the overall resulting CO_2 emissions are the same or probably worse. Although Western Europe has tried raising their electricity prices in recent years, they are now seeing the consequences, and have now even stated that they are not willing to push further unless the world does so.[417] Each added uncompetitive windmill and solar generating plant or biomass plant used for energy makes the country building them less competitive since someone has to pay the added costs, which other countries not doing so do not have to pay. The Obama

---

[415] Congressional Budget Office, 2010.

[416] Furchtgott-Roth, 2013

[417] Peiser, 2014.

Administration apparently has not learned this, and appears to be pushing harder than ever. The following table shows the extent of the disparity in the cost of electric energy between the US and Europe. Germany has one of the highest such costs in Europe, and appears to be realizing what damage that can do to its economy.

Europe's Energy Calamity

Average cost of energy by price per kilowatt hour in 2012 (U.S. cents)

	United States	European Union	European Union cost disadvantage
Industrial Firms	6.67	12.25	+5.58
Households	11.88	19.75	+7.78

Source: Eurostat, US Energy Information Administration

Comparative Cost of Electrical Energy in the US and the EU[418]

At the EU level Benny Peiser has characterized the economic problems resulting from the EU's pursuit of GWD as follows:[419]

> The EU's unilateral climate policy is absurd: first consumers are forced to pay ever increasing subsidies for costly wind and solar energy; secondly they are asked to subsidize nuclear energy too; then, thirdly, they are forced to pay increasingly uneconomic coal and gas plants to back up power needed by intermittent wind and solar energy; fourthly, consumers are additionally hit by multi-billion subsidies that become necessary to upgrade the national grids; fifthly, the cost of power is made even more expensive by adding a unilateral Emissions Trading

---

[418] Darwall, 2014.
[419] Peiser, 2014.

Scheme. Finally, because Europe has created such a foolish scheme that is crippling its heavy industries, consumers are forced to pay even more billions in subsidizing almost the entire manufacturing sector.

Growth: Why the US "Green" Stimulus Program Was Bound to Fail While Encouragement of Non-conventioanl Oil and Gas Production Could Have Paid Dividends

One of the partial casualties of Obama's GWD/GEF policies has been his attempts to stimulate the economy. If the US wants to maintain its current standard of living and solve its current economic problems, which appears to be generally accepted by both political parties, it has to find a way to stimulate its economy while halting, or better, reversing the runaway growth of US Government debt. The primary means tried since the beginning of the current recession in late 2007 has been to run large Federal budget deficits, to push down interest rates by the Federal Reserve, and to increase Federal spending for specified purposes to stimulate the economy. The results have been very slow. Paul Krugman has argued that the problem is that the fiscal stimulus has not been large enough—in other words, we should be running an even larger budget deficit while pursuing roughly the same policies.

I argue that the Obama Administration did not do a very effective job of targeting its stimulus program towards those purposes that would maximize its effects on increasing economic output. The major problem was caused by the loss of jobs in the home building industry when the "mechanism" used to finance their work imploded. One way to solve this would have been to finance work that would have employed similar skills through the stimulus. This would presumably have involved public construction projects with lasting value for the economy. Instead, the Obama Administration chose to devote about $90 billion of the $840 billion in stimulus funds to GWD-related purposes. The $90 billion included specialized manufacturing of high-tech renewable energy components. Although the evident purpose was to use taxpayer

465

funds to subsidize GWD-related private investments, various arguments were advanced, particularly the alleged need to build a US "green" industry which was supposed to be the wave of the future. They also argued that it would lead to "green jobs." This amounts to almost pure state capitalism and has been largely a complete failure from an economic and financial perspective given that many of those receiving the "investments" have gone bankrupt.

It is clear to me that the Obama Administration was trying to achieve more than economic recovery with its Federal stimulus funds. The obvious problem with this $90 billion expenditure is that besides not being very helpful in terms of its economic stimulus effects because it subsidized industries which were not tied to US production, it was bound to fail in its GWD goals both through lack of clear thinking as to how to achieve these goals and because it ignored the individual decisions made in the marketplace that would ultimately decide whether they would lead to a change in direction of these components of the economy as the authors appear to have hoped. The authors apparently believed that by directing Federal funds to various components of "clean" technology that GWD ideology approves of, including electric cars, solar, and wind technology, that somehow these environmentalist-supported expenditures would actually alter public behavior in the marketplace. Like many examples of state capitalism, the results have been largely failures. One of the not so amusing items funded by the stimulus was climate modelers—as if they did not already have more than enough Federal dollars, were facing high unemployment, and would do something more than waste more dollars on more misleading models.

Subsidies for electric cars and their components is a case in point. Even given the (unwarranted) assumption that burning fossil fuels is a bad idea, electric cars require large batteries, which are not very environmentally friendly to build. If, in addition, electric cars are generally powered in the US by electricity generated from fossil fuels, particularly coal, as is generally the case in the US except in the Pacific Northwest, little if any reductions in CO_2

emissions can be expected. The major possibility for reducing anthropogenic sources of CO_2 is from the substitution of natural gas for coal because of lower natural gas prices resulting from fracking and other new techniques. The environmental costs of substituting electric vehicles for oil-powered vehicles would appear to exceed the benefits.[420] Like many other environmental proposals, investing in electric vehicle technology shows little careful analysis but rather ideology, which should play little role in the formulation of sound public policy.

One of the areas where the US has a clear comparative advantage is fossil fuel energy development. Russia and Canada have similar opportunities and problems (high wage levels relative to many less developed countries but large fossil fuel deposits) and have long since understood that fossil fuel energy development represents one of the areas where they can usefully expand in a way that will at least maintain their current standard of living. The Obama Administration appears to believe that the US should not do this—at one point proclaiming loudly that the US cannot drill its way to energy independence. Subsequent large increases in shale oil production have proved this argument (which seems to have been disappeared completely from Presidential rhetoric) false. Substantial expansion of US drilling, especially using non-traditional techniques, can and has now been proven to be capable of greatly reducing our continuing high payments for imported oil and loss of potential government royalties for oil and natural gas. We could have pushed this expansion much further by actively leasing suitable public lands for this purpose. Unlike attempts to subsidize "clean" technology, which at best produce products few people want to buy, there is a ready worldwide market for oil and natural gas, especially if US Government restrictions on their exports were removed.

One of the few relatively easy opportunities to increase government revenue is from oil and gas on lands and waters it

---

[420] Hawkins et al., 2012.

owns. Radical environmentalists are strongly opposed to such development, but offer no realistic alternative. I find the energy development option an easy choice. It may not solve all or even most of our economic problems, but it could make a significant contribution. A further discussion of what might be involved in a serious effort to develop US fossil fuel energy resources can be found in the next subsection.

One thing is for certain, the increasing worldwide demand for energy will in the end be satisfied, and largely not using very expensive, unreliable "renewable" energy sources. The only issue is which countries will satisfy this demand. If the US continues to hold back on energy development on its extensive public lands and restrict production on suitable private lands (such as by prohibiting fracking in New York), other countries will develop their energy resources, almost certainly using much lower environmental standards. The environmental case for not developing these US resources as a whole is not supportable. The US will do a better job of protecting the environment than most alternative sources and energy development presents one of the growth industries for which the US has a strong comparative advantage.

In some cases governments can greatly stimulate the economy by making changes in existing regulations rather than making stimulus expenditures. They have no immediate budgetary cost but may result in substantial private investments. It is important to note that less expensive energy, which has occurred as a result of recent natural gas and oil development, can make an important contribution to job creation in user industries either by reducing the cost of their feedstocks or the energy they use. In industries where these costs are important components of total costs, locations with low energy costs are a powerful magnet to these industries. This is one of the many reasons why policies that result in high energy costs are very not compatible with job creation.

Growth: Developing America's Vast Hydrocarbon Reserves Presented Historic Opportunity for US to Solve Some of Its Economic Problems and Still Does

As just discussed, the radical environmentalist view of energy development is one of the important factors that has kept the US from using its vast hydrocarbon deposits to help solve some of its current economic problems. Texas employment, spurred by the use of fracking and other non-conventional drilling on private land, avoided the sharp drop in the early years of Great Recession that most of the rest of the US experienced, and has greatly exceeded employment growth in the rest of the country since then.[421] If such development had been encouraged elsewhere, particularly on government-owned land, it could have provided additional well-paid jobs much needed for reducing the high unemployment rate after the Great Recession, large new revenue sources from additional royalties paid to state and Federal governments much needed to balance severe budget deficits, significant improvements in the US current balance of payments, which has been strongly negative for many years, and greatly decreased dependence on oil from politically unstable sources such as the Middle East. At least some analysts believe that this approach would have generated enough revenue to pay off the national debt.[422] Even if it did not, the cost would have been much lower and the results superior to the "green" stimulus funding. We had a historic opportunity to stimulate oil and gas development using new techniques because of the very high price of oil in recent years until late 2014 which would have allowed the development of domestic resources that would not have been competitive at the much lower prices that prevailed earlier.

The big news in oil and gas development in recent years has been the development and use of a combination of hydraulic

---

[421] Perry, 2014.
[422] See Ferrara, 2012.

fracturing (fracking) and horizontal drilling to obtain both tight oil and natural gas from abundant shale deposits in the US. Some unconventional energy investments involving fracking on privately owned land have been profitable so that no government subsidies have been required for active development. The best examples are the Bakken, Eagle Ford, and Permian Basin tight oil developments in North Dakota and Texas, but there are indications that similar potential may exist in Southern California and Oklahoma. It is no accident that North Dakota and Texas have very low unemployment rates and large inflows of energy production taxes to bolster state finances. And the contribution of shale gas and oil appear to be making to the US economy is reported to be quite large in the last few years.[423] There is little question that the Federal Government has shale deposits under its vast lands as well as other non-reservoir sources of petroleum. Because of environmental opposition, however, New York State has imposed a moratorium on the use of fracking. California appears to have approved fracking but with a number of restrictions that may delay development. As of late 2014 the Obama Administration appears to publicly favor the use of fracking and the EU has avoided banning it, but neither one may be very enthusiastic about it.

Until recently, oil prices were at or near their peak and allowed higher cost non-traditional production techniques to be tried out and made more efficient through experience using them. These have proved so successful that US production using these techniques has soared in the last few years, and apparently as a result world oil prices fell precipitously in late 2014. This was a unique opportunity which may not come again soon. Fortunately, significant shale oil resources are located on private land in states that encourage oil and gas production, so government could not prevent such development. But the development of additional shale oil could have been even more rapid if states with significant shale oil resources and the Federal Government had actively encouraged

---

[423] Blackmon, 2013.

such development. They still can, although the lower prices would provide less incentives for the private sector to add non-conventional production.

Russia has carried oil and gas development to an extreme for a large country by basing most of its economy on exploiting its resources. During the period of high oil prices Russia made ends meet despite the absence of internationally competitive industries in many other areas (such as aircraft production) despite substantial investments over many years. The major thing that prevented this partial solution to the economic problems of the US in recent years has been the Obama Administration and its decision to adopt the restrictive policies of the CIC towards fossil fuel resource development.

Despite the rapid rise (and fall) of world oil prices in recent years, the rapid decrease of natural gas prices and increases in available reserves in the US means that human use of fossil fuel resources has a great future and that the US, with its extensive resources of coal, oil, and natural gas, is probably better situated than any other country in the world, including the Middle East and Russia, to greatly increase its contribution to meeting the growing world demand for additional sources of energy. Attempts to raise domestic prices of fossil fuels and waste resources on subsidized, inefficient, and environmentally damaging alternatives such as solar, wind, and biomass, and subsidized[424] and environmentally risky nuclear simply postponed using America's vast hydrocarbon resources to help solve its major economic and debt financing problems. A 2013 study suggested that shale gas and shale oil production represented the best opportunity to create jobs and boost GDP by 2020.[425] The major US liability is the continuing opposition to such development by the Obama Administration and environmental organizations.

For example, the construction of LNG export terminals and

---

[424] As a result of the Price Anderson Act.

[425] Lund *et al*, 2013; price drops in 2014 may limit development potential.

facilities to take advantage of the price differential between North America and the rest of the world for natural gas alone would result in tens of billions of dollars of investments if only the Administration had more rapidly approved export licenses and environmental organizations had refrained from delaying lawsuits. Or, much better would be to remove the requirement for natural gas export licenses, which only retard international trade, limit the ability of the market to operate efficiently, and slow US efforts to develop its natural gas economy and help solve its economic problems. Similar problems exist in the case of oil; most crude oil exports are banned under the 1975 Energy Policy and Conservation Act. This ban similarly needs to be removed if the US is to develop its oil supply economy and help to solve its economic problems.

With substantial encouragement, domestic fossil fuel sources could supply most of US oil needs, especially if unconventional sources are allowed to be used and the rest could be met through market substitution of natural gas for needs now satisfied by oil. Market forces have already made a start in this direction and all that is required is for state and Federal governments to get out of trying to control the energy markets and to open up state and Federally-owned hydrocarbons energy resources to development with careful attention to avoiding environmentally extremely risky sites and sources such as the current generation of nuclear energy.

According to the Institute for Energy Research:[426]

> *Technically recoverable resources total 1,194 billion barrels of oil and 2,150 trillion cubic feet of natural gas that is owned by the federal taxpayer...The value of the estimated oil resources is $119.4 trillion and the value of the estimated natural gas resources $8.6 trillion for a grand total of $128 trillion.[427] These numbers, however, are likely*

---

[426] Simmons, 2013.

[427] There is a significant difference between technically recoverable and proven resources, but historically technically recoverable oil resources are

to be low, since little is known, for example, about the offshore energy resources where a moratorium has been in place since 1981 on 85 percent of the waters in the lower 48 states and most of Alaska. The Obama Administration has effectively continued the moratorium lifted by Congress in 2008 through its 2012–2017 leasing plan.

Although an active program to develop these resources is probably the easiest and possibly the only realistic way to begin to pay off the current approximately $18 trillion US Federal debt, oil production per year on Federal lands has increased by only a small amount during the Obama Administration compared to George W. Bush's second term. The Institute also found that US GDP and jobs and state revenue would have substantially increased under a more active leasing program.[428]

Substantial royalties paid to the Federal and state governments on state and Federally-controlled economically competitive oil and natural gas resources could have allowed a significant down payment on bringing down both Federal and state budgetary deficits as well as hundreds of thousands or even millions of new jobs. The substitution of natural gas for coal and various energy conservation measures have already resulted in significant reductions of US CO_2 emissions, unlike most of the countries that have actually pursued mandatory government intervention to bring it about. Yes, you read that correctly, in the last few years the US has been reducing its energy-related CO_2 emissions, more than any other country,

---

a much better indication of potential future production over the long term than proven resources since proven resources have remained relatively static at about three years US use in recent decades despite substantial production, huge new finds, and significant improvements in technology. Proven resources indicate reserves that geological and engineering studies demonstrate can be recovered from known reservoirs under existing economic and operating conditions, and are not classified as such until these studies have been completed.

[428] Simmons, 2013.

primarily by using more lower-cost natural gas and less coal in power plants for economic reasons. And US CO_2 emissions per capita have been falling even more dramatically (see cut below).[429]

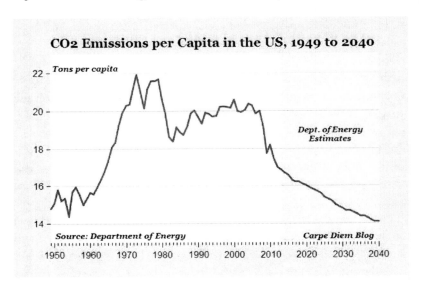

CO2 Emissions per Capita in the US, 1949 to 2040

Tons per capita

Dept. of Energy Estimates

Source: Department of Energy

Carpe Diem Blog

By pursuing its GWD/GEF approach to energy usage, the Obama Administration has only made the situation worse rather than taking advantage of the historic opportunity of high oil resulting from increasing world demand and the gradual depletion of many of the large, low cost oil reservoirs in the world to help solve many of the current economic problems, including unemployment, the need for lower energy prices, and growing Federal and state debt.

The decrease in CO_2 emissions in the US suggests that the radical environmentalists' approach of using mandatory government regulations and price changes to reduce CO_2 emissions and banning fracking may be less effective than allowing the energy markets to operate more freely in reducing CO_2 emissions. The reason is that

---

[429] Perry, 2013; source for cut: Mills, 2012.

emissions from burning coal are much higher than from burning natural gas per unit of energy obtained. Fracking makes possible much larger gas production from shale deposits, which decreases prices, which encourages substitution of gas for coal, which decreases CO_2 emissions. Many Western European nations ban fracking but heavily subsidize solar and wind. This has not resulted in reductions in CO_2 emissions but greatly increased energy prices and complaints about these prices. So the market approach appears to have been far more effective than the government intervention approach in reducing CO_2 emissions based on current experience in Western Europe and the US.

Although the environmentalist sustainability arguments have no rational basis, there is a legitimate concern about whether opening up additional oil resources to development may result in oil spills and other adverse shorter-term environmental effects. Some past developments have resulted in oil spills and it is likely that future ones will too even if the best available control technology is conscientiously used, which I strongly favor to the extent economically feasible. There are also concerns about impacts on wildlife habitat in some cases, which need to be carefully addressed. But the idea that the US can continue to prohibit development of so much of its oil resources and instead import oil from other nations at current high prices and still maintain its standard of living is no longer viable. It is obvious that if the US continues to impede oil development on unreasonable environmental grounds that the oil will be produced somewhere else, and probably under much weaker environmental standards than those that would be followed in the US. The US has a better opportunity than most other countries to increase oil output in an environmentally sound way if only the environmental views of the governments and environmental organizations in the US permit this to happen. Preventing increased output in the US may prevent a few accidental oil spills in the US but will increase them elsewhere at much higher environmental costs. And it will mean fewer well-paying jobs in the US and more elsewhere. Attention needs to be centered on preventing spills, not preventing production and distribution.

An interesting observation in this respect, however, is that in California and other oil rich states there have been oil seeps throughout recorded history. These seeps have the same adverse environmental effects as oil spills even though they are entirely natural in origin. Because adverse environmental effects are man-made does not make them any worse in terms of their effects. Further development of California's oil reserves in the areas of these seeps[430] would undoubtedly decrease them since it would relieve some of the pressure in the oil reservoirs from which they come. So some of the adverse environmental effects of California's oil resources will occur either with or without new development. I vote for careful further development with all reasonable safeguards in California and elsewhere where it is likely to have a significant reduction in oil seeps. For state and Federal Governments with severe budgetary problems this is really a no-brainer.

Very recently the Sierra Club has decided to oppose the use of natural gas to generate electrical power in addition to their long-standing opposition to the use of coal. In their view natural gas-based energy is no longer to be a bridge fuel to a more "self-sufficient" future but is now one of the fuels to be consistently opposed. I find this irresponsible, irrational, and damaging to US national interests, and I am happy that I am no longer a Club member. These decisions should be left to the market as long as environmentally responsible safeguards are in place.

By using faulty science and not understanding the role that the very extensive US energy resources can play in solving US economic problems, the Obama Administration has squandered a historic opportunity. Replacement of oil imports with domestic energy resources and a large expansion of hydrocarbon exports is now possible; it could be much more rapid if state and Federal Governments get out of restricting the development and distribution of hydrocarbon resources, lease their land holdings

---

[430] US Geological Survey, 2009; their map is based on Hodgson, 1987.

more rapidly but environmentally responsibly, and allow the markets to function more normally. The potential is illustrated by the economic results of exploiting the Bakken shale formations in North Dakota for oil and the Eagle Ford formation in Texas. The rapid development of shale oil in parts of Montana and North Dakota has resulted in a booming economy with much higher pay for both highly skilled and low income workers.[431] There is a strong contrast between economic activity in adjoining areas of Pennsylvania (which allows fracking) and New York (which does not).[432] Employment in the US oil and gas industry increased by 92% between 2003 and 2013 compared to a 3% increase in all jobs.[433] Under high pre-late 2014 prices, North America could have become the world's leading energy supplier.[434] The main obstacles were political rather than geological or technological or even economic. This may change somewhat with the lower oil prices in late 2014, however.

In addition to its failure to even maintain let alone expand leasing on Federal lands, the Obama Administration has imposed or is planning to impose numerous new regulations that will have the effect of decreasing hydrocarbon development through uncertainty, delay, prohibitions, and increasing costs. Canada has recently introduced a one-portal, one-permit Federal policy for all permits with tight deadlines for response by regulatory agencies.[435] A similar policy should help energy development in the US and avoid the endless delays and reviews that are now routine in the US. The proposed Keystone XL pipeline is a perfect example of this. Canada has clearly recognized the opportunity to use its hydrocarbon resources for its economic development; Russia has long done so; it is time for the US to do so as well. The US may no longer be

---

[431] Polzin and Whitsitt, 2014.

[432] *Ibid.*

[433] *Ibid.*

[434] Mills, 2012.

[435] *Ibid.*

competitive in low-tech manufacturing, but the US can compete in terms of hydrocarbon production at least at current world prices prior to late 2014.

A number of environmental organizations have taken a strong stand against the Keystone XL pipeline primarily because it would bring what they call "dirty oil" into the US and may increase CO_2 emissions. "Dirty oil" is oil that requires the release of additional CO_2 beyond that required for reservoir oil in order to make it available for use. In this case the tar sands have to be heated in order release the oil. As explained in earlier chapters additional CO_2 release is very likely to be good for the environment by possibly warding off the onset of a new ice age or little ice age (assuming it has much of any effect at all). Furthermore it would bring a supply of "ethical oil" into the US from a trusted neighbor and thereby reduce US dependence on other countries run by less than savory demagogues, dictators, and monarchs.

If the US refuses to allow the pipeline to cross the border, the oil is likely to be exported at higher cost in both CO_2 emissions and dollars using an all-Canadian pipeline to a West Coast port and the US will import more oil from much more dangerous and less ethical suppliers. As in so many other areas the US environmental organizations have gone mad. Some of them, including the Sierra Club, have even now resorted to civil disobedience to try to try to stop the pipeline. This is not evidence of a useful environmental movement but rather of a far left fringe group bent on preventing useful development. Their activities make no economic or national security sense and will harm their own cause as well as the US economy. There are newspaper reports that environmental opposition to new pipelines is spreading to other proposed pipelines.[436] This encourages the transport of oil by railroads, which have a much worse safety record, and slows the delivery of natural gas to power lower CO_2-emitting gas power plants and to heat homes and other buildings, which is particularly critical during

---

[436] Harder, 2014 and 2014a.

particularly cold weather.

Encouraging oil and gas development represents not only good economics but also good long term environmental policy. As we have recently seen, such development can impact high world oil prices and thereby help US consumers and thereby the non-oil and gas economy.

The reason why investment in shale or tight oil production is attractive needs some explanation. The US was one of the first countries to make major inroads into its reservoirs of cheap-to-pump petroleum, which were never as large or as inexpensive to tap as some of those subsequently found in the Middle East. Although the US provided most of its own oil until the 1970s, the US could not compete with these other sources of oil from large reservoirs. This led to our dependence on foreign sources in recent decades.

But more recently with the advent of OPEC and the expansion of oil demand, particularly in the developing world, relative to cheap supplies, the Middle Eastern owners of the large reservoirs began to charge far more than their cost of production for their cheap-to-produce oil. And they compounded this by using their profits in part to subsidize the living standards of their populations. This made it more difficult for these Middle Eastern countries to reduce their high prices nearer to their costs of production, even if they wanted to do so. This provided a golden opportunity for the US to use its somewhat higher cost oil resources to become an even larger supplier of world oil, or at the very least greatly reduce our oil imports. The Canadians worked hard to use their tar sands for exactly this purpose. The Russians have been exporting their large conventional oil and natural gas resources for years, and have done everything they can think of to discourage their customers in Western Europe from developing their own shale resources. This larger effort enabled them to maintain or improve their standard of living for several decades until late 2014 despite a shortage of competitive world-class industries.

The US needs to follow Canada's lead rather than to frustrate them by denying approval for the construction of the Keystone XL pipeline, which would bring their politically-safe oil to the US market and provide critically needed, low cost "on-ramps" for the Bakken oil along the way. Shipping oil by higher cost rail or road is more expensive in both dollars and CO_2 emissions, and probably more dangerous. Non-conventional oil development is already happening where the resources are privately owned in the Bakken and the Eagle Ford shale oil deposits in North Dakota and Texas. Relaxation of state and Federal restrictions on developing similar publicly-owned resources could have encouraged similar development of them, hopefully with every known economically sound environmental safeguard. The primary obstacle has been the environmental movement and their political supporters, not the availability of resources to develop. I cannot support their current viewpoints, which may have a devastating effect on the future of the US economy and result in little if any environmental benefits.

California is a splendid example. California could probably have solved or at least greatly alleviated their serious budgetary problems by doing more to encourage the development of their state-controlled oil resources and encouraging the use of new technologies for developing oil now in shale resources such as the huge Monterrey Shale formation in Central and Southern California, but has encountered considerable opposition by environmental groups. The US Energy Information Administration has estimated that the Monterey/Santos shale in California holds 64% of the technically recoverable shale oil resources in the on-shore lower 48 states, far more than the Bakken and Eagle Ford shales, which they estimate hold about 29%.[437] In 2013 California passed legislation allowing fracking if certain environmental safeguards are observed. This was bitterly opposed by the Sierra Club, but it remains to be seen whether the oil industry will be willing to use fracking under the safeguards finally adopted. I believe that the environmental groups would be more effective by

---

[437] US Energy Information Administration, 2011.

picking a few of the developments that would have the most adverse short-term non-GEF environmental effects for their opposition while supporting most environmentally responsible development as a whole.

The environmental movement points out that oil development may have some short-term environmental costs, particularly oil spills and encroachment on wildlife habitat. I believe that much needs to be and can be done to minimize these effects, particularly if that were the focus of the efforts by environmental organizations rather than opposition to any and all fossil fuel development. They also argue that fossil fuel use is "non-sustainable" and results in additional carbon dioxide emissions. I have argued that sustainability is not a relevant criterion and higher levels of CO_2 appear to be desirable from an environmental viewpoint given that the Holocene interglacial period appears to be coming to an end and a new ice age represents one of the worst possible environmental disasters. Finally, they are arguing that the new technologies developed by the oil industry for extracting oil from shale formations, particularly hydraulic fracturing or "fracking," will have adverse environmental effects[438] despite the fact that such effects have never been demonstrated over the long history of use by the oil industry.

I believe that the time has come to do everything possible to take advantage of our vast fossil fuel resources, particularly non-conventional and offshore conventional oil, to help solve the nation's larger economic problems. If the environmental movement does not wish to play a useful role in this development, perhaps it would be better to discount their views. I hope it does not come to this, but it currently looks likely given their ideologically-based intransigence, their continuing efforts to misrepresent the effects of energy use, and their continuing efforts to smear those who are attempting, often on a volunteer basis, to present an alternative and accurate assessment of the science and economics.

---

[438] US Senate, 2014c.

I have personally spent considerable time in the California areas where shale and offshore oil development may be promising, the Colorado plateau, and the Arctic, and value these ecosystems as much or more than anyone I know. But if the choice is between useless and uneconomic wind, solar, and biomass generation with their adverse but largely unrecognized adverse environmental effects and useful and very profitable oil and natural gas development and substitution of natural gas for other fossil fuels, I have no hesitation in advocating environmentally responsible oil and gas development. Obama's former standard argument (not heard in recent years) that we cannot drill our way out of oil importation is deliberately deceptive and simply wrong, just like his support for Spain's renewable energy development. The US has the knowledge and the resources to do very well at non-conventional oil and gas development and to minimize their adverse environmental effects, and passed up an opportunity to pursue it expeditiously while the recent economic pricing opportunity existed. Both state and Federal Governments could have played an important role in this. No serious effort was made, in part because of the environmentalist ideology of the Obama Administration and some states such as California.

Failure to follow this approach has resulted in a lower standard of living than would otherwise have been possible and thus much less concern for the environment in the long run, as well as a continuation of the economic problems we are now all too well acquainted with. Politicians would obviously prefer to avoid an open confrontation with the seriously misguided environmental movement, but the alternative of slow economic growth should ultimately leave them no choice. Hopefully the much needed confrontation will occur sooner rather than later. As the GWD continues to unravel hopefully more politicians will work up the courage.

In 2014 a new development underlines the importance of encouraging fracking and other innovative techniques for developing new sources of oil and natural gas. The US and many developed countries have been paying a high price for oil and gas in recent years because of a relatively tight supply-demand situation with regard to oil and gas from large reservoirs. This has cost those countries without access to large reserves of low cost oil and gas huge external transfers of wealth that would otherwise have largely gone to greater consumer demand for domestic goods and services. As shown by the rapid descent of oil prices in 2014, primarily as a result of the huge new supplies coming from non-conventional sources on private land in the US, the development of these new sources has resulted in much lower prices for oil importers throughout the world staring in 2014 and lower natural gas prices in the US in recent years. This can only be good news for US consumers, competitiveness, and the US balance of payments. It also shows how wrong the peak oil hypothesis has been and the argument that environmentalists have too often made that energy prices must inevitably go higher thus making renewables more competitive.

Avoiding the Effective Abandonment of Good Economics as an Imporatant Tool for Regulatory Policy

One of my major interests has always been cost-benefit analysis of natural resource and environmental projects, regulations, and policies. Cost-benefit analysis came into being primarily as a result of its application to evaluating the economic efficiency of US water resource development projects. Since much such development is done at the behest of Congress and groups that might benefit from such developments, cost-benefit analysis has not always had a favorable image in terms of its effectiveness in insuring that only the most economically efficient water projects are constructed. I tried to encourage EPA to apply such analyses to environmental regulations by sponsoring research as to how the economic benefits of EPA regulations could practically be determined.

I had mixed success prior to 1981 since not all the laws under which EPA operated required such analyses and many questioned the need for such analyses. The Reagan Executive Order of 1981, however, changed all that since it required such analyses on all proposed regulations which imposed substantial impacts on the economy. After the Order, the question was not whether such analyses would be done but rather how accurate and unbiased they would be. Prior to the Obama Administration I believe that EPA was reasonably responsible in how it carried them out although I believed that significant improvements could be made in how they were produced. Since the beginning of the Obama Administration, however, I believe that the analyses have become much more political in nature since the Administration already knew what regulations it wanted and has viewed the cost-benefit requirement as just one more procedural obstacle that needed to be satisfied to get what it had already determined it wanted. This sometimes occurred in previous administrations, but it has become much more pronounced under Obama. These are not good circumstances for useful such analyses, which achieve most of their benefits by encouraging users to reexamine their assumptions and proposals to make them more economically efficient. I had already had professional disagreements with economists with radical environmentalist views, which included one who turned out to be my second level supervisor under the Obama Administration.[439]

I continue to support cost-benefit analysis for environmental regulations but regret that as in the case of water projects it is all too easy to abuse such analyses since there are no independent arbiters of such analyses under current Federal practices other than OMB. So despite having spent much of my career at EPA promoting such analyses and sponsoring research to make them more accurate I believe that they are likely to be abused in the environmental case if their content is determined by those with a bureaucratic-interest in the outcome. It is long past time to move this function out of EPA into a less easily biased agency or at least out of the offices that are proposing and advocating new regulations within EPA. OMB has historically filled some of this role but in

---

[439] Carlin, 2005.

administrations with strong biases towards greater regulation there is evidence[440] that they may not be fulfilling this need adequately.

Unfortunately, the result of the collision of requirements for benefit-cost analysis of environmental regulations has been that many Federal regulatory agencies have bent their economic analyses to conform to the environmental policies they have been told to pursue either by Congress or the President. In other words, Federal regulatory analyses have become attempts to support the prevailing policies of those currently in power rather than independent attempts to make sure that proposed regulations are actually economically efficient. This is not unlike what has happened in the water development arena, where the same thing happened. This has been a source of great disappointment to me given my emphasis on this during many periods of my career. A renaissance of benefit-cost analysis will apparently require some significant changes in the current administrative arrangements that have led to this unfortunate state of affairs.

In a more ideal world, I would strongly favor changing the basis for environmental regulation to include economic efficiency where this is not already the case. Some Federal environmental laws already do this but some do not, particularly the Clean Air Act, under which economics cannot be used as the basis for the most important regulations. Such a change would hopefully result in more economically rational decisions rather than reflecting the environmental coloration of the current administration in power. The Clean Air Act is particularly problematic in this regard since it can be read as not allowing any adverse effects from designated air pollutants. This is not a rational policy since there will always be some residual effects, no matter how much is spent. The object should be to prevent pollution that can be prevented at a cost consistent with the benefits achieved rather than some abstract concept of no adverse effects.

---

[440] Gayer and Viscusi, 2013 and 2014.

Avoiding "Sustainability" and Other PR-related Goals Rather than Using Economic Efficiency Criteria or Allowing Free, Competitive Markets

As discussed in Chapter 10, the radical environmentalists are attempting to substitute "sustainability" for economic efficiency or market outcomes. Sustainability is not the only seemingly useful slogan now promoted by environmental organizations to gain acceptance of their agenda. Others include energy efficiency, renewability, recycling, and biodiversity. Each one suggests a worthwhile but vague goal that environmental organizations hope everyone will support. The results are well illustrated by the energy market disasters that have occurred where governments have chosen to follow "sustainability" or GWD rather than economics, including California, Great Britain, Germany, and Spain.

Avoiding Changing Behavior by Government Regulation/Incentives When Inconsistent with Environmental Ideology

One aspect that most modern radical environmentalists appear to share is the belief that when people's behavior is not consistent with their environmental ideology, these people's behavior can and should be changed by government action rather than allowing consumer preferences to determine the outcome through the actions of a relatively free market. They appear to believe that taxpayer funds should be used to spread their ideology, and since even that is often not sufficient, behavior must be subjected to government regulations and incentives if their ideologically determined outcome is to be achieved. This is roughly the opposite of a free market approach—if the consumers do not agree with the ideology then they must pay to hear government propaganda on the issue; if that is still insufficient, then the consumer is wrong and must have his/her behavior directly changed by government action to fit the ideology. This is a direct threat against one of the underlying ideas of what we call Western Civilization. Others have suggested that it has similarities to Eastern European Communism, where the Government decided how much of what was to be produced and consumers could buy what might sometimes be available rather than what they might want. The history of such non-market approaches to economic decisions has not been very favorable.

There is a very strong relationship between economic output and energy use in various economies. Academics supporting the radical environmentalists have been well aware of this and have constructed sophisticated arguments that the incentives created by their proposed governmental interventions in the energy markets as well as recommended large public R&D expenditures to increase energy efficiency will somehow make it possible to rapidly reduce energy use by dramatic percentages despite the obvious rapid increase in energy-using devices in the modern world. I believe that most of this is at best wishful thinking; others also have doubts.[441]

One of the most important goals of modern environmentalists in recent years has been to prevent additional fossil-fuel electric generating capacity from being built. The consequences of failing to maintain adequate capacity to meet demand are particularly dire in the case of electrical transmission networks, which must maintain a constant balance between demand and supply to avoid their collapse and thus work much better when they are operated with very high reliability and with ample reserve capacity. If radical environmentalists are successful in halting such development, this would further strain these networks and greatly increase reliability problems, which lead to enormous costs when networks fail. It appears much more likely to me that the only realistic way to rapidly and reliably decrease energy use to the extent they have proposed is by deliberately bringing on an ever-deepening depression. Voters might not find that particularly attractive, but may not realize this until electrical networks start repeatedly failing.

An amusing sidelight is that primarily as a result of the

---

[441] Galiana and Green, 2009.

operation of a relatively freer market, the US has done much "better" than most other developed countries in recent years in decreasing CO_2 emissions in significant part as a result of market substitution of natural gas for coal in electricity generation. This is a result of the operation of market forces as a result of the decrease in price of natural gas as a result of the increasing use of fracking and horizontal drilling. It is generally unrelated to governmental intervention to promote "renewable" energy sources or efforts to decrease coal use by making its use more expensive by government regulation. In other words, market forces have resulted in decreasing coal use more than the environmentalists' and the Obama Administration's efforts to punish coal use by government regulation or disincentives. Environmentalist support of fracking might produce similar reductions in CO_2 emissions elsewhere. Instead, many environmental groups are doing everything possible to outlaw fracking in the US and abroad. They have an unlikely ally in Russia, which wants to preserve its lucrative market for natural gas in Western Europe by supporting European environmentalists who oppose fracking.

Avoiding the Consequences of US Government Intervention in Auto Production

As a result of the Great Recession starting in 2007-8, two US auto companies, General Motors and Chrysler, went bankrupt. The US Government intervened and provided financial assistance. One of the objectives of this intervention appears to have been to reduce CO_2 emissions from future new cars produced by these firms. This took the form of encouragement of electric and small car production. At the same time the Government got industry "agreement" to much more strict auto mileage standards. They obviously hoped in this way to "persuade" the industry to adopt the Government's much broader GEF goals. The results have not exactly fulfilled these goals. As of late 2014 consumers have shown a strong preference for pickup trucks, SUVs, and luxury cars apparently as a result of declining gasoline prices, probably as a result of increased US oil production as a result of the use of fracking and horizontal drilling for oil.[442] The Obama Administration has thus illustrated the futility of state capitalism. The government is unable or in this case does not want to respond to—let alone understand—changing consumer preferences and other marketplace changes such as new oil supplies. Like their similarly futile and counterproductive GEF efforts under the stimulus program, this illustrates why the government should stay out of subsidizing/regulating/forcing the private sector to conform to its ideological preferences except in the rare instances where there is a good economic case for doing so.

Avoiding Use of Ultra-low Discount Rates

Many radical environmentalists believe that one of the cornerstones of economics, use of discount rates that have some reasonable relationship to the rate of return on capital found in the free market for similar risks should not apply to cost-benefit analyses of their proposed environmental policy proposals. The result is to assume that future (presumably richer) generations should gain at the expense of the current generation. When pushed to show the economic feasibility of their proposals, their supporters (think Stern[443] and Garnaut[444]) reduce the discount rates they use close to zero[445] so as to try to economically "justify" what they cannot do in any other way given their assumptions.[446]

---

[442] White, 2014.

[443] Stern, 2006.

[444] Garnaut, 2008.

[445] Carlin, 2011.

[446] My brief review of both the Stern and Garnaut reviews can be found in Section 5 of Carlin (2011). A recent detailed review of Stern (2006) can be found in Lilley, 2012.

Avoiding Government Intervention in Free Markets without Any Economic Basis

Many and perhaps most economists have long argued that government intervention in free markets is best limited to cases where there are clear market failures (in which case the free market may not provide optimal solutions), where government resources are involved, and where the governmental interventions proposed have economic benefits that exceed their costs (so as to avoid reducing the economic welfare of society). Such interventions can take many forms; a number were discussed in earlier in this chapter. Besides regulation, this can include denial of access to government-owned resources that the market would otherwise develop if they were privately owned, subsidies, taxes, and direct government investments. Government regulation is expensive for all concerned and almost invariably leads to a variety of unintended adverse consequences in addition to correcting whatever problem they were intended to correct. It should be emphasized that direct government investment is often not successful and ends up being a net loss to taxpayers due to the use of political or ideological rather than economic criteria.[447]

Some US radical environmentalists have even argued that cost-benefit analysis has inherent problems, was being administered with an anti-environmental bias by the US Office of Management and Budget prior to the Obama Administration, and that it largely only helps the regulated industries by delaying and weakening new regulations.[448] I have argued strongly against these allegations.[449] Many radical environmentalists appear to believe that the sensible "guidelines" on governmental action discussed above should not apply to their proposals. Despite comprehensive regulations which have long restricted emissions from coal-fired electric power plants

---

[447] Lerner, 2009.

[448] Ackerman *et al*, 2004

[449] Carlin, 2005.

in many countries, they believe such plants must be even further regulated despite the lack of evidence of a current market failure in the form of significant adverse effects of the pollutants actually proposed to be regulated.

I maintain that none of these modern radical environmentalist views on energy are compatible with the conclusions reached by applying neo-classical economics or with optimal economic growth. The result is that attempts to pursue many of the objectives pursued by radical environmentalists would result in making society economically worse off rather than "saving" it as they so fervently believe. This is almost certain to result in lower growth than otherwise would be the case, and an increased tendency towards economic stagnation. It is far from certain that the proposed environmentalist "solution," a major government-imposed reduction in CO_2 emissions, would achieve its stated objective, to reduce global temperatures to the extent that has been hypothesized.[450] Unfortunately, very few economists have taken a public stand against GWD. GWD is simply not compatible with the application of neo-classical economics and deserves to be opposed by economists as being incompatible with their economic worldview.

As discussed in Chapter 2 with regard to Indian economic development, central planning has not had a very good record in promoting economic growth. Regulations requiring a certain proportion of electricity be generated using certain fuels or that gasoline used be from certain sources or subsidies that promote one source of electric energy over others are no better than central planning in one sector. Energy efficiency standards are as well. The reason that central planning has had such bad economic results is that central planners have little knowledge of the unintended consequences of their decrees. They may not even care since they get paid whatever happens. The most efficient economic system is the one that places the decision-making responsibility with those

---

[450] Carlin, 2008.

who have both the most knowledge and the most incentive to make efficient decisions. That generally means with the industry involved, not the government.

The radical environmentalists have been pushing Western developed countries towards central planning for energy generation and use for several decades. The results are exactly what one would expect—economic inefficiency and lower growth. The best evidence for this is what has happened in those Western European countries that have imposed the strongest requirements for use of "renewable" energy. As detailed earlier in this Chapter, their energy situation is rapidly deteriorating, energy prices are roughly three times US prices, and energy-using industries are considering moving elsewhere. And one government intervention results in more interventions to correct the problems resulting from the first intervention. If the US follows further down the same road, as the Obama Administration wants, the result will be the same. The central planning involved will not even try to take account of the impacts on other sectors, energy prices, or economic growth, as other planned economies have sometimes tried to do. It is simply the imposition of government command and control (reduce CO_2 emissions by X% by year Y) on the energy sector without regard to what this will result in. The results will be no better than the energy policy disaster in Western Europe and might be worse.

Unintended Consequences of Unjustified Government Intervention in the Energy Market

Western Europe presents a real life example of what happens when governments make unjustifiable interventions in the energy market, in this case by providing incentives or regulations that result in greater investment in renewables. The Obama Administration initially held up Spain as the example that the US should follow. The problem was that Spain had introduced very expensive incentives for renewables but did not have a mechanism to pay their rapidly increasing cost, particularly when Spain began to experience problems with the EU common currency, the Euro.

More generally, as a result of EU requirements and support by radical environmentalists in individual countries, several other Western European governments also pushed renewables. One of the resulting problems has been that uneconomic investments in renewables have resulted in the loss of revenue for conventional generation, particularly natural gas plants, resulting in a lack of investment in such plants and some pressure by the owners to close them. But when cloudy, windless periods come there is a shortage of electrical power from conventional sources to supply the needed electrical energy. The result is another government intervention to keep open or even build new backup fossil fuel capacity at the expense of ratepayers and taxpayers. The result is that the cost of electricity is very high and rising rapidly, which discourages investment in industries using significant amounts of it, and encourages such industries to move elsewhere where power is less expensive. The ultimate change in CO_2 emissions is close to zero since the emissions and the jobs that go with them are simply exported to other countries. The futility of this needs to be understood by politicians and the voters that elect them.

In addition to the earlier discussions of the non-economic effects of increasing use of renewables, this appears to have had several unfortunate economic effects.[451] One is that energy is significantly more expensive, which makes countries using renewables much less competitive. Another is that renewables make the operation of natural gas plants much less profitable by providing power when sun and wind are available, with the result that there is a lack of investment needed to build and operate backup for renewables, which means that gas plants are less likely to be available when the sun does not shine or the wind does not blow. A third is that either ratepayers or taxpayers must pay for whatever incentives are offered to build and operate renewable sources and they are often not willing to do so. These problems account for many of the energy problems experienced in Western Europe in

---

[451] Peiser, 2014.

recent years and for a substantial pull back by Spain and Germany and other EU countries from renewables.[452] Spain can no longer afford the incentives it had promised renewables, and in 2014 decided to remove all of them for renewable sources built before 2004. The result is likely to be catastrophe for investors who had counted on them. German electricity rates are the second highest in Europe and roughly triple those in the US, which encourages industry to move elsewhere. Both households and industries are complaining.

In the US most states with renewable portfolio standards (RPSs) have significantly higher electricity rates than those without them.[453] Colorado electricity prices have risen 20% faster than the national average since the state enacted a RPS in 2004, and twice as fast since 2007 after stricter mandates were enacted.[454]

These economic distortions and problems can be expected anytime that government makes unjustified intrusions into energy (or other) markets. And they usually grow worse with time as additional government "fixes" are piled on the preceding ones. This is exactly what happens as incentives are created to use renewables rather than leaving energy generation and use to market forces. The Western European experience should be a warning of what will happen in the US if these incentives are not curbed, but there is no evidence of learning by the Obama Administration or a number of blue states that have imposed RPSs and renewable subsidies. The 2014 proposed EPA regulation to reduce CO_2 emissions from existing power plants actually proposes the use of RPSs as one control strategy states can use.

---

[452] *Ibid.*

[453] Bryce, 2012.

[454] Taylor, 2014a.

Some Rationalizations Often Used to Support GWD/GEF

I have heard a number of rationalizations for pursuing GWD/GEF when the scientific invalidity of CAGW is pointed out. One is that we should "play it safe" just in case the CIC science should be correct. The other is that the GWD/GEF push is in the "right direction" anyway, so why not pursue it?

Since there is simply no question that the CAGW and EWD scientific hypotheses are invalid, we would not be "playing it safe" by assuming that are. All we would be doing is spending resources on policies which the CIC alleges would counteract global warming/climate change/extreme weather. These policies would have no effect on extreme weather and no measurable effects on global warming or climate change. Spending resources on policies that have no or no measureable effects is not an insurance policy but rather a waste. It is far better to spend resources on the basis of sound science and economics than to waste them on alleged solutions that will not solve the alleged problems identified.

As I argued in Chapter 9 there is substantial evidence that Earth is descending into a new ice age over the next few thousand years. Under these circumstances, if indeed there is any basis for CAGW, what is needed is higher, not lower CO_2 emissions. If there is little basis for CAGW and none for EWD science, government-imposed reductions in human emissions of CO_2 harms humans by reducing the assistance they can obtain from using fossil fuels as substitutes for human labor and time and provides no benefits in terms of reducing catastrophic global warming or climate change or extreme weather. "Feel good" policies like reducing consumption, reducing fossil fuel use, and sustainable development may appeal to some people as wise but have no basis in economics or science, despite CIC statements to the contrary.

The "right direction" argument usually assumes that we know the future. Unfortunately, no one does. The market mechanism seems to do better at this than any other approach tried, so we should use it rather than trying to guess the future.

Compatibility of Environmental Improvement and Economic Development

Since one of the major effects of GWD/GEF is to decrease economic growth and development, the immediate question is whether this would be good or bad for the environment. I argue that economic development is not only compatible with improvements in environmental quality but ultimately promotes it. This is contrary to one of the underlying assumptions of radical environmentalism, which views development as an evil that must be stopped. Why do I hold this view? First of all, after years of effort the best argument that radical environmentalists have been able to come up with against economic growth and development is GWD/GEF/EWD. And as argued in this book, this is at worst a minor effect that is not worth worrying about.

Second, environmental economists have long hypothesized something called the Environmental Kuznets Curve (EKC), which argues that after an initial drop in environmental quality as standards of living start to increase, environmental quality starts to improve at higher income levels.[455] One of the reasons this presumably occurs is that at lower levels of income people have neither the means nor place a priority on improving their environmental quality as development and urbanization get underway. A second reason is that with higher incomes demand becomes less material-intensive and thus less polluting. As Beckerman[456] puts it, *"The strong correlation between incomes and the extent to which environmental protection measures are adopted, demonstrates that in the longer run, the surest way to improve your environment is to become rich."*

One of the exceptions to this Curve that has been noted in the literature is the continuing increase in atmospheric CO_2 levels which has occurred in the 20th Century at the same time as increases

---

[455] IBRD, 1992; Dinda, 2004.
[456] Beckerman, 1992.

in development and living standards. But as argued in this book CO_2 is not a pollutant and is not itself an indicator of environmental quality. No one dies or is made ill by current or foreseeable atmospheric CO_2 levels. Plants are helped, not hurt, by higher levels. Alarmists have tried to associate it with catastrophic increases in global temperatures and conventional pollutants from fossil fuel combustion, which are of concern environmentally. But CO_2 increases have very little impact on climate and may actually be a result of increasing global temperatures. Atmospheric levels of conventional pollutants from combustion do fall with increasing income after an initial increase, as the EKC predicts. CO_2 should not be considered an exception to the EKC hypothesis since it is not a measure of environmental quality.

Effects on National Security

The precipitous drop in world oil prices in late 2014 and the earlier drop in US natural gas prices was in significant part due to the success of US fracking and other innovative techniques on private land in greatly expanding US oil and natural gas production. The changes in world oil prices were quite rapid once they started in mid-2014. The change in world natural gas prices has been much slower because of the technical, legal, and political difficulties of exporting large volumes of US LNG by sea.

It is very much in the interest of the US that these lower world oil prices be sustained. The drop in prices has similar effects inside the US to a tax cut for lower income earners in that it results in expanded purchasing power for non-energy goods. Lower oil prices will primarily adversely affect US adversaries, particularly Russia, Iran, and Venezuela, all of whom are heavily dependent on oil revenue.

Oil producers are seeing less income, of course, which will decrease their outlays for exploration and development. But this can be partially offset by decreasing government interference with oil and gas production where there is no justification for it. This could include reducing unjustified Federal and state regulations and greatly expanding access to Federal and state lands for exploration and development. GWD/GEF-inspired regulations are an important example of unjustified interference. AFD-inspired regulations are particularly important in that they can and have closed off important oil and gas resources from contributing to oil and gas production increases.

In addition to all the other adverse effects of GWD/GEF/AFD, it is important to recognize its effects on national security, particularly for the US's Western European allies. In the end, what determined the outcome of the Cold War was not the number of strategic missiles or bombers[457] but the fundamental dependence of the Soviet Union on foreign currency from exports of oil and natural gas due to its lack of many world class industries. This vulnerability proved its undoing,[458] and could have similar effects on Russia as a result of the major drop in world oil prices in late 2014. If Russia is again forced to seek Western finance as they did in the early 1990s as a result of the precipitous drop in oil prices, the Western world may again have an opportunity to influence Russian behavior without going to war. In the previous experience, the Soviet Union voluntarily avoided using military force to prevent the downfall of their Eastern European satellite governments apparently in order to obtain Western financing. and this soon caused them to change their own form of government and their geographical boundaries.

If continued long enough, the world price drop might result in Russia finding it advantageous to show much greater respect for Ukrainian and other Eastern European borders if they want to obtain Western financing. The US legal ban on crude oil exports and the slow approval process for natural gas export facilities is not helping. The opposition of major environmental groups to the proposed LNG export terminal at Cove Point, MD, is a case in point.

---

[457] As discussed in Chapter 2.

[458] Gaidar, 2007.

Unfortunately, the Russians have strong counter-leverage—their control over the natural gas that they sell to Western Europe. Non-cooperative Western European countries may find their natural gas supplies cut off or only available at a higher price. It is thus important for those Western European countries now dependent on Russian gas to rapidly minimize their dependence so that the Russians will lose this major source of leverage they now have over them.

Currently the European Union is importing about 80% of its oil and 60% of its natural gas; about a third of each comes from Russia.[459] The obvious way for countries that purchase Russian natural gas to minimize their dependence is for them to develop their own non-conventional natural gas resources. This, however, would require fracking, which the CIC and Russia oppose, ostensibly for the usual AFD ideological reasons. As of late 2014 no Western European country has been willing to make a serious effort to promote the development of such gas resources although some governments support the idea and a few are even making some preliminary efforts. If Western European nations want to get the Russians out of Ukraine and assure that they will not attempt to annex other Eastern European nations, denying loans and pushing fracking appear to be the easiest approaches for them to take. The US could assist in this by encouraging fracking on public lands and expediting permits for export terminals for liquefied natural gas, which a number of environmental organizations also oppose. Or better yet, it could change the law so that Federal permits would no longer be required. This would allow the US to provide a back-up source of natural gas, which might be particularly important during a transition period of decreasing use of Russian gas by Western Europe. And it would help the US economy and balance of payments.

When the US buys a large share of a vital raw material, oil, from unstable or unfriendly countries it is risking the adverse effects

---

[459] Macaes, 2014.

that would occur if these countries cut off the supply. When it facilitates the movement of oil into the US from a friendly country (such as Canada and the proposed Keystone XL pipeline) it increases its national security because this decreases its need to import oil from unstable or unfriendly countries. Since the XL pipeline would be built by the private sector, and appears to be a much safer and less disruptive to the rail network than the current rail transport used to move most of the oil, there is no economic reason to oppose it and significant national security reasons to approve it. For the reasons discussed in Chapter 9 there are no major environmental reasons to oppose the XL pipeline. The opposition of many environmental groups to XL is still another reason I can no longer support such groups and instead question their objectives.

The US and Western Europe are likely to have much greater success in pursuing their own interests the more they are self-sufficient and in the US case even net exporters of oil and gas. GWD/GEF/EWD and especially AFD are very unlikely to be helpful in that regard.

Besides the obvious economic benefits if the US Government had really pushed shale gas and oil hard in recent years, the national security benefits of falling oil prices might have been achieved much earlier. The US might now actually be able to offer effective relief from new possible Russian cutoffs of natural gas to Western Europe, which has been frightened by environmentalists into generally avoiding fracking to increase shale gas and oil output. But better late than never. Many environmental organizations have gone on record as opposing fracking, and some even favor keeping "our nation's fossil fuel reserves in the ground."

The international situation has changed dramatically in 2014 and the US and Western Europe need to make equally dramatic changes in their energy production and distribution capabilities. The changes needed cannot be achieved overnight, but the sooner the EU and the US start the sooner Eastern European security threats are likely to be reduced.

Much lower oil prices also adversely affect a number of the US' other major opponents in the world besides Russia, including Venezuela and Iran.[460] Conversely, higher oil prices, which result from GWD/GEF/AFD, help these countries. It is in the national security interests of the US to keep oil and gas prices low; the way to do this is to promote non-conventional sources throughout the world and make available suitable US public lands for this purpose.

Summary of Economic Impacts of GWD/GEF

The adverse economic and national security effects of GWD/GEF are many and large. Besides reducing economic growth in general, they discourage investments in fossil fuel development, which has a ready market for its products. These costs are not offset with any measureable benefits, so are net costs.

GWD/GEF is strongly detrimental economically because it leads to slower economic growth, wastes very large resources on solutions that will have little or no effect on the problems identified, hurts primarily lower income groups throughout the world, and decreases the availability and reliability while increasing the price of energy, one of the primary sources of economic growth and development and human advancement.

Developed countries' attempts to implement GWD/GEF/EWD are wasting hundreds of billions of dollars each year building and operating uneconomic "renewable" energy generation sources/facilities which in most cases will have to sooner or later be abandoned when the government subsidies run out. Some developed countries are trying to convince/force less developed countries to follow similar approaches. If these countries should be so unwise as to agree to these policies, their development will be slowed significantly, with adverse impacts particularly on the poorest of the poor. Many people who would otherwise live as a result of gaining access to the modern uses of energy such as sewage

---

[460] Higgins, 2014.

treatment, refrigeration, and even purchased food—now made more expensive by competition with growing fuel—will not only be denied the benefits of increased energy use but die for the lack of access to these basic benefits of energy use. Thus GWD/GEF not only will not help the environment but will harm the most vulnerable of the poor around the world and slow the development of the less developed world. Essentially, what the CIC is arguing is that useless attempts to slow warming are more important than the economic development of the less developed world and continued growth of the developed world. I could not disagree more. Economic growth and development provide real gains for everyone and for the environment. GEF provides only negative benefits.

Even if the CAGW hypothesis were correct just the way the IPCC argues, the CIC "solution" is highly unlikely to be effective because the major and rapidly rising source of emissions, the less developed countries, would not and should not agree to the economically devastating effects of mandatory CO_2 reductions on their future development and the welfare of their people. Even if they should agree, the cost in terms of payments from the developed world would not be approved by developed country taxpayers. And even if they did it is highly unlikely that the less developed countries could or would honor their commitments.

12
How Did We Get into This Mess and How Do We Get Out and Prevent It from Happening Again?

How Did We Get into This Mess?

WE GOT INTO our current GWD/GEF mess because various components of the CIC saw an opportunity to achieve their private ends by hijacking an obscure and unsettled area of science for their mutual benefit. Some previously not well known scientists, who sometimes build expensive but so far highly inaccurate and thus not very useful climate computer models, saw this as an opportunity for major research grants from government agencies to sustain their very expensive hobby, for public stardom that they would otherwise never achieve, and to claim that they were "saving the world" in the process. The mainstream media saw this as an opportunity to attract readers that were fast moving to the internet and to support the liberal, "progressive" politics of many of their owners. The environmental groups saw this as an opportunity to extend the life of their organizations which had already achieved many of the gains needed for society to pay greater attention to the impact of humans on nature. Espousing GWD/GEF and more recently EWD would attract supporters and particularly funding to greatly enlarge their donations, grants, paid staff, and influence. The relevant parts of the UN were and are in the firm control of alarmists who have used it to try to legitimize the scientific case for CAGW, promote GWD/GEF at the expense of the developed world, and transfer resources to the less developed world. The larger organization may have seen this as an opportunity to show how world government could "solve" a global environmental problem, and thus demonstrate its importance and relevance. They may also have hoped that this would lead to a permanent stream of income through handling transfer payments from the developed to the less developed world. Margaret Thatcher saw an opportunity to decrease the power and influence of British coal miners, who were making life hard for her.

More recently, liberal politicians in the developed world appear to have seen this as an opportunity to get green votes, possibly to enhance their governments' tax base, and possibly a chance to portray themselves as supporters of all that is "clean" and public spirited, even though their predominantly lower income supporters would be hurt. Some business interests hoped that they could build profitable and protected markets for their products by influencing new laws and regulations or sought to disadvantage competitors with a greater adverse exposure to environmental ideology. Corn farmers and agro-processors have seized on GWD/GEF as a golden opportunity for higher prices, output, and, of course, income and land values. Less developed countries have seized on it to argue for financial assistance from developed countries for adaptation and mitigation. More recently they have seized on EWD to argue for "loss and damage" payments from the same source to "compensate" them for losses from extreme weather events,[461] supposedly caused by the developed countries' CO_2 emissions. The Russians have used the need for ratification of the

---

[461] United Nations Framework Convention for Climate Change, 2013.

Kyoto Protocol to obtain membership in the World Trade Organization.

The losers have been everyone who did not fall into one or more of these categories—mainly ordinary and not so wealthy taxpayers and ratepayers and businesses in the developed world and poor people in less developed countries whose food comes from agricultural products now used for energy production—who would end up paying the enormous bills (both in terms of human hunger and financial outlays) for GWD/GEF. This includes many business interests, who have nevertheless often adopted a public relations approach claiming how "green" they are. This last, however, has greatly handicapped skeptic fund raising despite the myths about bountiful polluter funding promoted by alarmists. All these groups, which I have characterized as the CIC, came together to promote a common ideology and political agenda.

The self-serving myths created by the CIC are easy to explain. They all appear to be those that would help their cause from a public relations viewpoint rather than accurately portray their views or evidence. Portraying skeptics as in the pay of evil polluters is one example of this. Calling skeptics ugly names, such as "deniers" as in holocaust deniers, is another. I question whether these underhanded public relations myths are accidental; they rather suggest a strong PR influence on the CIC enterprise.

The public has never really understood the alarmist/skeptic scientific battles, but appear to have been a little skeptical towards model builders and business interests by nature. Thanks to the internet and a number of successful skeptic bloggers, interested members of the public could get the skeptic side of the argument if interested, which in an earlier day would have been completely drowned out by the mainstream media. The more liberal, "progressive" politicians saw little to lose and something to gain by supporting the CIC. The more conservative politicians were opposed in principal to larger government and higher taxes/energy rates except in Western Europe, where they have largely gone along with the whole GWD program. Some conservative US politicians, such as Senator Inhofe, actually took the trouble to understand the CAGW science, realized that it was not valid, and publicly said so.[462]

Finally, nature has not cooperated with the alarmists, and decided to keep global temperatures fairly constant since at least 1998 after a minor increase during 1997-98. Some citizens have noticed the at least 17 year hiatus in temperatures. If Romney had won in 2012 the climate bubble might have burst then since that would probably have been the end of the high funding for CAGW research, and the alarmist scientists would have found it in their interests to backtrack rather than continuing to push an invalid hypothesis. But Romney did not win and many of the groups just mentioned have found it in their interest to continue pursuing GWD/GEF at least as long as Obama is President and the Republicans lack sufficient votes in the Senate to prevent EPA from implementing GWD/GEF using regulations.

One unfortunate aspect of the current situation is that hundreds of billions of dollars are being wasted every year in the world pursuing GWD/GEF despite many other urgent needs such as reducing unemployment and improving education and health. This will continue as long as the subsidies for inefficient "renewable" sources of power continue and as long as regulations putting fossil fuel use at a disadvantage and favoring "renewables" continue in place or, more likely, become stronger. If, as Obama wants, these regulations are made even more strict by preventing new or further discouraging existing fossil fuel power production, the waste of resources will escalate even further. The so-called "renewable portfolio standards" enacted in many US states have similar effects. EPA is now trying to make these standards mandatory in all states with the help of their new regulations on existing coal power plants.

Obama has now publicly stated that he intends to implement GWD/GEF through regulations promulgated by EPA and other

---

[462] Inhofe, 2012.

Federal agencies. This avoids, he hopes, the problem of obtaining Congressional approval and largely removes the issue from public discourse by relegating it to the *Federal Register*, newsletters, and websites. It is increasingly evident that carbon tax or cap and trade schemes, although they would generally be more efficient at implementing GWD/GEF, are politically unsalable because many of the supporters are willing to "spend" other people's money but not their own for such illusory benefits. This is the lesson of the 2009-10 cap and trade bill in the US Congress and the Australian carbon tax. Politicians who support such high profile action in support of the GWD/GEF rhetoric are likely to be voted out of office. So the Obama Administration now intends to try what they hope will be an under-the-radar regulatory approach. This is just what I feared in early 2009 when I decided to submit negative comments on the draft Technical Support Document for the Endangerment Finding. This is clearly the best approach they have to implement GWD/GEF, but the worst for the American people because it has the best chance of "success." It has all the costs of cap and trade plus large regulatory costs and still no (or more likely, negative) benefits.

Where Have Environmental Regulations Gone Wrong and What Can We Learn from These Failures?

One way to analyze where improvements could be made to environmental regulation is by examining where and why it has gone wrong in the US. The three major cases are probably the DDT ban, the Greenhouse Gas Endangerment Finding, and the ethanol requirement (known as the Renewable Fuel Standard or RFS). A fourth possible case is the ban on chlorofluorocarbons, but much more research is needed in this case before a defensible conclusion can be reached. In the first two cases EPA did not follow careful and sound science to reach its determination. As discussed in Chapter 8, in the case of the DDT ban EPA did gather the scientific information, but then ignored what it had gathered in reaching a decision, with disastrous results. In the case of the Greenhouse Gas

Endangerment Finding they largely copied out what the UN IPCC and various US Government review groups had said without any attempt to reach an independent, science-based decision as a result of White House pressure for immediate action.

In the case of the Endangerment Finding it is clear, as was discussed in Chapter 11, that large expenditures will result in no, or more likely, negative benefits if CO_2 emissions reduction regulations should be implemented as EPA proposes.

The ethanol program certainly must rank as one of the major boondoggles of recent decades. It increases world food prices, hurts the poor, results in increased soil erosion and water pollution from farmland, removes wetlands and other environmentally sensitive land from the Conservation Reserve Program,[463] costs more per mile than gasoline as of late 2014, and accomplishes little if any CO_2 emissions reductions, which are not desirable anyway. There are a number of villains here, particularly Congress, but also George W. Bush, who signed the legislation, the Obama Administration which has been more than supportive of it, and EPA, which has not used various options available to it to avoid implementing parts of it (quite possibly at the behest of the Administration). Congress is always looking for ways to subsidize farmers, perhaps because of their disproportionate representation, particularly in the Senate, and seized on GWD as another excuse to do so. Most environmental organizations do not support the program. But if the GWD rationale for reducing CO_2 emissions had not existed, Congress would have been less likely to seize on it. And if the Obama Administration came out in opposition to the program, it would hopefully have long since disappeared.

How Do We Get Out of This Mess?

There are two categories of changes needed to get the US out of its current climate/energy mess and avoid future similar problems.

---

[463] Secchi and Babcock, 2007.

The first is the changes needed to halt the Administration's multi-faceted approach towards implementing GWD/GEF/EWD. The second is the changes required to insure that similar attempts to hijack the regulatory and political system for the purpose of implementing natural resource/energy policies that are not supported by sound science and economics will be less likely to occur in the future. These changes will be discussed in the remainder of this chapter.

Changes Needed to Avoid Implementation of GWD/GEF/EWD

The environmental movement's one constant with regard to energy (as well as most other issues) is increasing government regulation and involvement. What is needed, however, to solve the current energy mess is exactly the opposite—to get government out of any involvement that is not necessitated by valid economic, environmental, or ownership considerations. This would end a large number of existing or proposed Federal and state programs in the US and in the EU and member countries.

Change Needed: Prevent Attempts to Resurrect Carbon Tax and Cap and Trade Proposals

Although the Cap and Trade Bill died in the US Senate in 2010 and will hopefully not be revived, support for carbon taxes and cap and trade bills continues to appear. Obama appears to have given up on Cap and Trade in terms of new legislation based on his June 25, 2013 speech. Although it may be a more economically efficient way to achieve CO_2 emissions reductions than the regulatory approach now being pursued by his Administration, there is no environmental justification for either of them since changes in CO_2 emissions would have no significant effect on global warming/climate change/extreme weather and minor warming would be good, not bad. Both of these approaches would impose major burdens on the American population, particularly the less affluent ones, for negative benefits. Australia enacted a carbon tax in 2012, but a new government repealed it in July, 2014. In June, 2014, however, the Obama Administration revived cap and trade through EPA regulations allowing it as an option that states can use to meet EPA's new GHG emissions reductions on existing coal-fired power plants. This deliberate end run around Congress deserves to be defeated just as much as the original Waxman-Markey bill, which had a number of the same features.

Cap and trade, if implemented without favoritism (a big if, of course), is one way to try to take into account the adverse environmental effects of energy use on climate change. But attempts to use this approach assume that there is a significant relationship, and that if there is, that we can accurately determine the exact "cap" that would avoid the alleged adverse effects of the resulting increased emissions of CO_2. The hypothesis that there is a significant relationship is not supported by available observations. Hence there is currently no basis for a cap on CO_2 emissions.

Similarly, since using fossil fuels and modest increases in temperatures helps the economy and improves human life there is no economic or environmental reason for imposing a carbon tax—just the opposite.

Change Needed: End Economically Unjustified Attempts to Rig Energy Market Decisions Regarding Energy Production and Use

Radical environmental groups are constantly urging governments to rig the energy markets to favor the environmentalists' preferred energy sources in a number of ways. Federal and state legislatures should not be involved with selecting energy sources. Such decisions should be based on economics and justifiable pollution standards rather than politics, assuming that market prices reflect any major market failures involved. No subsidies/taxes/preferences should be imposed unless so justified.

One of the mechanisms used for this purpose is renewable performance standards (RPSs), sometimes referred to as renewable energy standards, which require that specified percentages of electricity are generated by "renewable" sources. They have been enacted in over half the states, are included as an option for states in the EPA 2014 proposed standards on existing power plants, and amount to a tax on users' utility bills to subsidize much higher cost, unreliable sources that cannot compete in the open marketplace. RPS requirements do nothing to increase the efficiency of US energy production[464] and a great deal with increasing the cost of energy. In addition, wind turbines have a devastating impact on bird and bat populations including many rare and endangered species.

In addition to RPSs, there are also widespread subsidies for wind and solar power production and use. These simply substitute tax payer dollars for rate payer dollars. Production and investment tax credits are subsidies from taxpayers to wind/solar producers; feed-in tariffs increase the prices paid by other users to the extent that the payments exceed the wholesale price of electricity at the times provided to the utility since the utility overhead, distribution, and backup costs are generally not subtracted from the amounts paid to those providing electricity.[465] Solar/wind are highly unlikely to ever be economically viable because of the highly diffuse nature of their energy sources, their high original costs, the relatively high cost of transmitting the power they generate in high solar/wind areas with few existing transmission lines to load centers, their extreme unreliability in an electric grid requiring very high reliability, the necessity to keep fossil fuel reserve capacity continuously available to compensate for the random non-availability of renewables, and their major and unsolvable environmental problems.

Even the strongly pro-GEF Google Corporation has given up on alternatives with the following statement by two of its employees who worked on a project (RE<C) to generate

---

[464] Additional discussion can be found in Chapter 11 and Carlin, 2009c.
[465] Tanton, 2014.

renewable electricity more cheaply than coal-fired power plants do:[466]

> At the start of RE<C, we had shared the attitude of many stalwart environmentalists: We felt that with steady improvements to today's renewable energy technologies, our society could stave off catastrophic climate change. We now know that to be a false hope....Renewable energy technologies simply won't work; we need a fundamentally different approach.

There are simply no economic or environmental arguments that can justify such monumental wastes of resources and destruction of wildlife. If and when the subsidies/RPSs are gone, most wind/solar installations are very likely to be abandoned, as in the past, which will then often result in huge cleanup problems, but happily greatly improved life expectancy for birds and bats.

Wind and other renewable subsidies need to be ended since they create a genuine environmental problem resulting from government interference in the marketplace, but could be solved by ending all of them. The Federal renewable Production Tax Credit (PTC) is perhaps the most important one, although some states also offer such subsidies. Unfortunately, it is very popular in Congress, like many subsidies. Even more difficult but essential is to require that the owners of all wind/solar sites pay for dismantling, removing, and disposing of them when they are no longer economic to operate. Since many firms operating wind/solar may go bankrupt when and if subsidies end, performance bonds may be one way to enforce such requirements.

The general problem is that the CIC has succeeded in altering energy production methods at the expense of taxpayers and ratepayers who have not been sufficiently organized or even

---

[466] Koningstein and Fork, 2014.

informed to resist them. One of the main battlegrounds is currently at the state level, which makes coordinated resistance efforts difficult and expensive but nevertheless vital if rationality is to be restored to the energy markets. The Federal PTC needs to be permanently ended and not resurrected as has repeatedly happened in recent years. Wind/solar preferences are actively supported by most environmental organizations, the Obama Administration, many European countries, and some US states. There is simply no environmental or economic case for them, and a strong case against them. If wind/solar investments were economically justified, government preferences, subsidies, and environmental exemptions would not be needed to make them viable. The continued strong push for them by the CIC suggests that the subsidies are "needed" to make them economically viable and that seems likely to always be the case.

One of the major problems with attempts by government to select winners among energy sources is that they are very likely to select inefficient and quite possibly even environmentally inferior choices compared to leaving these choices to the workings of the market. The choice of energy sources is complicated, region specific, and subject to being influenced by the many special interests with something to gain by the choice made. Industries that make solar panels or windmill parts are more than likely to lobby for their particular solutions without regard for the larger national interest.

The national interest is not to satisfy various interest groups but to supply reliable energy at the lowest possible cost after fully taking into account the adverse environmental effects of each alternative through subsidies or taxes or pollutant regulations. Energy is an input to many products and services; higher cost energy increases the cost of these products and services, and makes them less competitive in the marketplace.

Some politicians may regard renewable performance standards and other attempts to rig energy markets as a "free" consolation prize for environmentalists. Unfortunately, it is far from free. It is leads to significantly higher energy costs for households and industries for many years and quite possibly inferior environmental choices. In the case of households that means that there will be less money available for other needs. For lower income households this means that other necessary expenditures including healthcare and education will have to be reduced. Where the energy source is intermittent, such as wind and solar, the cost of building substitute sources to insure availability needs to be taken into account, as they would be in private markets.

Change Needed: Prevent/Rescind Other Indirect GWD/GEF-Inspired Regulations/Subsidies

It is sometimes not realized just how widespread GWD/GEF-inspired regulations/subsidies are or will be in the US if current EPA proposed regulations are implemented. They are found at both the Federal and the state levels. In some cases, they are not even advocated by environmental organizations, but are instead intended to help various special interests using GWD/GEF as a cover for their attempt to hijack government policy for their benefit. It seems possible, however, that if GWD/GEF should be discredited it would make it more difficult to maintain these efforts.

Any attempt to enumerate all these regulations/subsidies would be incomplete. But just listing a few will illustrate just how widespread they are. They include ethanol requirements for gasoline, mercury emissions reductions from power plants; saline effluent reductions from coal mining; dunes sagebrush lizard preservation under the Endangered Species Act to reduce oil production and energy efficiency standards. The mercury standard does not target CO_2 from power plants, but the effect is very similar because of the stringent requirements imposed on coal power plants for little or no valid purpose.

Ethanol requirements for gasoline are an example of requirements not supported by most environmental organizations but promoted using GWD/GEF ideology by special interests—mainly corn farmers and agribusiness—for their own benefit.

Without government endorsement of GWD/GEF these requirements would hopefully soon wither away for lack of ideological "cover." But the current result is higher prices for gasoline, lower miles per gallon, and the possibility of damage to some types of engines, particularly small ones, all of which hurts primarily lower income users. Without government regulation, primarily a result of Congressional pressure, a free market would not add ethanol in gasoline, and everyone except the special interests involved would be better off for it. There is some dispute whether the requirement reduces CO_2 emissions, but the effect is small if it exists at all. As argued in this book, that is not an important consideration anyway due to the minor and likely beneficial effects of increased CO_2, particularly at current levels.

Until consumers buy automobiles or major appliances or light bulbs they may not be aware that for a number of years the Federal Government has gone well beyond the original purpose of such regulations to supply information so that consumers can make better choices with regard to the relative energy characteristics of these items to outright regulations as to what the energy characteristics of such purchases must be. The government claims that the major benefits of these regulations are to save purchasers' money by reducing energy usage over the lifetime of the vehicle/appliance. If so, there is no economic justification for these regulations since consumers can make much better choices on their own, and with much more information as to their own preferences and budgets.

These regulations are now quite widespread. The best known are the automobile fuel efficiency standards, which the Obama Administration has pushed very hard (but not transparently) to tighten. The largely ignored non-financial costs in reduced safety for vehicle occupants. One way to get increased fuel economy is to reduce vehicle weight. This reduces fuel use per mile but at the cost of reduced safety for occupants when involved in accidents, particularly with larger vehicles. Since it is difficult to say exactly what the added costs of any vehicle accident are because of changes in vehicle weight, and the victims are largely unorganized, this has become an easy method for GWD/GEF supporters to reduce energy use. And with some large vehicle manufacturers beholden to government for their recent very survival, the manufacturers have chosen not to mount an effective resistance. As usual, lower income groups are the chief losers because of increased accident losses and increased prices for vehicles. Even some higher income households will be affected, however. Those with a strong interest in safety, such as for their children, will either not be able to purchase this added safety or have to pay dearly for it.

Efficiency standards for light bulbs are another example of the inefficiency of GWD/GEF-inspired regulations. In this case, the legislation was initiated in Congress rather than environmental organizations, but is now defended by the Obama Administration, which is preventing any repeal effort in Congress along with Liberal Democrats. As usual, the winners are the well-to-do and light bulb manufacturers, and the losers are consumers, particularly lower income ones. The fluorescent bulbs now required make some sense for lights that are used for long periods of time but not for those used only briefly or infrequently since their cost is far higher and need to be used enough to compensate for their higher initial cost before they burn out. There is also an important environmental safety problem with the fluorescent bulbs because they contain mercury which can harm those in areas contaminated if the bulbs are not handled with extreme care. But consumers are no longer allowed to exercise any judgment, but should be allowed to.

One possibility might be to require by statute that Federal regulatory decisions demonstrate a market failure and be based on applying the scientific method using independent analysts. The requirement to demonstrate a market failure is part of existing Executive Orders but has not been very effective since the regulators are also usually the authors of the analyses.

Although it may be hard to influence proposed Federal regulations, there is also a need for non-government research into the various attempts to impose non-related regulations whose real purpose is to increase the costs of using coal and other fossil fuels. An example may be the proposed mercury emission regulations for power plants that serve no useful purpose according to some skeptics.[467]

The Obama Administration is attempting to achieve a significant increase in energy efficiency through a drastic tightening of energy efficiency standards, particularly for motor vehicles. They strongly believe that this should be done by government fiat rather than by consumer choice. Presumably they believe people are too dumb or careless to follow their own self-interest by reducing their energy bills and must be told exactly what they are to use in order to promote the Administration's GEF/EWD doctrine whether citizens want it or not. This shows a complete lack of faith in the choices of individual consumers, who may desire one type of bulb or type of car over others but are not allowed to exercise their own judgments based on market prices. This decreases their quality of life as well as insulting their intelligence.

Change Needed: Rescind EPA Endangerment Finding and All Regulations Based on It

At the behest of the Obama Administration, EPA has become reckless and appears to be almost totally in the control of the CIC. Since the Endangerment Finding is based on invalid science, as discussed in Chapter 9 and Appendix C, the most straightforward approach is to rescind it as well as any regulations dependent on it. This would have the major advantage that EPA would no longer be subject to environmental lawsuits to expand its GEF activities beyond what even it wants to do, as explained in Chapter 10. Withdrawing the Endangerment Finding might take a while, but is the cleanest way to solve the energy policy problems created by this Administration. If so, hopefully due consideration would be given to independently gathering and analyzing the available evidence.

---

[467] Soon and Driessen, 2011.

This would slow down the process, of course, but would be well worth it to set the record straight and show how such an analysis should be done. Unfortunately, this is unlikely to occur during the remaining years of the Obama Administration.

Most of the other straight forward approaches that Congress might take, like reversing the 2007 Supreme Court decision, are very likely to vetoed by Obama. Another possibility would be vetoes of individual regulations under the Congressional Review Act, but this would also require Presidential signatures. The most viable effort would probably be for Congress to deny funding to implement any CO_2 related regulations, but even that would leave the Endangerment Finding on the books to cause trouble later.

Change Needed: Encourage Environmentally Responsible Development of Energy Resources on Federal Lands and Remove Restrictions on Oil and Gas Exports

Private enterprise has discovered very extensive, widely distributed source of natural gas on private land underlain with shale formations. Access to it is available through horizontal drilling and fracking, which involves fracturing the rock around the horizontal bore hole to facilitate the flow of the gas into the well. As a result, US natural gas prices have dropped so much that gas can compete with coal in many uses for energy production and could be exported if only the US Government would allow it. The price may go even lower since shale gas is partly a by-product of the much more valuable shale "tight oil." The US was already the Saudi Arabia of coal, so it can truly be said that the US currently appears to be the best endowed country in the world in terms of energy resources.

There is no shortage of fossil fuels now or in the indefinite future, despite the claims to the contrary by GEF adherents. The only question is how much it would cost to extract and bring to market each of the sources using non-conventional extraction techniques. There is considerable dispute over this, but there can be no doubt that there are substantial sources of state and Federally-owned reservoir oil which are currently off limits for political and environmental reasons. There can also be no doubt that there are immense quantities of non-conventional oil under state and particularly Federal control.

Green ideology emphasizing energy scarcity/energy saving is clearly not in accord with reality or consumer preferences. But still GEF activists attempt to deny access/use through legal harassment and promote resource use restrictions on particular energy resources such as coal and oil.

US energy resources are abundant, but US conventional oil has not been able to compete in recent decades because of the availability elsewhere of very inexpensive oil in large reservoirs, particularly in the Middle East. But in recent years increasing use in developing countries and collaboration by the owners of these reservoirs to maintain much higher prices in the world market have resulted in escalating prices. The result is that the US has had the opportunity to develop its somewhat more expensive sources of oil as well as many other energy resources. Wind and solar, the energy sources so strongly favored by environmental organizations are hopelessly expensive and unreliable at the current time, and probably for the indefinite future. And at recent lower prices for oil and natural gas, wind and solar are even less competitive. In many cases they do not even reduce CO_2 emissions since additional fossil fueled "spinning reserves" are needed to balance their erratic production in order to avoid very costly electrical network crashes or intentional blackouts. The principal exception appears to be areas where there are abundant hydroelectric sources that can also be varied minute by minute to balance the supply and demand for electricity where substantial wind and solar sources are used. Only the Pacific Northwest has the abundant hydroelectric resources required for this purpose in the continental US.

Although there is considerable shale gas and oil on private land in the US, there is an obvious need to open more energy resources owned or controlled by Federal and state governments where markets will use them, and this would help to maintain the very recent low world prices for oil and for gas in the US. There is also a need to greatly reduce legal harassment by environmental organizations to restrict the use of domestic fossil fuel resources, particularly coal. This harassment is both costly and time consuming in terms of bringing new energy supplies to market.

There has long been increasing concern about USG budget deficits, the US balance of payments, and US unemployment. Development of US coal, oil, and natural gas resources on public lands could make a significant contribution to all three of these problems, but is being shunned at every step of the way by the Obama Administration and its environmental supporters. Such resource development results in a net inflow into the US Treasury rather than the direct outflow of subsidies required to develop wind and solar alternatives. The substitution of domestic sources of fossil fuels for imported ones improves the US balance of payments and national security. The development of these added resources provides real jobs (as opposed to imaginary and heavily subsidized "green jobs") for Americans which cannot be outsourced to other countries since the energy sources are located in the US. There is no better, cheaper, or more effective way to achieve all these objectives than by encouraging development of fossil fuel energy resources on suitable Federal lands. Such development, of course, requires careful attention to effective environmental safeguards by both private developers and government overseers.

One of the curiosities of oil and gas policy in the US is the Federal ban on export of both natural gas and crude oil. The US now has a surplus in natural gas and is increasing its oil output substantially. The unnecessary and unproductive bans on exports need to be removed.

Change Needed: End Research Predominantly Supporting Only CAGW

As was discussed in Chapter 11, the US had already spent over $39 billion for climate research, predominantly for CAGW-oriented research, through 2013. For the reasons discussed in Chapter 9 this has been a failure since it has not resulted in a valid justification for CAGW as it was intended to do and also did not make significant advances in understanding Earth's climate. Now I do not rule out that reforms could be instituted that resulted in a more balanced research effort including funding for both CAGW and skeptic research and use of the SM. But this did not prove to be the case even during the eight years of the George W. Bush Administration and appears even less likely under the Obama Administration. So the easiest solution is to avoid funding CAGW research until major reforms have been made that will insure a balanced approach. Perhaps the minimum that should be done is to add funding for a comprehensive program of skeptic research.

Change Needed: End US Support of IPCC and US NCAs and Aid to Less Developed Countries Based on Climate Change Criteria

The IPCC and US National Climate Assessments are little more than CIC propaganda activities rather than objective scientific assessments.[468] Since reform appears unlikely and difficult, the time has come to end their US Government support by zeroing out their budgets from the US Government. Since the IPCC was formed the US has been the largest funder—over $43 million.[469]

This would probably be one of the more effective reforms. It is becoming increasingly clear the UN is not capable of preparing impartial assessment reports that use the scientific method to determine scientific validity. So the best course would seem to be to terminate all US support for this activity rather than try to reform it. It is always possible that funding would be continued by private funding sources, but at least it would not have the implied US Government label. The UN's latest effort in its 2012 Rio+20 meeting in Rio de Janeiro appears to be even less based on science

---

[468] See Chapter 9; also Easterbrook, 2013, and Booker, 2013.
[469] Arnold, 2014.

and more defuse than their earlier climate efforts and does not deserve support either.

Change Needed: End US Government Big Brother GEF Indoctrination and Suppression of Free Speech by Public Employees on Scientific Issues

It is not widely known, but taxpayer funds have been used to support GWD/GEF/EWD and particularly CAGW indoctrination during the Obama Administration. The amounts involved are sometimes relatively minor compared to the expenditures on wind turbines and solar power, but I strongly object to the principle. I do not see why taxpayers should have to pay for their own indoctrination with Green/GWD/GEF/EWD ideology. Most of these programs are hidden in obscure parts of the Federal budget, but that just makes them harder to find, not more acceptable.

The Obama Administration has been trying to indoctrinate the public with its climate ideology in many ways and through a variety of agencies. This includes material on agency websites, advocacy of climate "education,"[470] exhibits in National Parks,[471] and grants by the National Science Foundation. One example is the $700,000 NSF grant to The Civilians, a New York theatre company, to finance the production of a show entitled "The Great Immensity,"[472] "a play and media project about our environmental challenges."[473] A second example is a $5.7 million grant to Columbia University to record "voicemails from the future" that paint a picture of an Earth destroyed due to climate change.[474] A third example is a $4.9 million grant to the University of Wisconsin-Madison to create scenarios based on America's climate actions on climate change

---

[470] McMahon, 2014.

[471] Carlin, 2010.

[472] Their website can be found at http://theGreatImmensity.org as of 2013.

[473] Harris, 2010.

[474] Watts, 2014.

including a utopian future where everyone rides bicycles and courts forcibly take property from the wealthy.[475]

The general approach pursued by the Administration for arts and education-related climate propaganda appears to be very similar to the similar propaganda campaigns by Soviet and Eastern European governments to promote their political ends. Direct legislative prohibition is probably the best approach towards ending such attempts to influence public opinion using taxpayer dollars in the US. Although it is not clear whether GWD/GEF/EWD is an ideology or a religion, it is certainly not science, the purview of NSF. Continued funding for this purpose is outrageous and all too reminiscent of the book *1984*. Why should taxpayers pay for their own indoctrination with GWD/GEF/EWD ideology/religion and invalid science? Why should Government carry out indoctrination of its own citizens? We are much closer to the situation portrayed in *1984* than most people realize.

Equally reprehensible is the Obama Administration's attempts to prevent the expression of free speech on purely scientific issues by public employees and to discourage public employment of those who hold other viewpoints. This was discussed earlier in Chapter 6.

Change Needed: Encourage Fracking and Other Non-conventional Techniques for Oil and Gas Extraction

As a result primarily of private enterprise the US oil and gas industry has developed and applied a number of new techniques that result in substantially increased output, particularly from "tight" oil and gas located in shale formations. The use of these techniques has been regulated primarily at the state level in the US and by national governments in Western Europe. In areas with favorable geolocy in the use of these techniques has developed rapidly in the last decade except in New York and California, but is being fought aggressively by the environmental movement on the

---

[475] Harrington, 2014.

grounds that fracking is dangerous, even though no documented cases of adverse effects have been found. Since underground resources are owned by the government in Western Europe, very little exploration and almost no production has taken place there. Both the US and state governments as well as the Western European countries need to encourage the development of these techniques on an urgent basis. The Obama Administration says that it supports such use but has not been aggressive in encouraging it.

Change Needed: Avoid US Ratification or Support for a New International Climate Protocol or Understanding

Since there is no real possibility that the Earth will undergo CAGW, that governments could do anything about it if there were, or that the less developed countries will agree to meaningful CO_2 emissions reductions, there is no use or need for a new treaty or understanding. Without US support, the effort will inevitably fail.

Summary: How Can We Best Get Out of This Energy Mess?

The best way out of the energy mess created by GWD/GEF in Western Europe and now actively being promoted by the Obama Administration in the US is simply for governments to get out of energy production and use decisions except where their own resources are involved or there are uncompensated external effects. In the US there is a great need for government to do a better job of providing access to fossil fuel resources on public lands. The market has done a far better job on private lands and is highly likely to continue doing so if allowed to do so by government.

Longer-Term Reforms Required to Avoid Repetition

The Obama Administration has based its proposed GHG regulations on bad science and economics despite extensive efforts over the last 35 years to build a Federal regulatory system that would prevent this from happening. Unfortunately, the authors of the basic US environmental laws did not really contemplate that EPA would be captured by environmentalists.

Longer-term reforms are even more difficult than the short-term reforms discussed in the first half of this Chapter, but must be faced. Given the huge expenditure of very scarce citizen skeptic effort in recent years it would be a shame if the same effort had to be mounted the next time a very expensive regulatory control scheme is proposed with no real scientific, economic, or legal basis. Even if we should be so lucky as to escape huge damage this time, we might not be so lucky next time. It is hard to raise longer-term problems in the heat of a short-term battle, but it is certainly time to start. The rest of this chapter explains what I believe needs to be done.

The longer-term problems are deep-seated and difficult to understand let alone resolve. The problem is that big government science has lost touch with the very essence of science—the comparison of hypotheses to explain the real world with data which would verify or deny their validity. Several commentators have discussed aspects of this problem, including Richard Lindzen[476] and Arthur Robinson.[477] Lindzen has summarized the problem as follows:[478]

> In brief, we have the new paradigm where simulation and programs have replaced theory and observation, where government largely determines the nature of scientific activity, and where the primary role of professional societies is the lobbying of the government for special advantage.

One of the things shown by the climate caper is just how easily the extensive safeguards developed over the last 35 years to try to insure that good science and economics are used in US Federal

---

[476] Lindzen, 2008.
[477] Robinson, 2010.
[478] Lindzen, 2008, p. 4.

regulations can be sidestepped by a President determined to use bad science and economics. I am particularly concerned by this since many of my efforts at EPA were to promote and improve these safeguards. Similarly, the legal safeguards built into the laws and Constitution of the US are also being seriously compromised. Citizens and taxpayers need to be concerned since it is they that will suffer the consequences if these abuses are not curbed. There can be little doubt that at least parts of the EPA air office were captured by the CIC during the George W. Bush Administration, and that all of EPA and parts of other agencies such as Interior were captured by the CIC starting early in the Obama Administration. This must not be allowed to happen in the future if the public is to be well served by the EPA or other regulatory agencies. Agencies captured by the interest groups which they are supposed to regulate are incapable of taking into account the broader public interest inherent in their responsibilities. Obviously much stronger safeguards need to be put in place so that this can never happen again.

Under current rules, there is simply no forum in which opponents to proposed regulations can effectively present the scientific, economic, or even the legal aspects of an agency decision. The courts defer to the agencies on the science and try to avoid any consideration of it. Congress is too divided to effectively address the merits of particular proposed regulations. And the Executive Branch can and has effectively prevented the opposition from receiving a real hearing and then implemented what it wanted to do in the first place. The results are what we now see with regard to EPA's proposed climate regulations as well as other air pollution regulations in recent years. The problem is how to keep this from being repeated in the future. The remainder of this Chapter will present some ideas for accomplishing this.

Longer-Term Reform: Insure that Wide Spectrum of Hypotheses Receive Funding in Government Funding of Research and Scientific Assessment

One of the fundamentals of good science is that a variety of viewpoints be heard and investigated. Probably because of the widespread view within government that the GWD was the only science that deserved funding, virtually all of the over $39 billion in research funding for climate/global warming over the last two decades has been devoted to GWD-inspired research. As long as this continues to be the case, the outcome will remain the same—more GWD-inspired research. The only way out at this point is either to stop government-sponsored research or to somehow mandate that multiple lines of research be followed. Given that the Obama Administration is unlikely to move off its GWD/GEF/EWD ideology, the only source for such a mandate is Congress.

One of the most important effects of such a change would be that the strong incentives that have existed for many years in the academic world to support the GWD would come to an end. During all this time if one wanted USG financing it paid to be a GWD supporter. Very few if any grants were made to those espousing alternative viewpoints. This continues to this day. If this economic incentive were broken much of the support for GWD-related research in the academic world would almost certainly disappear almost overnight. Grant-seeking academics would realize that their non-GWD ideas could be funded regardless of whether they supported GWD and would be free to support whichever side they believed had the most scientific merit rather than which side controlled the grant money. So this might be the most important change that could be made in terms of changing the intellectual climate for GWD in the US. Unfortunately, instituting such a change would be far from easy. The question of how to judge which side a proposal should be judged as supporting is far from clear and would require careful drafting and oversight by Congress.

527

Longer-Term Reform: Insuring Independence and Objectivity of EPA

Congress has granted sweeping regulatory powers to EPA on the assumption that it would be responsibly used. This has taken many years and with some exceptions such as DDT and recent air regulations was based on a long record of what I believe has been widely perceived to be generally careful regulation where needed. And as a result of how EPA was formed in 1970 by President Nixon, EPA reports directly to the President and thus has no independence from the Executive Branch like some regulatory agencies. This means that if an Administration wants to use its power to determine regulations, it can impose exactly what it wishes to do subject only to the Congressional Review Act and Congress' powers of appropriations, both of which have proved ineffective so far in preventing Obama from doing what he wants with regard to EPA. Nixon may have created this situation in order to avoid asking Congress to approve a more complicated reorganization and perhaps because he did not want to dilute his authority over regulations. Until the Obama Administration, EPA tended to be more often prevented from following some of its regulatory preferences by the Administration in power rather than being told to do more than was dictated by science and economics. The Obama Administration reversed this by urging EPA to do much more than it might otherwise have done. This is probably legal, but has greatly increased the already substantial tensions between the political parties. Some may recall that Obama promised to bring the parties together but at least in the case of EPA he has done much to create a new area of intense discord.

I am personally convinced that the Obama Administration will not voluntarily give up its leverage through fiat regulation by EPA, perhaps because of its commitment to radical environmentalism, and has done anything short of endangering its own reelection to promote the worldwide GWD/GEF cause. The political risks of doing so appeared to me to be large, particularly in the 2010, 2012, and 2014 elections, but were apparently regarded as secondary by the Administration to their GWD/GEF goals except in 2012. It seems obvious to me that they tried to postpone most of the more radical CO_2 and other regulations until after the 2012 elections were over while carefully building the regulatory groundwork for moving very rapidly afterwards. For a long while I could not understand why Obama was endangering his own reelection in 2012 and those of his party in Congress by pushing such obviously unpopular measures as cap and trade and EPA GHG regulations. It is now clear that it was a calculated gamble that he could get reelected as long as his intentions were hazy and most of the worst proposed EPA regulations had not been publicly announced or put in place.

The 2012 proposed CO_2 power plant regulations are a case in point. According to *Politico*,[479] the draft EPA regulation included existing power plants when it was reviewed by OMB, but OMB removed this and everyone has since denied that this was their intent. Obviously it was; but the Administration did not want to raise this unpopular measure during a presidential campaign. The President's party did lose the 2010 elections to the House of Representatives which was indeed a major loss in terms of Obama's ability to enact his program. It is clear that some of these losses were due to his environmental proposals, but some were probably also due to the weak economy.

EPA is under the direct control of the Administration, unlike some other regulatory agencies, which are under a variety of bi-partisan approaches even though they often deal with much less partisan issues. During the Obama Administration EPA has been compromised by its political masters who in turn have been primarily responsive to the demands of environmental groups, with the result that it has lost its credibility as an independent arbiter of science and regulations. Insulating it from direct control by the Administration would help to correct this at the cost of making the

---

[479] Martinson, 2012.

President less accountable for EPA's actions. But increased insulation needs to be seriously considered to avoid a repeat of the problems during the Obama Administration. One approach might be to create a bi-partisan commission to run EPA. The result would hopefully be that only those regulations agreed to be necessary and useful by representatives of both parties would be adopted. It would also result in greatly increased transparency, which has been so much needed during the Obama Administration despite their claims to the contrary.

Longer-Term Reform: Prohibit EPA from Using "Secret Science" and Speculative "Co-benefits"

The EPA Air Office has increasingly used "secret science" and speculative "co-benefits" as part of its "playbook"[480] to justify ever more stringent, inefficient regulations that increasingly represent regulatory overkill. They use these approaches in order to increase the alleged environmental and economic benefits of proposed regulations and the chances that these regulations will be adopted. For reasons discussed in Chapter 11 and elsewhere, there is no excuse for using research results which cannot be independently verified by outside reviewers or which are speculative in nature. Health "co-benefits" should not be used to justify regulations unless it can be shown that they meet the same requirements for solid science that the primary benefits do and that all data and research used to determine these "co-benefits" are publicly available. In other words, the time has long since come to abandon this and other parts of the EPA air "playbook."[481] Previous regulations that were dependent on "co-benefits" for their economic justifications need to be reviewed and reanalyzed in light of this proposed

---

[480] US Senate, 2014; the EPA air "playbook" is discussed in Chapter 11.
[481] US Senate, 2014.

requirement, particularly if they used alleged benefits from controlling PM$_{2.5}$.[482]

Longer-Term Reform: Revise Clean Air Act to Allow Consideration of Costs

Perhaps the major shortcoming of the Clean Air Act is that many of the major provisions of it do not allow consideration of the economic costs involved. As the regulations become ever more stringent and expensive, and the benefits ever more speculative and hard to prove, these restrictions need to be revised or even removed so that costs of control can be taken into account.[483] Pursuing minor and improbable risk reductions at huge costs is not reasonable or a prudent use of resources.

Longer-Term Reform: Analyze Proposed Regulations Using a Hormetic Dose Response Model

As discussed in Chapter 11, EPA as a matter of long standing policy routinely analyzes most dose-response functions using a linear no-threshold model. This exaggerates the benefits from pollution control and is an important part of the EPA air "playbook" used to justify ever more stringent air regulations. Now that the cost of new controls is rapidly increasing, it is long since time to reexamine this assumption. My recommendation would be to require that such analyses should also be done using a hermetic dose response model as well as other appropriate ones such as a threshold model, as discussed in Chapter 11. EPA should then be required to compare the results from the various models, including the economic costs involved, and analyze why one is more likely to correspond with reality for each pollutant considered for regulation.

---

[482] A list of these regulations can be found in *ibid.*, Appendix A. PM$_{2.5}$ is discussed in more detail in Chapter 11.

[483] These issues were explored in more detail in Chapter 11. The Utility MACT (mercury) rule is an illustration of the need for this change.

Longer Terms Reform: Require that EPA MUST Carry Out Independent Analyses

EPA has reached its GHG Endangerment Finding by relying primarily on the UN IPCC reports and other reports based on them without any independent review thereof. This saved considerable time and probably controversy at the time, but resulted in no real analysis of the scientific issues. Primary responsibility for scientific assessments should not be assigned to non-EPA groups and particularly non-US groups for determining the basis for EPA regulations in my view. The actions of EPA in largely adopting the IPCC conclusions and US reports based on the them is basically an abdication of its responsibilities to independently review the science and economics of proposed regulations.

Longer-Term Reform: Reduce Incentives for EPA Managers to Follow Administration

Besides the normal bureaucratic controls, the pay of all EPA executives and senior analysts are directly determined by EPA management within certain boundaries determined by Congress and the President. This is unlikely to lead to independent action or thought by these crucial civil service employees. Yet independent analysis is desperately needed if EPA is to reflect good science and economics rather than science determined by their political masters.

Longer-Term Reform: Periodically Review and Evaluate Existing Major EPA Regulations

Although ambient air quality regulations are subject to periodic review, many programs do not currently have a process in place to review the continuing scientific justifications for some major classes of pollutants. Given the changing science for a number of these regulations, this results in the possibility that regulatory actions will continue in place when they may no longer be needed or when stronger action is justified. Perhaps the place to start is the regulations concerning stratospheric chlorofluorocarbon control. Some existing legislation requires periodic reviews of pollution control levels, but the need is for program reviews at a much broader level to determine what is being accomplished and whether it is worth continuing. Such reviews may have to done by other agencies.

Longer-Term Reform: Require that EPA Base Its Decisions on Using the Scientific Method

EPA claims to select alarmist GHG science primarily on the basis of alleged peer review of assessments.[484] Yet one thing that is clear is that there are numerous cases where the IPCC peer review requirement was not implemented as written.[485] EPA says that these deviations are of no importance because nothing crucial to the CAGW hypothesis has been questioned. So EPA wants it both ways. It wants to select on the basis of peer review, then argue that peer review failures are of no importance because substance was not compromised by the failures. Obviously EPA should have done its own analysis, selected which studies to use on the basis of which correspond with observable reality and the scientific method, and performed peer review using reviewers from all spectra of opinion, not just those that support their desired outcome. It is evident that EPA was under orders to move much too rapidly for such an independent review to take place, although even a careful review may well have resulted in the same decision since the political policy makers would still have pushed their views. These decisions have now been approved by the Supreme Court and need to be avoided in the future by legislative action to make EPA more independent. Careful gathering of the evidence might have provided

---

[484] Knappenberger, 2010.
[485] Laframboise, 2010; Knappenberger, 2010

independent evidence concerning the Endangerment Finding, but cannot insure that politicians will consider it.

Given the death of cap and trade legislation in the US Senate in 2010, the short-term outcome of the US debate on actions that allegedly might reduce climate change may rest primarily on what the USEPA manages to actually do. So it is of some importance what criteria EPA claims to be using in determining the scientific merits of its Endangerment Finding. One criteria that should not be used is consensus, which has no useful role in science.

Longer-Term Reform: Transfer Responsibility for Preparing Economic Analyses Out of Offices Which Prepare Regulations

The responsibility for economic analyses of proposed regulations has always been with the EPA program offices (air, water, toxic substances, etc.) that have the responsibility for preparing new regulations. Although these analyses are reviewed by the economic office in EPA's policy office this review has apparently not been sufficient to prevent biased benefit-cost analyses from being prepared, particularly in cases where the regulations are being prepared at the direction of the White House. If these benefit-cost analyses are to be of any real effectiveness, it is vital that the responsibility for the analyses rest with as independent an organization as possible from those preparing the regulations. The obvious first step to take is to move the primary responsibility to the central economics office. This is still within the jurisdiction of the Administrator but would allow those preparing the analyses to at least express reservations concerning the economics of proposed regulations.

Longer-term Reform: Require Congressional Approval of All Proposed Regulations above a Specified Cost

The House of Representatives passed the Regulations from the Executive in Need of Scrutiny (REINS) bill in 2013.[486] This approach would require that any Executive Branch regulation costing more than $100 million obtain explicit House and Senate approval before becoming effective. In the case of the Senate it might even require 60 votes out of 100. This would result in a huge decrease in regulations and the speed with which they are approved. It would effectively greatly strengthen the Congressional Review Act, which allows expedited rejection of proposed regulations but requires an unlikely Presidential signature since the regulation would not have been prepared if the President opposed it. This bill would prevent future regulatory overkill but at the cost of a much slower regulatory process. Although not an ideal solution it may be the best available realistic solution to restoring sanity to the US regulatory regime.

Longer-term Reform: Split Responsibility for Data Gathering and Data Interpretation

There is no other approach that will remove the risk that data interpreters might try to change the data to fit their interpretations, as has repeatedly happened with climate data.[487] The latest report by Anthony Watts shows what can happen if there is no such separation.[488] The proposal by the Obama Administration to create a National Climate Service would only further decrease the diversity of thinking required to better understand our complicated climate and alter data to fit the ideology of those in power.

---

[486] US House of Representatives, 2013.
[487] D'Aleo and Watts, 2010.
[488] Watts *et al.*, 2012.

Longer-term Reform: Abandon Precautionary Principle

Many radical environmentalists support the use of the precautionary principle, which argues that when an activity raises suspicions of harm to the environment or human health, precautionary measures should be taken even if the relevant cause and effect relationships have not been established scientifically. In their view this justifies making assumptions as to the risks posed by various alleged environmental problems even though there is no real data to judge what the risks might be. They believe that the "Principle" exempts any need to await the development of such data and justifies their rush to governmental intervention. Suspicion of an environmental problem is enough to justify governmental action to correct such alleged problems even though detailed study might show that there is no need for the proposed action. This shows mainly arrogance rather than rationality.[489]

Longer-term Reform: Get Government Out of the Choice of Energy Sources

Federal and state governments should not be involved with selecting energy sources. Such decisions should be based on economics and justifiable pollution standards rather than politics, assuming that market prices reflect any major market failures involved. No subsidies/taxes/preferences should be imposed unless so justified.

The national interest is not to satisfy various interest groups but to supply reliable energy at the lowest possible cost after fully taking into account the adverse environmental effects of each alternative. Energy is an input to many products and services; higher cost energy increases the cost of these products and services, and makes them less competitive in the marketplace.

With some significant exceptions, the choice of energy sources

---

[489] Bailey, 1999.

to be built and used has traditionally been primarily decided in the United States by the market rather than by the Federal Government. In order to insure that the environmental effects of each source are fully taken into account it is necessary that the full social costs of these adverse effects be taken into account by those making the choices. This is ideally done by including these environmental costs in the price of energy from these sources.

Contrary to current efforts to impose arbitrary "renewable energy standards" and Federally mandated CO_2 emissions controls on power plants, the most economically justifiable approach would be to adjust existing taxes on various energy sources to account for the adverse environmental effects that we know exist and allow the market to work its will. In the ideal this effort might start by bringing up to date the findings by me and several colleagues[490] and adding the adverse environmental effects of solar, wind, and biomass sources.

Longer-term Reform: Rethink How Federal R&D Decisions Are Made

In hindsight it is increasingly clear that the warmist GHG control effort has been largely funded by government itself, particularly the US Government, during both Democratic and Republican administrations. Government often uses a peer review approach sometimes even involving current grantees or those sympathetic to current funding trends to decide how to direct new research funding in the same area. Rapid progress in the sciences is dependent on the availability and testing of a wide variety of hypotheses, which is less likely to happen when research proposals are funded on the basis of this type of "insider" peer reviews. What is needed is a broad set of hypotheses to be explored. This may be one reason that the climate research program became trapped in a narrow, unproductive area. I believe that most of the over $39

---

[490] Viscusi *et al.*, 1994.

billion spent by the US Government for AGW-related climate research was wasted. This should have been evident to at least one of the grant-making agencies over recent decades.

Longer-term Reform: Provide Forum Where Stakeholders Can Present Their Scientific and Economic Cases

One of the problems during EPA's consideration of climate and other recent air pollution regulations is that opposition stakeholders have not been given a public forum to present a well-developed case. Yes, there have been public hearings, but even these have been politicized by EPA and environmental groups to crowd out any serious discussion of the issues. In addition to such hearings I believe that there needs to be a separate forum in which detailed presentations of the science and economics of proposed regulations by credentialed specialists representing major stakeholder groups can be made; hopefully this would allow a meaningful discussion of the issues that is simply not possible in three minutes at a crowded public hearing that no one listens to. Since the courts are not willing to seriously consider either the science or the economics of regulations, there needs to be a forum for these topics involving major stakeholders.

13
In Summary

I SUPPORTED THE environmental movement in the 1960s and early 1970s in their successful efforts to prevent construction of two dams in the Grand Canyon and in other often largely land use campaigns of the day and by serving as the Chairman of the second largest chapter of the Sierra Club. I continue to support them in many of their efforts to reduce real pollution and other environmental problems where they advocate real solutions to real problems. I supported them because I believed that their objectives would improve both the environment and the economy. As a result of their success in the Grand Canyon and other campaigns and the public environmental enthusiasm of the 1970s, the environmental movement attracted considerable public support and contributed to creating many new laws and environmental regulatory agencies. In a surprisingly short time, many of the major pollution problems were substantially reduced or at least greatly improved in the US.

I am not arguing for abandonment of the pre-Brundtland Report ideals of the environmental movement but rather abandonment of goals that are not supportable on sound economic, scientific, and legal grounds, such as their current climate campaign. What is needed is not an end of the movement to improve the environment but rather a major course correction to bring it back to reality.

The radical environmentalists have built a fantasy world to support their claim that the world's climate will change disastrously unless fossil fuel energy use and production is immediately greatly curtailed by government fiat. When someone challenges their ideology, they brand the challenger a "denier." When they proclaim that CO_2 emissions can and must be reduced by 80%, they conveniently forget that the current standard of living in the developed world is based on the use of fossil fuels. As a result of Congress' rejection of their cap and trade "solution" in 2009-10, they and the Obama Administration proposed that EPA should impose many of the provisions of their cap and trade bill through regulatory fiat despite the lack of a scientific, economic, legal, or constitutional basis for doing so. When surface temperature records fail to support global warming, friendly governments "adjust" the data so that they (somewhat) do. When skeptics present data showing that the alarmists' science is invalid, they are attacked for being in the pay of polluters. When less developed countries say that they do not want to reduce fossil fuels use and lose their chance to escape poverty through economic development made possible by using more fossil fuels, they are told that hundreds of billions of dollars will be given to them by the developed nations if they just play along. The president of the US and the Administrator of EPA call emissions of a trace atmospheric gas (CO_2) essential to life on Earth "carbon pollution" even though it was at vastly higher levels during most of Earth's history with no evidence of adverse effects.

When all this unreality is pointed out, the cooperative mass media attack the authors of the heresy, Democratic senators and representatives demand that anyone employing such individuals supply communications from and information on funding received by the heretics, and the President sanctions mass public attacks on elected officials who question his science and energy policies. Living in a world of unreality is a symptom of madness. Radical environmentalism has gone very, very mad.

In Summary

The radical energy environmentalists who have come to dominate most environmental organizations in recent years have used their earlier widespread support to promote a "solution" to climate problems that cannot be achieved in the way proposed. This "solution" would actually make the world worse off both economically and environmentally (as discussed in Chapters 9, 10, and 11).

They believe that fossil fuel use must be greatly reduced primarily because of three potential drawbacks from its use— conventional pollution, alleged climate effects, and the fact that fossil fuel resources are not replaced by nature over human life spans. Fortunately, none of these drawbacks pose a major problem to either humans or the environment and can be overcome without remaking Western society to greatly reduce fossil fuel as radical environmentalists insist is necessary.

Conventional air pollution can and is being controlled in many developed countries—and over-controlled in the US. Happily, there is just no credible evidence that increasing human-caused CO_2 emissions are anything but beneficial to humans and the environment, and especially to plants, which can make good use of all they can get. The key hypothesis that alarmists use to make their case for the alleged adverse effects of increasing CO_2 on climate is invalid according to the scientific method. And the non-renewable characteristic of fossil fuels is not a serious drawback if human ingenuity is allowed to operate through relatively free markets and not curtailed by unjustified government regulations. The huge increase in oil and natural gas output as a result of the recent expansion of the use of fracking and horizontal drilling has again shown this to be the case.

In the last few years the radical environmentalists' efforts have gone well beyond opposition to carefully selected energy use projects in the US which have particularly adverse environmental effects to active efforts to reduce emissions from whole classes of energy facilities, particularly coal-fueled power plants, and more recently natural gas production and the building of pipelines and natural gas export terminals. Their methods now include using civil disobedience, and some affiliated groups oppose capitalism as a system despite its obvious success where it has been used.

The anti-fossil fuel objectives of the radical environmental movement are promoted by a climate-industrial complex (CIC) composed of the principal groups that would benefit from bringing this about. The CIC is a very large enterprise with scientific, propaganda, governmental, and other arms funded mainly by taxpayers, wealthy radical environmentalists, and suppliers of renewable energy systems.

Despite widespread CIC propaganda to the contrary, the CIC is far better funded than the skeptics, who lack any funding by government and lack the tight internal coordination that characterizes the radical environmentalists' activities. The CIC funding by wealthy radical environmentalists is very tightly controlled by multiple interlocking foundations, which serve the purpose of hiding its sources and providing wealthy radical environmentalists with tax deductions for their contributions even though substantial resources appear to be going towards influencing legislation and election campaigns.

Although great progress has been made in solving the more visible conventional US pollution problems, more remains to be done in selected, usually less visible, areas such as non-point water pollution control. Instead of concentrating on these real pollution problems, the radical energy environmentalists have jumped far ahead of the science, which refutes their catastrophic anthropogenic global warming (CAGW) hypothesis, and advocate remaking energy supply system at immense cost by drastically reducing CO_2 emissions. The Obama Administration with a little help from the radical environmentalists, has now attempted to rewrite the Clean Air Act through imaginative but illegal interpretations of it which would allow EPA to require states to implement CO_2 reductions by fuel switching in the electric power sector or by enacting legislative changes outside plant fences that would reduce CO_2 emissions using a number of the approaches rejected by Congress in 2009-10. The

In Summary

Administration is also reportedly trying to circumvent the Constitutional requirements for Senate ratification of a new global climate treaty.

The CIC initially claimed that their "solution" of reducing CO_2 emissions would reduce global warming, then when there was no significant warming, prevent climate change, and when there was nothing more than normal climate change (climate has been changing for much of Earth's history and will undoubtedly continue to do so), to reduce extreme weather. One advantage of their latest objective from their viewpoint is that they can try to point to every unusual storm as "proof" of the need to reduce CO_2 emissions, despite the lack of any objective basis for this.

Unless the President and EPA are stopped by the courts or Congress, much worse than what EPA has so far proposed to reduce CO_2 emissions from power plants is very likely to follow because EPA has opened up a legal hornet's nest which will allow radical environmental organizations to achieve a stranglehold on the US economy by forcing EPA to restrict the use of fossil fuel energy to any extent they desire. The President has already promised China further US CO_2 emissions reductions beyond those proposed by EPA by 2025. Whether this increment, if it should occur, would be taken entirely from the electric generation system or other areas is uncertain. Their next target appears to be the oil and gas industry, with proposed Federal regulations on methane emissions from oil and gas production.

So in response to unvalidated and much too warm computer models (discussed in Chapter 9) and more recently unsupported assertions as to the effects of increasing CO_2 on extreme weather events (as discussed in Chapter 6), the CIC and the Obama Administration have acted in ways that would give radical energy environmental groups effective control of energy generation and use, the meaning of the Clean Air Act, and some of the Senate's treaty approval rights through unilateral and in many cases unconstitutional Executive Branch decisions (as discussed in Chapters 6 and 8). The real danger is that freedom of speech and the rule of law will be seriously compromised, not the alleged adverse (but actually positive) effects of increasing CO_2 levels that are being used to argue for rebuilding the Western energy supply system at huge expense in dollars and greatly decreased reliability.

Although I continue to support economically, scientifically, and legally justified pollution control, the CIC's radical energy-related efforts will have the effect of reducing economic growth and development and hurting the financially less well-off worldwide rather than reducing measurable, scientifically verifiable, and damaging pollution. In the unlikely event that their prescriptions actually resulted in less global warming, this would harm the environment, the economy, and the human population by slightly increasing the major real climate risk—a new Little or full Ice Age.

So like many radical political movements, the environmentalists have become fanatics, in this case left-wing fanatics. Like most such groups, they have now exceeded the limits imposed by US laws and the Constitution and have resorted to trying to rewrite laws and even the Constitution's separation of powers, undermining the cooperative federalism with the states which is the basis of most Federal environmental laws, and trying to circumvent the Constitution's treaty approval requirements. What started out as an obsession with reducing fossil fuel energy production and use has ended up creating what is likely to be lengthy legal and political battles over the meaning of the Clean Air Act and the Constitution. Even if these climate proposals represented good economics and science, which they do not, I believe that they need to be defeated in order to reduce the increasing threat posed by an imperial Presidency. The laws should be written and treaties approved by Congress, not an increasingly all-powerful Chief Executive intent on imposing his/her will over that of elected representatives in Congress.

The public does not appear to be generally aware of what the Obama Administration is trying to do in its climate/energy policies in terms of what John Boehner has called "aggressive unilateralism" but I would characterize as dictatorial behavior. They need to be since this represents a serious threat to American democracy and the rule of law. The public is somewhat more aware of Obama's related efforts in health insurance and immigration, where he is trying to impose his policies by attempting to override laws passed by Congress and the framework imposed by the Constitution. If the Obama Administration is allowed to get away with all this, future presidents may follow their example. The longer it takes for the public to push back against Obama's imperial executive orders, the more extreme he is likely to become to take advantage of what he apparently believes are loopholes in the Constitution that allow him to rule directly without interference by Congress or the laws it enacts. It is important that the Constitution be upheld even in the face of alleged global environmental threats; in this case the threats are bogus anyway. Yet the mainstream press has almost never even discussed the problem. They need to.

The CIC includes sympathetic Western governments, much of the Western climate science establishment, the liberal mass media, left-of-center politicians, and producers of "CO_2 emission-reducing" products in addition to most environmental organizations. Using a sophisticated, massive, and sometimes even government-financed propaganda campaign akin to that portrayed in the novel *1984*, the CIC has proposed to vastly reduce energy use, which is one of the major requirements for economic development and growth, and emissions of a trace gas necessary to life on our planet, carbon dioxide. They claim that their agenda is based on science and sometimes even that science "demands" it. The evidence they have offered for these claims ignores the scientific method, the basis for determining what is and is not valid science. As discussed in Chapter 9, their "science" would have been shown to be invalid if the scientific method had been applied, but it has not. Even if significant reductions in CO_2 emissions could actually be achieved in the Western developed world, they would have little if any effects on climate, particularly since emissions in less developed countries are increasing rapidly.

Current Risks Posed by Radical Energy Environmentalism in the US

Sixty percent of Americans regard themselves as active environmentalists or sympathizers, down from about 70% in 2000. What many of these 60% may not realize is that the environmental movement has changed radically since the 1970s when it achieved popular acceptance. I experienced the bipartisan support enjoyed by the movement in the 1960s and 1970s as an activist and Sierra Club leader, spent 38 years as a senior EPA employee, and in 2009 argued against the bad science being promoted by the Obama EPA to justify EPA regulations on CO_2 emissions. EPA's 2009 Endangerment Finding led in 2014 to an EPA attempt to unilaterally rewrite the Clean Air Act to allow much of the failed Waxman-Markey (Cap and Trade) Bill in Congress to be imposed directly as EPA regulations. Even if this effort should be successful, it will not change the climate or extreme weather in any measureable way even though Obama has proclaimed it will. It will simply increase the rates paid for less reliable energy, with lower income Americans bearing most of the burden along with the slow recovery of the US economy.

This book explains how the environmental movement of the 1970s has been hijacked by a radical fringe who are attempting to change the way of life for all Americans through EPA regulations drafted in response to a blueprint prepared by the movement itself and aimed particularly at forcing those who do not accept their radical ideology to abide by it by Federal fiat.

EPA's new proposed power plant regulations on CO_2 emissions will happen unless a way is found to stop them. There are only three possible ways this could happen: A president who will withdraw or modify them, aggressive action by both houses of Congress to prevent their implementation, or rejection by the courts. Obama appears unlikely to withdraw the regulations before he leaves office in 2017. Rejection by the courts has not proved a dependable strategy for skeptics to date, but the Obama

Administration proposals are becoming increasingly outrageous legally. Congress is the only somewhat dependable avenue in the near term, but it appears to be paralyzed by its extreme divisions. A number of environmental groups tried largely unsuccessfully to influence the outcome of the 2014 mid-term elections, probably for fear that Republican control could lead to rejection or a reduction of recent proposed CO_2 regulations.

Costs, Benefits, and Contradictions of What the CIC Proposes

The CIC in general and the radical energy environmentalists in particular have apparently lost perspective on the effects of what they have proposed with regard to energy production and use. This is evident from examining what the major effects have been and will be if CO_2 emissions were controlled as the CIC and the Obama Administration propose, separated into costs and benefits:

Costs:

1. *Adversely affect lower income people in the developed world by substantially raising the cost of the energy they use since it makes up a greater proportion of their outlays compared to higher income people.*

2. *Greatly decrease the chances that lower income people in the less developed world can improve their standard of living and instead increase the risk of their deaths through malnutrition and continued poverty as a result of higher prices they pay for energy produced from renewables and food used as fuel.*

3. *Large transfers of wealth by the developed world to the less developed world in order to bribe the less developed world to accept (2), which they would otherwise not accept, and probably would not anyway even if such bribes were paid.*

4. *Reduced national security of the US and its allies*

because of restrictions on fracking, particularly in Western Europe, and on US exports of natural gas.[491]

5. *Very large subsidies paid by taxpayers and ratepayers in the developed world for building "renewable" energy sources that are economically inefficient and are highly likely to be scrapped when the subsidies end.*

6. *Undermining the rule of law and usurping Congress' constitutional role by attempting to rewrite the Clean Air Act and imposing regulations that Congress has already rejected in order to bulldoze their "War on Coal" into effect.*

7. *Publicly attacking and harassing those who differ.*

8. *Misdirecting attention to the possibility of future warming rather than the much greater risk of global cooling in coming centuries/millennia.*

9. *Endangering the public health and welfare, which EPA is charged with protecting, by requiring the shutdown of coal fired power plants necessary to preserve the stability of the US electric grid and to supply the electric power needed under adverse weather conditions.*

Benefits:

1. *The alleged possibility that global warming/climate change/extreme weather might be reduced by an unmeasurably small amount at an unknown time in the future by government-mandated reductions in CO_2 emissions based on unvalidated climate models that have consistently substantially overestimated future temperatures, failed to predict the 17 or more year halt in increases in global temperatures, and contain major assumptions based on invalid science as determined by*

---

[491] Technically, the cost (4) imposed by prohibiting fracking in much of Western Europe applies only to the CIC and not to the Obama Administration since the Administration publicly supports fracking, although not exactly enthusiastically.

the application of the scientific method. Emissions reductions by developed countries would require the active support of the less developed world to have any measureable effect, but this is contrary to their desires for economic development.

2. *The alleged health "co-benefits" in the US from CO$_2$ reductions from implementing the EPA proposed regulation on existing coal power plants which are grossly overstated because they make extreme assumptions concerning the effects of pollutants at very low concentrations and are based on questionable EPA-sponsored research that EPA cannot or will not release to the public. Western European countries have not alleged such benefits, and some have actually increased their use of coal for producing electrical power.*

I find the trade-offs involved to be overwhelmingly negative; in fact, I find the CO$_2$ emissions reduction proposals to be reprehensible in terms of their adverse impacts on mankind, the economy, the rule of law, and the environment. From every perspective government-imposed CO$_2$ emissions reductions in the world as a whole and the two proposed EPA CO$_2$ regulations on power plants in the US should be abandoned, along with other unjustified EPA regulations such as mercury from power plants and further tightening of the ozone standard.

Radical energy environmentalism is a series of contradictions. Even if their objectives could be achieved, the contradictions guarantee that very little if anything would be achieved. These contradictions include the following:

- *Any reductions in CO$_2$ emissions by Western countries will be more than offset by increased emissions from less developed countries such as China and India. The CIC/UN strategy is for the developed countries to pay (bribe) the less developed countries to accept substantial* *reductions, but this is fundamentally unacceptable to them and is likely to fail even if funding could be found.*

- *The only realistic way to significantly reduce global CO_2 emissions would be to substitute natural gas for coal wherever possible, which would preserve the vital economic gains from substituting fossil fuel energy for human time and effort while reducing CO_2 emissions because of the chemical composition of the fuels. Doing this by government edict may result in minor reductions in some Western nations but will ultimately fail as long as coal's cost is lower in most areas.*

- *The only way around this is to make energy from natural gas lower cost than from coal, but this would require pushing fracking, which the radical environmental movement and many sympathetic governments strongly oppose.*

- *Every proposal for reducing CO_2 emissions by radical energy environmentalists involves higher energy costs through either added taxes or higher energy prices and/or lower grid reliability. The left of center political parties that support reducing CO_2 emissions often generally claim that they represent the interests of lower income citizens who will be most hurt by these higher taxes and prices.*

- *The major energy sources that the environmental movement approves of, wind and solar, have a significant adverse effect on birds and some other wildlife which the environmentalists normally strongly oppose, but now largely try to ignore.*

- *Decreased living standards resulting from increasing energy costs will decrease the capability of society to withstand global warming/climate change/extreme weather when and where they occur.*

In Summary

- *The radical environmentalists' favored energy and food production policies result in an expansion of cultivated areas at the expense of reduced wilderness, often involving the use of more easily damaged marginal land, increased hazards for birds and bats, and construction of unsightly, low density energy production and transmission facilities over large areas.*

These and other contradictions inherent in the climate movement's doctrines will ultimately prove to be its undoing, but it appears likely that considerable damage to the economy, the environment, the rule of law, and even the Constitution may be done in the meantime.

In brief, the best response to the minor warming, minor climate change, and normal to lower-than-normal extreme weather is to have the courage to do nothing except careful, unbiased scientific research until humans have a much better understanding of weather and climate, and to leave energy use and generation decisions as much as possible to the markets to decide. Government intervention and regulations have only made matters worse and will continue to do so. Germany, Spain, and some other Western European countries provide an unfortunate example of what is likely to happen.

To date there has been comparatively little public debate concerning the energy fantasy promoted by the CIC, probably because most citizens have little interest in it. This may change starting in 2015 as a result of the domination of the US Senate by the Republican Party as a result of the 2014 midterm election. The new Senate Majority Leader was reelected in part based on his support for solving the problems created by the CIC in his state of Kentucky. The Obama Administration is pressing ahead with what appears to be an all-out campaign to "save the World" from the effects of one of the most essential building blocks of life on Earth, carbon dioxide, which it deliberately mischaracterizes as "carbon pollution" for propaganda purposes.

More Detailed Picture

The CIC's proposed "solution" to the alleged climate problems created by human activity—government-imposed CO_2 emissions reductions—have proved astronomically expensive, not achievable in the real world, have no measureable effect in lowering global temperatures (which it would actually be better to increase moderately for the resulting benefits and decreased chances of disastrous global cooling), but instead have created an energy mess and have proved detrimental to the environment, the economy, and national security where they have been tried, especially in Western Europe. The Obama Administration now proposes to bring all these "benefits" to the US using blatantly illegal EPA reinterpretations/rewriting of the Clean Air Act which have little or no basis in law or the principle of separation of powers enshrined in the Constitution. The CIC is are also trying to sell these efforts by increasingly misleading statements if not deliberate falsehoods about what their proposed regulations would accomplish if implemented.

As explained in Chapter 11, the EPA proposal on CO_2 emissions from existing power plants would be very expensive, and have no or even negative climate benefits, all for the possibility of minor and largely speculative health "co-benefits" based largely on "secret science" which EPA cannot or will not release to the public. It would also result in a much less stable electric grid, particularly during severe warm and cold episodes, with increased risks of blackouts and/or load shedding with the possibility of enormous economic losses in a society very dependent on reliable electric power. In non-academic terms, EPA is proposing a monumental illegal boondoggle at the expense largely of lower income groups and at the direct behest of environmental organizations and economic beneficiaries.

It follows that the world would be better off without these radical energy environmental organizations and the other components of the CIC which have promoted bad science, bad law, and unworkable "solutions" to non-existent problems. They are doing much more harm than good and do not deserve the public, foundation, Obama Administration, Western European government, and EU support they are receiving.. By pursuing goals which have no basis in science, economics, or reality they set back, not advance, the environmental cause. This book explains why I believe their present agenda is so detrimental and how I came to change my views concerning the movement as a result as they increasingly shifted towards supporting their radical energy agenda in recent decades.

As discussed in Chapters 9 and 10, their agenda is detrimental environmentally because it diverts attention from and, if anything, would make the major longer-term environmental problem facing the world (catastrophic global cooling) worse in order to address alleged human-caused warming and extreme weather, which government can do nothing about in the way proposed. In addition the CIC advocates use of very high cost, unreliable energy sources that kill very large numbers of birds and bats, help the insects they pray upon, and in the case of wind turbines, impose high noise pollution, and damage landscapes where they are built. Their attempts to block fracking are detrimental to their goal of lower CO_2 emissions since lower cost natural gas resulting from its use results in its substitution for coal-based energy and hence significantly lowers CO_2 emissions. I do not believe CO_2 emission reduction itself (without reduction in other actual pollutants) is practical or useful, but they claim it is.

As discussed in Chapter 11, their radical environmental energy agenda is detrimental economically because it is economically inefficient even under the best technical assumptions, leads to slower economic growth, wastes very large resources on solutions that will have no effect on extreme weather, climate change, or on global warming, hurts primarily lower income groups throughout the world, and decreases the availability and reliability while increasing the cost of energy, one of the primary determinants of economic growth and development. And in the unlikely event that it were ever successful it would reduce moderate global warming that is beneficial rather than harmful.

Hundreds of billions of dollars are currently being wasted each year in the developed world on renewable energy facilities destined to be abandoned sooner or later when the subsidies run out since they cannot compete economically with fossil fuels in most circumstances. The CIC is trying to force their agenda on the less developed world, where it is already harming the poorest of the poor. Their agenda results in the death of people who would have lived if they had access to the benefits of fossil fuel use, and higher food costs which result from use of land that could be used to growing the food needed to survive but are now devoted to growing fuels. Hopefully the less developed countries will not be so foolish as to adopt the CIC agenda, but if they should, these damages can be expected to grow enormously. Continued pursuit of the CIC agenda has negative benefits and enormous costs.

Finally, even if the CIC "solution" could be achieved in the way they have proposed, significant reductions in global CO_2 emissions are practically unachievable given the understandable desire of less developed nations, the major and rapidly increasing source of increased CO_2 emissions, to substitute energy use for human labor for purposes of economic development, and their resulting refusal to try to reduce emissions. The idea that further reductions of US CO_2 emissions beyond what the US has already "achieved" and the decimation of the US coal industry, as the Obama Administration proposes, would somehow lead the less developed world to renounce their commitment to rapid economic growth, low energy costs, industrialization, and avoiding social unrest to follow our proposed lead is pure fantasy. Yet this is one of the crucial false assumptions under which the Obama Administration is proceeding with its climate campaign. They claim that the US will show the way towards government-imposed reductions in CO_2 emissions and the less developed countries will forget their well-founded opposition and approve a new protocol which the US would be unlikely to ratify—all because EPA has imposed burdensome CO_2 emission reduction regulations on the US energy industry. This is utter nonsense on top of the nonsense about the need for reducing CO_2 emissions in the first place. Among other problems is the strong possibility that even if the less developed countries agreed to reduce their emissions, they just might not carry out such commitments as they might make, and there would be little the developed countries could do about it.

The US private sector has (inadvertently) shown the way to actually reducing CO_2 emissions even though it would probably be better not to reduce them: Encourage fracking to increase natural gas production, which lowers prices, which leads to substitution of gas for coal, which greatly reduces CO_2 emissions. The great beauty of fracking is that its development and use came about not through government planning or subsidies but by the operation of a relatively free market on private land. And it is government, particularly in some US blue states and Western Europe that is attempting to stop it. Most other nations are not following our lead on fracking and other technologies to obtain natural gas from shale deposits, and are foregoing not only a major new source of oil and gas but also a large reduction in CO_2 emissions that would actually work. But instead of emphasizing these new technologies as the way forward (assuming that reductions in CO_2 emissions are actually needed, of course) for less developed countries, the Obama Administration is trying to reduce US emissions by reducing the use of coal and substituting "renewables" in the US and promising vast payments to less developed countries that go along with their scheme of government-imposed CO_2 reductions.

An alternative policy would be to strongly encourage fracking and other new and useful technologies, particularly on Federal lands and in foreign countries with suitable shale resources, so as to drive natural gas prices lower and hasten substitution for coal, and then encourage less developed countries to follow our lead for their own profit and reduced emissions. Working with market forces rather than imposing government solutions contrary to market forces has a far better record of success. Better yet would be for government to just worry about other problems that it must solve, such as national security. Providing abundant natural gas to Western Europe could pay major dividends to the national security of our European allies.

The radical environmental organizations are endangering national security by advocating reductions in fossil fuel energy production in the US and the European Union and impeding exports of US energy resources to Europe, which reduces the ability of Western Europe to become independent of Russian fossil fuels. Such advocacy has resulted in a reduced ability by Western Europe to use more effective economic sanctions against Russia when they invade and annex parts of neighboring countries, as they did in 2014. Unless the US or others can rapidly replace gas imports from Russia, the EU cannot avoid the possibility that Russia will use the continued flow of its currently critically needed supplies to blackmail the EU countries into agreeing to further annexations. The EU countries need to move rapidly to develop their natural gas resources and the US needs to facilitate the export of some of its new natural gas and oil to Europe in the interim as replacements for Russian supplies. Russia appears likely to try to further squeeze Ukraine's natural gas supplies; the environmental groups are effectively assisting them in this. Many environmental groups oppose natural gas development, especially fracking, and exports but apparently fail to understand the importance of such actions as one of the most important steps that can be taken to fight Russian expansionism in Eastern Europe.

Since 2006 the US has led the world in reducing CO_2 emissions and the UN's attempts to negotiate new international climate treaties have led to little more than an increasing crescendo of demands for payments from the developed to the less developed world and no serious commitments to reduce emissions by the less developed world, which is the only way that meaningful reductions in CO_2 emissions could possibly be made through international negotiations. This basic problem has existed since the Kyoto Protocol was negotiated (which exempted the less developed countries from reducing emissions) in 1997, and there is no realistic solution to it in sight as of 2014. The idea that the developed world can persuade the less developed world to accept permanent second class economic status by agreeing to meaningful, verifiable, and enforceable reductions of their CO_2 emissions shows a total lack of understanding of these nations and their aspirations. Of course these countries will insist on very large developed country payments, which developed country taxpayers will (correctly in my view) refuse to pay. And even if they did, most less developed countries would still not (correctly in my view) meaningfully reduce CO_2 emissions. The proposed payments for extreme weather damage are even less justified (if such is possible) and are even less likely to be funded by Western taxpayers.

A very simplified version of the central question is whether it is more important to encourage less developed countries to reduce CO_2 emissions, as the IPCC and CIC maintain, or to encourage them to greatly increase their economic development, which would necessarily involve greater use of fossil fuel energy. With such development the less developed countries would be better able to adapt to extreme weather events and climate change and provide more adequate food and medical care for their people. And with such development, there would be greatly increased interest in and concern about the environment as economic research has long shown. I have always supported such development, particularly now that it is clear that decreasing CO_2 emissions will make very little if any change in warming/climate/extreme weather and that moderate warming would be beneficial.

So the environmental "sustainability" ideology does not provide a rationale for reducing fossil fuel use, IPCC science is invalid, the CIC "solution" damages the environment, economy, national security, and poor people, and the Obama Administration's implementation of CO_2 reductions is undermining the rule of law and the Constitution. Their "solution" will not measurably reduce CO_2 emissions because of opposition by the less developed world, and would not have a measurable effect on climate or weather even if it could be carried out. The alternative of urging other countries to follow our example by allowing fracking to increase their natural gas supplies (and lower energy prices and CO_2 emissions) has been given some lip service but largely ignored by the Obama Administration. The even better alternative of ignoring the CO_2 non-problem and leaving it to the market to decide most energy issues has not even been considered, but should be.

And this fantasy CIC "solution" is far from free. Studies put the current global expenditures to promote the CIC climate agenda at about $1 billion per day; the environmental organizations want to expand this greatly. The Obama Administration has already given billions to foreign countries for climate mitigation and adaptation and promises many more. It has even agreed at the November, 2013 UN meeting in Warsaw to a "loss and damage" mechanism to reimburse less developed countries for damages incurred from extreme weather events. The Obama Administration is doing everything it can to assist the CIC in their efforts. Without its support the campaign would probably have died by now given that Australia, Russia, and Canada, and to some extent Japan and some Eastern European countries have all backed away from some future UN climate efforts.

For all these reasons I can only describe the radical environmental energy agenda as mad since it is not subject to reason or judgment. The environmental movement has lost all contact with reality as it has pursued the fantasies of government control of global warming/climate change/extreme weather and the peak oil hypothesis. A similar statement can be made concerning the Obama Administration's climate policies, although it has publicly supported fracking by trying to claim credit for it, which many environmental organizations do not.

The CIC bases their efforts on trying to scare people about problems that governments can do little or nothing to solve in the ways they have proposed, such as global warming, climate change, and extreme weather. Their advocacy has also disregarded the fundamentals of civil discourse and scientific integrity since they apparently believe that their dubious and changing "cause" is more important than scientific discourse based on observations and hypotheses based on them. They continue to make major factually inaccurate statements concerning their proposed policies and dismiss anyone who expresses contrary views with disparaging epithets and *ad hominem* attacks.

All this would be primarily of academic interest except that the CIC has obtained political backing in many Western countries from left-of-center political parties and used that to change these countries' energy policies in ways that will result in all the adverse environmental, economic, and national security outcomes outlined in this book. Although I have spent most of my career in the service of environmental improvement, the time has come to find better guides to an improved environmental future rather than following the energy fantasies now being promoted by most environmental organizations and the rest of the CIC.

One of the problems with the CIC agenda is that it assumes that government-dictated energy policies will be superior to market determined ones. That this is not the case is illustrated by the experience of a number of Western European countries that have adopted the CIC agenda. Instead of solving problems they have greatly increased energy prices, reduced the availability and reliability of electricity, and in the case of Germany and Britain have ended up forcing the building of new coal plants and diesel backup generators, respectively, which do not satisfy even the CIC's objectives. If governments simply stayed out of energy decisions not involving government-owned resources, urgent national security objectives, or actual proven pollution problems and let the market decide how to meet energy needs, everyone except the CIC would be much better off, including the environment. This is what has largely happened in the US until recent years, but is now greatly endangered by the radical environmental-inspired actions of the Obama Administration to use EPA regulations and continuing subsidies to distort the energy markets as well as the continuing effects of preferences enacted in some largely blue states for renewable energy sources. Given the now evident very adverse European experience it is hard to understand why the Obama Administration would want to repeat these failed European experiments or why some blue states would want to continue down their current energy-interventionist path. It is paramount that the government get out of most energy decisions in order to restore economic efficiency and improve the environment. Continuing the downward spiral of growing government intervention now so evident in Western Europe will only make the situation worse.

The UN IPCC claims that it cannot think of a natural explanation for the very limited global warming in the late 20th Century. In Chapter 9 I suggest some natural phenomena that would provide such a natural explanation for not only the late 20th Century but possibly even the last 10,000 years. One of the real problems is that the IPCC was directed to only look at human-caused climate change and has all too faithfully avoided taking a broader look at other possible causes. The results can best be described as biased and disastrous. Bad science and economics lead to bad policy and a monumental waste of resources.

At Best a False Alarm

Concern over catastrophic anthropogenic global warming (CAGW) is increasingly being seen for what it was and is—at best a false alarm, or if done on purpose, which it probably was, a hoax, or even a scam since many of the perpetrators are evidently gaining financially from their hoax. It probably represents the worst scientific scandal of the current generation because of the widespread attempt to corrupt the scientific method, the support of most of the scientific establishment and many Western governments in doing so, and the scurrilous methods used to denounce those who point out that observations refute CAGW.

None of the IPCC's scary climate predictions have been confirmed after 23 years and five IPCC reports, and there is no reason to believe that they will anytime soon since the current cooling phase of the 60 year Pacific Decadal Oscillation cycle is likely to last until the 2030s. The Canadian, Australian, Russian, and some European governments are pulling out of the false alarm/hoax/scam, in some cases because of the unpopularity of the very rapid rise in energy prices resulting from increasing reliance on wind and solar, but the US is actually getting more deeply into it thanks to the Obama Administration's misguided efforts. Western Europe, Japan, Canada, and Australia all drank the cool aid much longer and more deeply, but their parliamentary systems of government can reverse course much more rapidly than the US, and appear to be moving in that direction. Surprisingly, the US has reduced CO_2 emissions more than any other country in recent years through the operation of market forces rather than the Government-imposed emissions reductions advocated by the CIC. There would seem to be a lesson from this.

The aggressive commitment of the Obama Administration to the climate false alarm/hoax/scam and the electoral weakness of the Republican opposition in the Senate prior to 2015 in opposing Obama on this issue meant that the false alarm/scam/hoax had continued life in the US through the 2014 election. So while the conservative US political system has saved the US from the worst of the false alarm/scam/hoax so far, the Obama Administration hopes to impose its radical environmentalism using EPA regulations before its end in 2017. The economic costs of the green agenda in Western Europe have been enormous and seem likely to mount in the US over the next few years given the relatively free reign that the Administration has set for itself on this issue.

In order to implement its EPA fossil fuel regulatory scheme, the Obama EPA is trying to use the Clean Air Act as a vehicle to implement its radical energy policies without obtaining the approval of Congress, which Obama knows he cannot get. In doing so, they propose to reinterpret parts of the Act in ways that are illegal and challenge the separation of powers written into the US constitution. Unless these attempts are struck down by the courts, which may take some time, the US appears headed for a legal and even a Constitutional debate, particularly since the Obama Administration appears to be pursuing related "aggressive unilateral" approaches (to use John Boehner's term) in other policy areas such as immigration and health insurance.

A Misreading of Geologic History

The CIC may have done serious harm to the future of humanity and many other life forms on Earth. They have misdirected the world's attention from the most important longer-term environmental problem we face—a new Little and/or full Ice Age—to alleged problems which would not be significantly affected by their proposed "solution," or might even make it worse. Little thought and no preparations have been made to address the major environmental problem. Unfortunately, through ignorance or disregard of the science or self-interest the alarmists appear to have gotten their science exactly backwards. Fortunately, we may have several centuries or possibly even as much as several millennia to solve the major problem but this is very uncertain, the solutions are not easy, are made worse by the alarmists' ideology, and may require all the time we have left to solve.

Unfortunately, surprisingly few understand this risk. Even if we should escape during this millennium, as we may have barely done in the last one, the risk appears to be even greater in the following millennium, and for each succeeding millennium as long as the Milankovitch Cycles continue in their downward or cooling phase, which appears certain.

Barring new knowledge, the only hope for avoiding this fate is to increase Earth's heat retention so as to decrease the cooling trend evident for over 3,000 years, consistent with experience with Earth's interglacial periods over the last million years. The possibilities for doing this appear to be two-fold: increase atmospheric carbon dioxide levels or use geoengineering when and if necessary to increase temperatures, particularly in the northern latitudes, during critical future periods of low temperatures. Increasing CO_2 levels is not likely to be very effective because of the very small and uncertain effect CO_2 actually has on global temperatures, but might at least move climate change control in the right direction.

Since the current hiatus or downtrend in global temperatures will probably continue until the 2030s, and the next upturn may not be evident until the 2040s, the CIC will somehow have to keep the public interested longer than they probably can, particularly if the public realizes that the proposed US CO_2 reductions will end up personally costing them dearly. So although the alarm should by all logic have long since disappeared, surely it will at least gradually die out over the next two decades, or hopefully very much sooner. The CIC are trying to rebrand their goal in the US as extreme weather control, which they may hope will provide endless opportunities to claim benefits from CO_2 reductions even when there is no warming.

The new science discussed in Chapter 9 and Appendix C is based largely on the rediscovery that Earth's very complicated climate is primarily determined by natural variations in the orbital properties of the Earth, in the sun and other astronomical bodies, and in natural temperature-regulating physical processes, particularly in the tropics, and ocean oscillations rather than by the relatively puny activities of man singled out by radical environmentalists. What the rediscovered science implies is that we need to be doing just the opposite of what has been recommended by the CIC. It is better both environmentally and economically to increase carbon dioxide emissions (assuming that the emissions do not or can be controlled so as not to significantly increase conventional pollutants that are actually damaging), not decrease them as all these groups urge. There is very little if any risk of runaway warming (the basic fear being promoted by the alarmists), which is not known to have ever occurred during the last half million years at our current stage of the ice age cycle.

Good Environmental Policy Requires Good Science, Economics, Law, Transparency, and Scientific Integrity

One of the basic messages of this book is that science should not be dictated by government fiat to achieve political agendas; instead, great care needs to be taken to insure open dialogue between and within government, the scientific community, and the public so that scientific issues are freely and openly discussed, are not subject to government-sponsored propaganda, and that anyone in or out of government is encouraged to freely express their views on scientific issues. As we have seen in the climate issue, science is easily corrupted by large financial incentives offered by government to support one side of a scientific issue, and strong safeguards are needed to prevent this from happening on other issues. Scientific openness and integrity are prerequisites for avoiding future scandals similar to CAGW.

Another basic message is that environmental issues should be left to the economic marketplace rather than the political/regulatory process whenever they do not involve significant market failures requiring government intervention or involve government-owned resources. A third is that good economics and science, especially the scientific method, should be the basis for regulatory policy, not environmental, scientific, or political ideology. A fourth is that regulatory agencies should carry out careful and independent analyses of the science and economics on which their regulations are based, and not depend on analyses done by others.

Unfortunately, despite many virtuous sounding statements of good intentions, the Obama Administration has violated each of these precepts as a result of their attempts to impose their radical environmental ideology on the American public. This has led to scientific scandals similar in kind (although not so far with as much loss of life in the developed countries) to those in the Soviet Union and Nazi Germany, the destruction of EPA's scientific credibility, the waste of hundreds of billions of taxpayer and ratepayer dollars, slower recovery from the Great Recession, increasing unemployment in the coal sector of the economy, and decreased reliability of the electric grid. The attempts by the Administration to dictate scientific "truth" have reached alarming proportions and threaten the future of scientific inquiry by the US Government.

The use of incorrect, faulty, or speculative science and ignoring or distorting the economic efficiency of regulations only makes the situation worse by leading to remedies that decrease efficiency, will not remedy the problems addressed, and may cause major environmental problems as well. As the Obama Administration claimed in 2009, good science requires transparency, open discussion, and scientific integrity since there is a scientifically correct answer which is highly unlikely to be found otherwise, but just the opposite of what the Administration has actually done. Scientific issues should be decided by the use of the scientific method, not opinion surveys or a "consensus" of experts or the President. Using consensus, in fact, is the antithesis of science, which is based on skepticism and comparisons of hypotheses with the real world. When a scientific hypothesis fails to satisfy the scientific method it should be either discarded or modified, which the climate believers, including the Obama Administration, are unwilling to do with regard to their climate ideology. Economic analyses should be carried out in a non-partisan setting, not by government departments responsible for developing environmental regulations that have already been decided upon before the economic analyses are undertaken, which has long been the case despite the proliferation of economic analyses of proposed government regulations over the last 35 years.

Obama Administration's Radical Environmentalism Resulted in Missing Major Economic Opportunities

The Obama Administration's radical environmentalism has helped to keep it from what one might have thought would have been its primary preoccupation—creating jobs and economic recovery. Instead of actively promoting new technology for oil and gas production, which has resulted in avoiding serious adverse effects from the Great Recession in states which have pursued such development, the Obama Administration has promoted money-losing "green" stimulus spending. Encouraging oil and gas development could have been done fairly inexpensively with a high degree of confidence in a number of ways ranging from greatly increasing leasing of Federal oil and gas resources, encouraging the use of fracking and horizontal drilling to secure access to oil and natural gas long trapped in many shale formations, and rapidly approving natural gas export terminals and oil pipelines. With the much lower world oil prices in late 2014, such policy changes have become less effective but still worthwhile. Such development could, however, be used to supply natural gas to Western Europe to make Western Europe more independent of Russian supplies and discourage Russian expansionist plans. The primary opposition would undoubtedly come from radical environmental groups. And lower world oil prices resulting from the use of fracking will substantially reduce the US balance of payments imbalance and stimulate the non-oil production parts of the US economy.

In the Long Run Economic Development Is Helpful to Improving Environmental Quality

I have long believed that in the longer term economic development is in general supportive of environmental improvement, not its enemy. Prosperous countries are more likely to safeguard their environment than impoverished ones. So I believe environmentalists would do better to support economic growth in general while opposing selected ill-advised and particularly environmentally damaging development projects and policies based on careful analyses of their environmental effects, science, and economics.

In Summary

Impact on USEPA

Three types of major reforms would solve a number of problems at EPA; abolishing it is an alternative, but one I would not advocate. The first type of reform concerns governance; the second concerns its current agenda; and the third involves more closely involving Congress in the regulatory process.

Governance changes should encourage EPA to pursue regulations based on good science and economics rather than ideology and executive fiat. This would require increasing the independence of EPA from the White House so that EPA would be able to ignore the dictates of the White House concerning what the science and economics are. One possibility would be a bipartisan commission to run EPA with staggered terms with the chairperson coming from the President's party, somewhat similar to other, much less controversial agencies.

The second major change is to halt EPA's current determination to regulate emissions of CO_2, which would accomplish the opposite of what is needed at huge expense, and to rethink EPA's longstanding linear no threshold policy assumption concerning the effects of low dose pollutants so that its regulations improve health, not make it worse due to the hormesis effect. This should greatly decrease EPA's current efforts to regulate air pollutants now already at very low levels.

The third major change would be to affirmatively involve Congress more directly in major regulations. It currently has the authority to review and reject expensive regulations formulated by the Executive Branch, but this has rarely ever been used. A more expansive approach would be to require positive approval of major regulations before they could be implemented.

In Perspective

This has been a personal account of how I learned all this, why the mainstream press has managed to get virtually everything wrong and still does, how the US Environmental Protection Agency

567

attempted to hide the truth, and how the Obama Administration now proposes a tsunami of regulations to impose on the American economy to solve a non-problem that it cannot solve in the way it proposes, the science that makes this the case, and what I learned along the way about American government, politics, journalism, and environmentalism. These proposed government-imposed restrictions on energy production and use are not only not worth their huge cost, which would amount to a rate increase on anyone who uses fossil fuel-generated electric power and other selected targets, but would actually damage the environment.

Human use of the energy resources of the Earth was not planned or researched ahead of time; it was rather the "natural" and wonderful outcome of the operation of the free market over several hundred years. It turns out, however, that environmentally responsible use of energy is a source of hope and promise, not doom and gloom or a pretext to abandon a market-driven economy or limited government involvement with the economy. It is these last two policies which have brought the developed world the vast improvement in living standards it now enjoys and has made possible the improvements in the human and natural environment found in the developed world.

I believe I have been consistent—favoring economic growth and development as well as environmental protection. The radical environmental movement claims it is saving the world with policies that will not improve the environment in any perceptible way but will rather slow or even stop economic growth and development. In the less developed world this means continued poverty, disease, hunger, and deprivation. In the developed world it means lower growth in income and welfare, particularly among the less well to do. Making unmeasurable and likely imaginary reductions in global warming, "climate change," or extreme weather events is seemingly more important than making real reductions in poverty and malnutrition. It is the environmental movement that has changed, and no longer deserves my support or anyone else's as long as they support their current radical energy goals.

568

References

Ackerman, Frank, Lisa Heinzerling, and Raymond Massey, 2004, *Applying Cost-Benefit Analysis to Past Decisions*, Tufts University, July, available online at:
http://ase.tufts.edu/gdae/Pubs/rp/CPRRetrospectiveCBAJuly04.pdf

Agin, Dan, 2007, *Junk Science: An Overdue Indictment of Government, Industry, and Faith Groups that Twist Science for Their Own Gain*, Macmillan

Adler, Jonathan H., and Michael F. Cannon, 2014, Reining in Care—and the President, *Wall Street Journal*, July 23

Allen, Stephen J, and Julia A. Seymour, August, 2014, Who Watches the Watchmen? As Global Warming Theory Collapses, the Ignorant News Media Resort to Censorship and Name-calling, *Green Watch*, available online at:
http://capitalresearch.org/wp-content/uploads/2014/08/GW1408-final-for-posting-140725-140804.pdf

Alley, R.B., 2000, The Younger Dryas Cold Interval as Viewed from Central Greenland, *Journal of Quaternary Science Reviews*, 19, 213-26

Ambrose, Stanley H., 1998, Late Pleistocene Human Population Bottlenecks, Volcanic Winter, and Differentiation of Modern Humans, *Journal of Human Evolution*, 34, 623-51

American Association for Clean Coal Electricity, 2014, *Recent Electricity Price Increases and Reliability Issues due to Coal Plant Retirements*, February, available online at:
http://www.americaspower.org/sites/default/files/Electricity-price-spikes_Feb_2014.pdf

Andreen, William L., 2004, Water Quality Today—Has the Clean Water Act Been a Success? *Alabama Law Review*, 55, pp. 537-93, available online at:
http://papers.ssrn.com/sol3/Delivery.cfm/SSRN_ID556024_code2711 73.pdf?abstractid=554803&mirid=1

Andrews, Roger, 2014, Perception Trumps Reality—the IPCC Report on the Impacts of Climate Change, December 5, available online at: http://euanmearns.com/perception-trumps-reality-the-ipcc-report-on-the-impacts-of-climate-change/#more-5856

Archibald, David, 2011, David Archibald on Climate and Energy Security, February 12, available online at: http://wattsupwiththat.com/2011/02/12/david-archibald-on-climate-and-energy-security

Archibald, David, 2013, *Our Cooling Climate*, Washington, DC, September 16, available online at: http://www.iwp.edu/docLib/20130911_OurCoolingClimateCapitolHil l16thSeptember2013.pdf

Arnold, Ron, 2014, House Panel Takes Hard Look at UN Climate Change Process, *Washington Examiner*, May 27, available online at: http://washingtonexaminer.com/house-panel-takes-hard-look-at-un-climate-change-process/article/2548950

Arnold, Ron, 2014a, Tribes's Opinion that 'EPA's Clean Power Plan Is Unconstitutional Means More than You Think, December 23, available online at: http://blog.heartland.org/2014/12/tribes-opinion-that-epas-clean-power-plan-is-unconstitutional-means-more-than-you-think/

Arnold, Ron, and Alan M. Gottlieb, 1994, *Trashing the Economy: How Runaway Environmentalism is Wrecking America*, Free Enterprise Press.

Aristotle, 2007, *On Sophistical Refutations*, Translated by W. A. Pickard-Cambridge, eBooks, Adelaide, available online at: http://ebooks.adelaide.edu.au/a/aristotle/sophistical/

Bailey, Ronald, 1999, Precautionary Tale, *Reason*, April, available online at: http://www.reason.com/archives/1999/04/01/precautionary-tale

Bailey, Ronald, 2012, The Paradox of Energy Efficiency, *Reason*, November

References

Ball, Tim, 2013, Effects of Environmentalist and Climate Alarmist Crying Wolf Begin to Appear, December 10, available online at: http://drtimball.com/2013/2122/

Ball, Tim, 2014, IPCC Scientists Knew Data and Science Inadequacies Contradicted Certainties Presented to Media, Public and Politicians, but Remained Silent, March 21, available online at: http://wattsupwiththat.com/2014/03/21/ipcc-scientists-knew-data-and-science-inadequacies-contradicted-certainties-presented-to-media-public-and-politicians-but-remained-silent/

Barron-Lopez, Laura, 2014, Obama Pitches $1B Climate Change "Resilience Fund," February 14, available online at: http://thehill.com/blogs/e2-wire/e2-wire/198394-obama-to-announce-1b-climate-change-resilience-fund

Bastasch, Michael, 2013, Global Warming Gets Nearly Twice as Much Taxpayer Money as Border Security, *Daily Caller News Foundation*, October 28, available online at: http://dailycaller.com/2013/10/28/global-warming-gets-nearly-twice-as-much-taxpayer-money-as-border-security/

Bastasch, Michael, 2014, CO_2 Emissions Have Increased Since 2011 Despite Germany's $140 Billion Green Energy Plan, *Daily Caller*, April 9, available online at: http://dailycaller.com/2014/04/09/germanys-140-billion-green-energy-plan-increased-co2-emissions/

Beatty, W.A., 2012, The Religion of Global Warming, *American Thinker*, April 15, available online at: http://www.americanthinker.com/2012/04/the_religion_of_global_warming.html

Beckerman, W., 1992, Economic Growth and the Environment: Whose Growth? Whose Environment?, *World Development*, 20(1), April, 481-496

Bezdek, Roger, and Frank Clemente, 2014, *Protect The American People: Moratorium On Coal Plant Closures Essential*, Institute for Energy Research, available online at:
http://instituteforenergyresearch.org/wp-content/uploads/2014/06/Protect-the-American-People.-Moratorium-on-Coal-Plant-Closures-Essential.pdf

Bird, Joan, and John Bird, 2011, *Penetrating the Iron Curtain: Resolving the Missile Gap with Technology*, September 22, available online at:
https://www.cia.gov/library/publications/historical-collection-publications/resolving-the-missile-gap-with-technology/missile-gap.pdf

Blackmon, David, 2013, Behold the Bounty that Shale Oil and Natural Gas Have Wrought, *Forbes*, September 4,available online at:
http://www.forbes.com/sites/davidblackmon/2013/09/04/behold-the-bounty-that-shale-oil-and-natural-gas-have-wrought/

Bloomberg New Energy Finance, 2014, *Global Trends in Clean Energy Investment*, January 15, available online at:
http://about.bnef.com/files/2014/01/BNEF_PR_FactPack_Q4_Clean EnergyInvestment_2014-01-15.pdf

Booker, Christopher, 2013, Climate Change 'Scientists' Are Just Another Pressure Group: The IPCC and Its Reports Have Been Shaped by a Close-Knit Group of Scientists, All Dedicated to the Cause, *The Telegraph*, available online at:
http://www.telegraph.co.uk/earth/environment/climatechange/10356 276/Climate-change-scientists-are-just-another-pressure-group.html

Booker, Christopher, 2014, Lima Climate Talks: The Same Old Farce, *The Telegraph*, available online at:
http://www.telegraph.co.uk/comment/11291389/Lima-climate-talks-The-same-old-farce.html

Boone, Jon, 2010, Overblown: Where's the Empirical Proof? (Part IV), available online at:
http://www.masterresource.org/2010/09/overblown-part-iv/

References

Bradley, Robert L, 2000, *Julian Simon and the Triumph of Energy Sustainability*, American Legislative Council.

British Petroleum, 2013, *Statistical Review of World Energy June 2013 Workbook, Carbon Emissions Worksheet*, available online at:
http://www.bp.com/en/global/corporate/about-bp/statistical-review-of-world-energy-2013/statistical-review-downloads.html

Broeker, W. S., 2003, Does the Trigger for Abrupt Climate Change Reside in the Ocean or in the Atmosphere? *Science*, 300 (5625), 1519-22, June 6.

Brook, Barry, 2011, Lessons about Nuclear Energy from the Japanese Quake and Tsunami, April 11, available online at:
http://bravenewclimate.com/2011/04/07/lessons-nuclear-quake-tsunami/

Brown, Robert G., 2014, Real Science Debates Are Not Rare, October 6, available online at:
http://wattsupwiththat.com/2014/10/06/real-science-debates-are-not-rare.

Bryce, Robert, 2012, *The High Costs of Renewable-Electricity Mandates*, Manhattan Institute, February 10, available online at:
http://manhattan-institute.org/pdf/eper_10.pdf

Buchner, Barbara, et al., 2013, *The Global Landscape of Climate Finance 2013*, Climate Policy Institute, available online at:
http://climatepolicyinitiative.org/wp-content/uploads/2013/10/The-Global-Landscape-of-Climate-Finance-2013.pdf

Burnett, H. Sterling, 2015, Memo Reveals Bogus EPA Climate Strategy, January 28, available online at:
http://blog.heartland.org/2015/01/memo-reveals-bogus-epa-climate-strategy/

Calder, Nigel, 1974, Arithmetic of Ice Ages, *Nature*, 252(5480), November 15, 216-8

Caprara, Robert J., 2014, Confessions of a Computer Modeler, *Wall Street Journal*, June 8, available online at: http://online.wsj.com/news/article_email/confessions-of-a-computer-modeler-1404861351-lMyQjAxMTA0MDAwOTEwNDkyWj

Carlin, Alan, 1962, Letter from India, July 17, available online at: http://www.rand.org/about/history/wohlstetter/DL10703/DL10703. html

Carlin, Alan, and William E. Hoehn, 1966, Is the Marble Canyon Project Economically Justified?, printed in U.S. Congress, House, Committee on Interior and Insular Affairs, Lower Colorado River Basin Project, Hearings before Subcommittee, May 9-18, Part II, pp. 1497-1512.

Carlin, Alan, and William E. Hoehn, 1966a, Mr. Udall's 'Analysis': An Unrepentant Rejoinder, *ibid.*, pp. 1521-1535.

Carlin, Alan, and William E. Hoehn, 1967, The Grand Canyon Controversy–1967: Further Economic Comparisons of Nuclear Alternatives, House Committee on Interior and Insular Affairs, *Colorado River Basin Project*, Hearings before Subcommittee, May 2-5, 90th Congress, 1st Session, March 13-17, 1967, pp. 619-625. Also in U.S. Congress, Senate, Committee on Interior and Insular Affairs, *Central Arizona Project*, Hearings before Subcommittee, 90th Congress, 1st Session, pp. 489-497.

Carlin, Alan, 1968, The Grand Canyon Controversy: Lessons for Federal Cost-Benefit Practices, *Land Economics*, 44(2): 219-227, May. Reprinted in Charles J. Meyers and A. Dan Tarlock (eds.), *Water Resource Management*, Foundation Press, Mineola, New York, 1971, pp. 459-468. Earlier version printed in U.S. Congress, Senate, Committee on Interior and Insular Affairs, *Central Arizona Project*, Hearings before Subcommittee, 90th Congress, 1st Session, May 2-5, 1967, pp. 507-514. Also in House Committee on Interior and Insular Affairs, *Colorado River Basin Project*, Hearings before Subcommittee, 90th Congress, 1st Session, March 13-17, 1967, pp. 611-618.

References

Carlin, Alan, and Martin Wohl, 1968, An Economic Re-Evaluation of the Proposed Los Angeles Rapid Transit System, Paper P-3918, The RAND Corporation, Santa Monica, CA. Available online at: http://www.rand.org/pubs/papers/P3918.html

Carlin, Alan and R.E. Park, 1970, Marginal Cost Pricing of Airport Runway Capacity, 1970, *American Economic Review*, 60(3): 310-9, June. Reprinted in Peter Forsyth, Kenneth Button, and Peter Nijkamp (editors), *Air Transport Classics in Transport Analysis*. Edward Elgar Publishing, Northampton, MA, 2002, pp. 491-500.

Carlin, Alan, 2005, The New Challenge to Cost-Benefit Analysis: How Sound Is the Opponents' Empirical Case?, *Regulation*, Fall, pp. 18-23, available online at: http://object.cato.org/sites/cato.org/files/serials/files/regulation/200 5/9/v28n3-3.pdf

Carlin, Alan, 2007, Implementation and Utilization of Geoengineering for Global Climate Change Control, *Sustainable Development Law and Policy*, 7(2): 56-58, Winter, available online at: http://www.wcl.american.edu/org/sustainabledevelopment/documents /v7_2ClimateLawReporter.pdf.

Carlin, Alan, 2007a, Global Climate Change Control: Is There a Better Strategy than Reducing Greenhouse Gas Emissions?, *University of Pennsylvania Law Review*, 155(6), June, pp. 1401-97, available online at: http://www.pennlawreview.com/print/old/Carlin.pdf.

Carlin, Alan, 2008, Why a Different Approach Is Required if Global Climate Change Is to Be Controlled Efficiently or Even at All, *Environmental Law and Policy Review*, 32(3): 685-757, Spring, available online at: http://scholarship.law.wm.edu/cgi/viewcontent.cgi?article=1050&cont ext=wmelpr.

Carlin, Alan, 2009, *Comments on Draft Technical Support Document for Endangerment Analysis for Greenhouse Gas Emissions under the Clean Air Act*, March 16, unpublished report prepared for the US Environmental Protection Agency, available online at:

http://www.carlineconomics.com/files/pdf/end_comments_7b1.pdf, and additionally at:
http://www.openmarket.org/wp-content/uploads/2009/01/Carlin-Final-Report.pdf.

Carlin, Alan, 2009c, Why the Choice of Energy Sources Should Be Left to the Market after Externalities Are Taken into Account, September 13, available online at: http://www.carlineconomics.com/archives/483

Carlin, Alan, 2010, How Big Brother Is Using the national Parks and Other Agencies to Promote His Climate Religion Using Your Tax Dollars, available online at:
at http://www.carlineconomics.com/archives/947

Carlin, Alan, 2011, A Multidisciplinary, Science-Based Approach to the Economics of Climate Change, *International Journal of Environmental Research and Public Health*, 8(4): 985-1031, April 1, available online at: http://www.mdpi.com/1660-4601/8/4/985

Carlin, A., 2012, The Increasing Need for Research on Geoengineering Approaches to Reducing Potential Global Cooling, in Yu. A. Izrael, A. G. Ryaboshapko, and S. A. Gromov, editors, *Investigation of Possibilities of Climate Stabilization Using New Technologies, Proceedings of International Scientific Conference, Problems of Adaptation to Climate Change* (Moscow, 7–9 November 2011), Russian Academy of Sciences, Moscow, pp. 24-34.

Carlin, Alan, 2013, Modern Environmentalism: A Longer Term Threat to Western Civilization, *Energy and Environment*, 24(6), pp. 1063-72. Abstract available online at: http://multi-science.metapress.com/content/u7j410654k5876h8/?p=bd7c5322714a48c4995cdfd607580922&pi=0

Carlyle, Ryan, 2013, Should Other Nations Follow Germany's Lead on Promoting Solar Power?, *Forbes*, October 4, available online at: http://www.forbes.com/sites/quora/2013/10/04/should-other-nations-follow-germanys-lead-on-promoting-solar-power/

Catto, Neil, 2014, Volcano Activity, Temperature and Response Times, March 10, available online at

References

http://wattsupwiththat.com/2014/03/10/volcano-activity-temperature-and-response-times.

Chauhan, Chetan, 2015, India Unwilling to Be Treated on a Par with China on CO$_2$ Emissions, *Hindustan Times*, January 27, available online at: http://www.hindustantimes.com/india-news/no-indo-us-climate-deal-india-not-willing-to-treated-at-par-with-china-on-carbon-emissions/article1-1310900.aspx

Cheetham, Alan, 2012, 60 Year Climate Cycle, in *Global Warming Science*, October 10, available online at: http://appinsys.com/GlobalWarming/SixtyYearCycle.htm

Chen, Xiuhong, Xianglei Huang Xu Liu, July 16, 2013, Non-negligible Effects of Cloud Vertical Overlapping Assumptions on Longwave Spectral Fingerprinting Studies, *Journal of Geophysical Research*: Atmospheres, 118(13), 7309-20.

Chovanec, Patrick, 2012, China's Solyndra Economy, available online at: https://chovanec.wordpress.com/2012/09/13/wsj-chinas-solyndra-economy/

Coats, Dan, 2014, What Obama Could Learn from Germany's Failed Experiment with Green Energy, Fox News, June 5, available online at: http://www.foxnews.com/opinion/2014/06/05/what-obama-could-learn-from-germany-failed-experiment-with-green-energy

Conca, James, 2013, Like We've Been Saying—Radiation is Not a Big Deal, *Forbes*, January 11, available online at: http://www.forbes.com/sites/jamesconca/2013/01/11/like-weve-been-saying-radiation-is-not-a-big-deal/

Congressional Budget Office, 2010, *How Policies to Reduce Greenhouse Gas Emissions Could Affect Employment*, May 5, available online at: http://www.cbo.gov/sites/default/files/cbofiles/ftpdocs/105xx/doc10564/05-05-capandtrade_brief.pdf

Cook, John, Dana Nuccitelli, Sarah A. Green, Mark Richardson, Barbel Winkler, Bob Painting, Robert Way, Peter Jacobs, and Andrew Skuce,

2013, Quantifying the Consensus on Anthropogenic Global Warming in the Scientific Literature, *Environmental Research Letters*, 8(2), available online at: http://iopscience.iop.org/1748-9326/8/2/024024/article.

Cook, Russell, 2014, *Merchants of Smear*, Heartland Institute, September, available online at: http://heartland.org/sites/default/files/10-06-14_cook_merchants_of_smear.pdf

Corsi, Jerome P., 2014, Global Warming Threat? Now It's Asthma: White House Playing Heartstrings to Impose Onerous EPA Regulation of Carbon Dioxide, June 10, available online at:
http://mobile.wnd.com/2014/06/global-warming-threat-now-its-asthma

Cox, Anthony and Jo Nova, 2013, Has Man-made Global Warming Been Disproved? A Review of Recent Papers, The Hockey Schtick, October 26, available online at:
http://hockeyschtick.blogspot.com/2012/10/seven-recent-papers-that-disprove-man.html

Coyoteblog.com, 2014, Computer Modeling as "Evidence", available online at:
http://www.coyoteblog.com/coyote_blog/2014/07/computer-modeling-as-evidence.html

Crelinsten, Jeffrey, 2006, *Einstein's Jury: The Race to Test Relativity*, Princeton University Press.

Crichton, Michael, 2003, *Environmentalism as Religion*, presentation at the Commonwealth Club, September 15, San Francisco, CA, available online at:
http://scienceandpublicpolicy.org/images/stories/papers/commentaries/crichton_3.pdf

Crok, Marcel, 2013, [UK] Energy and Climate Change Committee Inquiry into AR5: Written Submission, December, available online at:
http://www.staatvanhetklimaat.nl/wp-content/uploads/2013/12/Marcel-Crok-submission-def.pdf

References

D'Aleo, Joseph, 2008, US Temperatures and Climate Factors since 1895, available online at:
http://icecap.us/images/uploads/US_Temperatures_and_Climate_Fact ors_since_1895.pdf

D'Aleo, Joseph, and Anthony Watts, 2010, *Surface Temperature Records: Policy-Driven Deception?*, available online at:
http://scienceandpublicpolicy.org/images/stories/papers/originals/surf ace_temp.pdf

D'Aleo, Joseph, 2014, Wikipropaganda on Global Warming, April 19, available online at:
http://www.cbsnews.com/news/wikipropaganda-on-global-warming/

D'Aleo, Joseph S., *et al.*, 2014, Scientists Respond to the Obama Administration's 2014 National Climate Assessment, May 19, available online at:
http://wattsupwiththat.com/2014/05/19/scientists-respond-to-the-obama-administrations-2014-national-climate-assessment/

Darwall, Rupert, 2013, *The Age of Global Warming: A History*, Quartet Books.

Darwall, Rupert, 2014, Europe's Stark Renewables Lesson, *Wall Street Journal*, January 28, available online at:
http://online.wsj.com/news/articles/SB100014240527023035 5320457 9348470302047850

Darwall, Rupert, 2014a, An Unsettling Debate: Global-Warming Proponents Betray Science by Shutting Down Debate, Summer, available online at:
http://www.city-journal.org/2014/24_3_global-warming.html

Davenport, Carol, 2014, Taking Oil Industry Cue, Environmentalists Drew Emissions Blueprint, *New York Times*, July 6, available online at:
http://www.nytimes.com/2014/07/07/us/how-environmentalists-drew-blueprint-for-obama-emissions-rule.html?

Davenport, Carol, 2014a, Obama Pursuing Climate Accord in Lieu of Treaty, *New York Times*, August 27, available online at:
http://www.nytimes.com/2014/08/27/us/politics/obama-pursuing-climate-accord-in-lieu-of-treaty.html

Day, Dwayne A, 2006, Of Myths and Missiles: The Truth about John F. Kennedy and the Missile Gap, *The Space Review*, available online at: http://www.thespacereview.com/article/523/1

Dayaratna, Kevin, and David W. Kreutzer, 2013, *Loaded DICE: An EPA Model Not Ready for the Big Game*, Heritage Foundation, November 21, available online at:
http://www.heritage.org/research/reports/2013/11/loaded-dice-an-epa-model-not-ready-for-the-big-game.

Dayaratna, Kevin, and David W. Kreutzer, 2014, *Unfounded FUND: Yet Another EPA Model Not Ready for the Big Game*, Heritage Foundation, April 29, available online at:
http://www.heritage.org/research/reports/2014/04/unfounded-fund-yet-another-epa-model-not-ready-for-the-big-game.

Delingpole, James, 2009, Climategate Goes SERIAL: Now the Russians Confirm that UK Climate Scientists Manipulated Data to Exaggerate Global Warming, December 26, available online at: http://blogs.telegraph.co.uk/news/jamesdelingpole/100020126/climat egate-goes-serial-now-the-russians-confirm-that-uk-climate-scientists-manipulated-data-to-exaggerate-global-warming/

Delingpole, James, and Kit Eastman, 2014, Global Warming 'Fabricated' by NASA and NOAA, Breitbart, June 23, available online at: http://www.breitbart.com/Breitbart-London/2014/06/23/Global-warming-Fabricated-by-NASA-and-NOAA

Delingpole, James, 2014, Household Recycling Is State-Endorsed Slavery, Breitbart, July 5, available online at:
http://www.breitbart.com/Breitbart-London/2014/07/05/Household-recycling-is-state-endorsed-slavery.

References

Delingpole, James, 2014a, Australian Bureau of Meteorology Accused of Criminally Adjusting Global Warming, Breitbart, August 25, available online at:
http://www.breitbart.com/Breitbart-London/2014/08/25/Australian-Bureau-of-Meteorology-accused-of-Criminally-Adjusted-Global-Warming

Delingpole, James, 2015, Forget Climategate. This 'Global Warming' Scandal Is Much Bigger, January, available online at:
http://jamesdelingpole.com/2015/01/forget-climategate-this-global-warming-scandal-is-much-bigger/

de Gouw, J.A. D.D. Parrish, G.J. Frost, and M. Trainer, 2014, *Reduced Emissions of CO_2, NOx and SO_2 from U.S. Power Plants Due to the Switch from Coal to Natural Gas with Combined Cycle Technology*, Earth's Future, available online at:
http://heartland.org/policy-documents/reduced-emissions-co2-nox-and-so2-us-power-plants-due-switch-coal-natural-gas-combi.

Deutche Welle, 2013, Germany's economics minister Gabriel seeks reform for renewable energy transition, available online at:
http://www.dw.de/germanys-economics-minister-gabriel-seeks-reform-for-renewable-energy-transition/a-17329275.

Dinda, Soumyananda, 2004, Environmental Kuznets Curve Hypothesis: A Survey, *Ecological Economics*, 49, 431-55, available online at:
ftp://ftp.soc.uoc.gr/students/aslanidis/My%20documents/papers/Dinda%20(2004).pdf

Dini, Jack, 2014, Is It the Sun?, *Canada Free Press*, August 6, available online at: http://canadafreepress.com/index.php/article/65111

DiPuccio, William, 2009, Have Changes in Ocean Heat Falsified the Global Warming Hypothesis? May 5, available online at:
http://pielkeclimatesci.wordpress.com/2009/05/05/have-changes-in-ocean-heat-falsified-the-global-warming-hypothesis-a-guest-weblog-by-william-dipuccio/

DiscovertheNetworks.org, 2009, Malaria Victims: How Environmentalist Ban on DDT Caused 50 Million Deaths, April, available online at:

http://www.discoverthenetworks.org/viewSubCategory.asp?id=1259

Douglass, David H., and John R. Christy, 2008, Limits on CO_2 Climate Forcing from Recent Temperature Data of Earth, *Energy and Environment*, August, available online at:
http://arxiv.org/ftp/arxiv/papers/0809/0809.0581.pdf

Douglass, David H., John R. Christy, Benjamin D. Pearson, and S. Fred Singer, 2007, A Comparison of Tropical Temperature Trends with Model Predictions, International Journal of Climatology, available online at:
http://icecap.us/images/uploads/DOUGLASPAPER.pdf

Driessen, Paul, 2014, Risking Lives to Promote Climate Change Hype, January 9, available online at:
http://www.wattsupwiththat.com/2014/01/09/risking-lives-to-promote-climate-change-hype/#more-100875.

DuHamel, Jonathon, 2012, NOAA temperature record "adjustments" could account for almost all "warming" since 1973, April 16, available online at
http://wryheat.wordpress.com/2012/04/16/noaa-temperature-record-adjustments-could-account-for-almost-all-warming-since-1973/.

Easterbrook, Don J., 2013, The 2013 IPCC AR5 Report: Facts -vs-Fictions, available online at:
http://wattsupwiththat.com/2013/10/03/the-2013-ipcc-ar5-report-facts-vs-fictions/

Easterbrook, Don J., 2014, Sanity Checking the National Climate Assessment Report Against Real Data Reveals Major Discrepancies, WattsUpWithThat.com, May 13, available online at:
http://wattsupwiththat.com/2014/05/13/checking-the-nca-report-against-real-data-reveals-major-discrepancies

Eenews.net, 2015, Legal Challenges—Overview and Documents [undated but appears to be early 2015], available online at:
http://www.eenews.net/interactive/clean_power_plan/fact_sheets/legal

References

Ehrenfreund, Max, 2014, Your Complete Guide to Obama's Immigration Executive Action, *Washington Post*, November 20, available online at: http://www.washingtonpost.com/blogs/wonkblog/wp/2014/11/19/your-complete-guide-to-obamas-immigration-order/

Eilperin, Juliet, 2013, White House Delayed Enacting Rules Ahead of 2012 Election to Avoid Controversy, *Washington Post*, December 15, available online at: http://www.washingtonpost.com/politics/white-house-delayed-enacting-rules-ahead-of-2012-election-to-avoid-controversy/2013/12/14/7885a494-561a-11e3-ba82-16ed03681809_story.html.

Eisenhower, Dwight D, 1961, *Farewell Address to the Nation*, January 17, available online at: http://mcadams.posc.mu.edu/ike.htm

Energy Ventures Analysis, 2014, *Energy Market Impacts of Recent Federal Regulations on the Electric Power Sector*, November, available online at http://evainc.com/wp-content/uploads/2014/10/Nov-2014.-EVA-Energy-Market-Impacts-of-Recent-Federal-Regulations-on-the-Electric-Power-Sector.pdf

E.ON Netz, 2004, *Wind Report 2004*, Excerpts from full report, available online at: http://aweo.org/windEon2004.html

Epstein, Alex, 2014, *The Moral Case for Fossil Fuels*, Portfolio Hardcover.

Eschenbach, Willis 2010, The Thunderstorm Hypothesis: How Clouds and Thunderstorms Control the Earth's Temperature, *Energy and Environment*, 21(4), August, pp. 201-16. Available online in a WattsUpWithThat.com version at: http://wattsupwiththat.com/2009/06/14/the-thermostat-hypothesis/

Eschenbach, Willis, 2011, The Empire Strikes Out, March 4, available online at: http://wattsupwiththat.com/2011/03/04/the-empire-strikes-out/

Eschenbach, Willis, 2013, The Tao of El Nino, January 28, available online at http://wattsupwiththat.com/2013/01/28/the-tao-of-el-nino/

Eschenbach, Willis, 2013a, Emergent Climate Phenomena, February 7, available online at:
http://www.wattsupwiththat.com/2013/02/07/emergent-climate-phenomena

Eschenbach, Willis, 2013b, Evidence that Clouds Actively Regulate the Temperature, October 6, available online at:
http://wattsupwiththat.com/2013/10/06/evidence-that-clouds-actively-regulate-the-temperature/

Eschenbach, Willis, 2014, Volcanoes Erupt Again, February 14, available online at http://wattsupwiththat.com/2014/02/24/volcanoes-erupt-again.

Ferrara, Peter, 2012, The Economic Implications of High Cost Energy, presentation at the Heartland ICCC7 Conference, available online at http://climateconferences.heartland.org/peter-ferrara-iccc7/

Feynman, Richard, 1965, *The Character of Natural Law*, MIT Press.

Friends of Science, 2014, *97% Consensus? No! Global Warming Math, Myths & Social Proofs*, February 3, available online at:
http://www.friendsofscience.org/assets/documents/97_Consensus_My th.pdf

Furchtgott-Roth, Diana, 2013, If Climate Change is Happening Now, What Do We Do?, Testimony before the Senate Environment and Public Works Committee, July 18, available at
http://www.epw.senate.gov/public/index.cfm?FuseAction=Files.View &FileStore_id=335ba07e-b4ce-4b39-a326-13bb582588e1

Gaidar, Yegor, 2007, *The Soviet Collapse: Grain and Oil*, American Enterprise Institute for Public Policy Research, April, available online at:
http://www.aei.org/wp-content/uploads/2011/10/20070419_Gaidar.pdf

Galiana, I. and C. Green, 2009, An Analysis of Technology-Led Climate Policy as a Response to Climate Change; Copenhagen Consensus on Climate: Copenhagen, Denmark; available online at:

References

http://thebreakthrough.org/blog/AP_Technology_Galiana_Green_v.6.0.pdf

Garfinkel, C. I., D. W. Waugh, L. D. Oman, L. Wang, and M. M. Hurwitz, 2013, Temperature Trends in the Tropical Upper Troposphere and Lower Stratosphere: Connections with Sea Surface Temperatures and Implications for Water Vapor and Ozone. *Journal of Geophysical Research: Atmospheres*, 118 (17).

Garnaut, R., 2008, *The Garnaut Climate Change Review: Final Report*; Cambridge University Press: Port Melbourne, Australia.

Gayer, Ted, and W. Kip Viscusi, 2013, Overriding Consumer Preferences with Energy Regulations, *Journal of Regulatory Economics,* June 30.

Gayer, Ted, and W. Kip Viscusi, 2014, Determining the Proper Scope of Climate Change Benefits, June 3, available online at: http://papers.ssrn.com/sol3/papers.cfm?abstract_id=2446522

Gillard, Julia, 2013, Julia Gillard Writes on Power, Purpose and Labor's Future, *The Guardian*, September 13, available online at: http://www.theguardian.com/world/2013/sep/13/julia-gillard-labor-purpose-future.

Gillis, Justin, and John Schwartz, 2015, Deeper Ties to Corporate for Doubtful Climate Researcher, *New York Times*, February 20, available online at: http://www.nytimes.com/2015/02/22/us/ties-to-corporate-cash-for-climate-change-researcher-Wei-Hock-Soon.html?_r=0

Glaser, Peter S., Carroll W. McGuffey, III, and Hahnah Williams Gaines, 2014, EPA's Section 111(d) Carbon Rule: What if States Just Said No, November 7, available online at http://www.insideronline.org/summary.cfm?id=23304

Goddard, Stephen, 2014, NOAA/NASA Dramatically Altered US Temperatures After the Year 2000, Real Science, June 23, available online at: http://stevengoddard.wordpress.com/2014/06/23/noaanasa-dramatically-altered-us-temperatures-after-the-year-2000/

Goddard, Stephen, 2015, Spectacular Climate Fraud from the White House, January 30, available online at:
https://stevengoddard.wordpress.com/2015/01/30/spectacular-climate-fraud-from-the-white-house/

Goddard, Stephen, 2015a, Barack Obama Goes Full Stalin, Real Science, February 24, available online at:
https://stevengoddard.wordpress.com/2015/02/24/barack-obama-goes-full-stalin/

Goklany, Indur M., 2011, *Wealth and Safety: The Amazing Decline in Deaths from Extreme Weather in an Era of Global Warming, 1900-2010*, Reason Foundation, Policy Study 393, September, available online at:
https://reason.org/files/deaths_from_extreme_weather_1900_2010.pdf

Goklany, Indur M., 2012, *Humanity Unbound: How Fossil Fuels Saved Humanity from Nature and Nature from Humanity*, Cato Institute, Policy Analysis No. 715, December 30, available online at:
http://www.cato.org/publications/policy-analysis/humanity-unbound-how-fossil-fuels-saved-humanity-nature-nature-humanity

Gold, Russell, 2013. Rise in Gas Production Fuels Unexpected Plunge in Emissions, *Wall Street Journal*, April 18, available online at:
http://online.wsj.com/news/articles/SB10001424127887324763404578430751849503848

Goldenberg, Suzanne, 2014, Al Gore Says Use of Geo-engineering to Head Off Climate Disaster Is Insane, *The Guardian*, January 15, available online at: http://www.theguardian.com/world/climate-consensus-97-per-cent/2014/jan/15/geo-al-gore-engineering-climate-disaster-instant-solutio

Goodenough, Patrick, 2014, Americans Spent $7.45B in 3 Years Helping Other Countries Deal With Climate Change, January 3, available online at: http://cnsnews.com/news/article/patrick-goodenough/americans-spent-745b-3-years-helping-other-countries-deal-climate

References

Goodman, Julie E., 2012, Testimony on EPA's Assessment of Health Benefits Associated with PM$_{2.5}$ Reductions for the Final Mercury and Air Toxics Standards, Subcommittee on Energy and Power, Committee on Energy and Commerce, February 8, available online at: http://energycommerce.house.gov/sites/republicans.energycommerce. house.gov/files/Hearings/EP/20120208/HHRG-112-IF03-WState-JGoodman-20120208.pdf.

Gosselin, P., 2011, Weed-Covered, Neglected Solar Park: 20 Acres, $11 Million, Only One and Half Years Old, July 4, available online at: http://notrickszone.com/2011/07/04/weed-covered-solar-park-20-acres-11-million-only-one-and-half-years-old/#sthash.UAOG315x.dpbs

Gosselin, P., 2015, Long List of Warmist Organizations, Scientists Haul in Huge Money from BIG OIL and Heavy Industry, February 9, available online at:
http://notrickszone.com/2015/02/09/long-list-of-warmist-organizations-scientists-haul-in-huge-money-from-big-oil-and-heavy-industry/#sthash.37UfnLFX.velEIIVJ.dpbs

Green, Kenneth P., 2011, *The Myth of Green Energy Jobs: The European Experience*, American Enterprise Institute, February 11, available online at: http://www.aei.org/article/energy-and-the-environment/the-myth-of-green-energy-jobs-the-european-experience/

Green, LC, and Armstrong, SR, 2003, Particulate Matter in Ambient Air And Mortality: Toxicologic Perspectives," *Regul Toxicol Pharmacol.* Dec., 38(3):326-35, http://www.ncbi.nlm.nih.gov/pubmed/14623483

Greenpeace, 2010, *GM Crops and World Hunger*, November 4, available online at:
http://www.greenpeace.org.uk/about/greenpeace-gm-crops-and-world-hunger.

Gregory, Ken, 2009, The Saturated Greenhouse Effect, July, available online at:
http://www.friendsofscience.org/assets/documents/The_Saturated_Greenhouse_Effect.htm.

Gregory, Ken, 2013, NASA Data Shows Decline in Water Vapor, WattsUpWithThat.com, March 6, available online at: http://wattsupwiththat.com/2013/03/06/nasa-satellite-data-shows-a-decline-in-water-vapor/

Gunderson, Bill, 2012, Why Solar Panels (and Stocks) Don't Work, *MSN Money*, November 2, available online at: http://money.msn.com/technology-investment/post.aspx?post=1caf21ea-2252-440d-8273-89a35459fcb0

Haapala, Ken, 2011, The Week That Was, Science and Environmental Policy Project (SEPP), May 21, available online at: http://www.sepp.org/twtwfiles/2011/TWTW%202011-5-21.pdf

Haapala, Ken, 2013, Federal Funding, The Week That Was, Science and Environmental Policy Project (SEPP), November 30, available online at: http://www.sepp.org/twtwfiles/2013/TWTW%2011-30-13.pdf

Hahn, Robert W., Randall W. Lutter, W. Kip Viscusi, 2000, *Do Federal Regulations Reduce Mortality*, AEI-Brookings Joint Center for Regulatory Studies, available online at: http://law.vanderbilt.edu/files/archive/011_Do-Federal-Regulations-Reduce-Mortality.pdf

Hammer, Michael, 2013, The NOAA Outgoing Long Wave Radiation Data Appears to Be Incompatible With the Theory of Anthropogenic Global Warming, December 10, available online at: http://jennifermarohasy.com/wp-content/uploads/2013/12/AGW_Falsified_Michael_Hammer.pdf

Harder, Amy, 2014, Protests Slow Pipeline Projects Across U.S., Canada, *Wall Street Journal*, December 9, available online at: http://www.wsj.com/articles/protests-slow-pipeline-projects-across-u-s-canada-1418173235

Harder, Amy, 2014a, Anti-Keystone Alliance Got Start in D.C. Dive Bar, *Wall Street Journal*, December 9, available online at: http://blogs.wsj.com/washwire/2014/12/09/anti-keystone-alliance-got-start-in-d-c-dive-bar/?mod=djem_EnergyJournal

References

Harrington, Elizabeth, 2014, Feds Paid $4.9 million to Create Hypothetical Utopian Climate Future: UN 'Youth Movement' Brings about 'New World Order' by 2070, *Washington Free Beacon*, available online at: http://freebeacon.com/issues/feds-paid-4-9-million-to-create-hypothetical-utopian-climate-change-future/

Harris, Rachel Lee, 2010, Science Foundation Backs Climate-Change Play, *New York Times*, October 3, available online at: http://www.nytimes.com/2010/10/04/theater/04arts-SCIENCEFOUND_BRF.html.

Haseler, Mike, 2014, Mike Haseler's Survey of Sceptics, February 25, available online at: http://www.bishop-hill.net/blog/2014/2/25/mike-haselers-survey-of-sceptics.html

Haun, William J., 2013, The Clean Air Act as an Obstacle to the Environmental Protection Agency's Anticipated Attempt to Regulate Greenhouse Gas Emissions from Existing Power Plants, March 11, available online at: http://www.fed-soc.org/publications/detail/the-clean-air-act-as-an-obstacle-to-the-environmental-protection-agencys-anticipated-attempt-to-regulate-greenhouse-gas-emissions-from-existing-power-plants

Hayward, Steven F., and Kenneth P. Green, 2009, Waxman-Markey: An Exercise in Unreality, *AEI Energy and Envtl. Outlook*, July, p. 1 and 3.

Hawkins, Troy R., Bhawna Singh, Guillaume Majeau-Bettez, and Anders Hammer Strømman, 2012, Comparative Environmental Life Cycle Assessment of Conventional and Electric Vehicles, *Journal of Industrial Ecology*, October 4, available online at: http://onlinelibrary.wiley.com/doi/10.1111/j.1530-9290.2012.00532.x/full

Heartland Institute, 2014, Alan Carlin, 2014 Winner of the Climate Change Whistleblower Award, available online at: http://climatechangeawards.org/alan-carlin/

Higgins, Andrew, 2014, Oil's Swift Fall Raises Fortunes Abroad, *New York Times*, December 24, available online at: http://www.nytimes.com/2014/12/25/world/europe/oils-swift-fall-raises-fortunes-of-us-abroad.html?_r=0

Hockey Schtick, 2011, New Paper Shows Global Warming Decreases Storm Activity, July 30, available online at: http://www.hockeyschtick.blogspot.com/search?q=warming+storm+activity.

Hockey Schtick, 2013a, Climate Scientist Dr. Murry Salby Explains Why Man-Made CO_2 Does Not Drive Climate Change, available online at: http://hockeyschtick.blogspot.com/2013/06/climate-scientist-dr-murry-salby.html

Hockey Schtick, 2013b, Swedish Scientist Replicates Dr. Murry Salby's Work, Finding Man-Made CO_2 Does Not Drive Climate Change, available online at: http://hockeyschtick.blogspot.com/2013/07/swedish-scientist-replicates-dr-murry.html

Hockey Schtick, 2014, New Paper Shows IPCC Underestimates Global Cooling from Man-Made Aerosols by Factor of 27 Times, June 10, available online at: http://hockeyschtick.blogspot.com/2014/06/new-paper-shows-ipcc-underestimates.html

Hockey Schtick, 2015, Over 200 Peer-Reviewed Papers Demonstrating Solar Control of Climate Published since 2010, February 22, available online at: http://hockeyschtick.blogspot.com/2015/02/over-200-peer-reviewed-papers.html

Hodgson, S.F., 1987, *Onshore Oil and Gas Seeps in California, California Department of Conservation*, Division of Oil and Gas, Publication No. TR26.

Homewood, Paul, 2015, All of Paraguay's Temperature Record Has Been Tampered with, January 26, available online at: https://notalotofpeopleknowthat.wordpress.com/2015/01/26/all-of-paraguays-temperature-record-has-been-tampered-with/#more-12774

References

Hoyt, Douglas V., and Kenneth H. Schatten, 1997, *The Role of the Sun in Climate Change*, Oxford University Press.

Hu, Zuliu, and Mohsin S. Khan, 1997, *Why is China Growing So Fast*, Economic issues No. 8, International Monetary Fund, available online at: https://www.imf.org/EXTERNAL/PUBS/FT/ISSUES8/INDEX.HTM

Hughes, Gordon, 2012, *The Impact of Wind Power on Household Energy Bills*, The Global Warming Policy Foundation, June 8, available online at: http://www.thegwpf.org/images/stories/gwpf-reports/hughes-evidence.pdf

Hughes, Gordon, 2012a, *The Performance of Wind Farms in the United Kingdom and Denmark*, Renewable Energy Foundation, available online at: http://ref.org.uk/attachments/article/280/ref.hughes.19.12.12.pdf.

Idso, Craig D., 2014, *Extreme Weather Events: Are They Influenced by Rising Atmospheric CO_2?* Center for the Study of Carbon Dioxide and Global Change, Tempe, AZ, available online at: http://www.co2science.org/education/reports/extremewx/extremewx.pdf

Idso, Craig, and S. Fred Singer, 2009 *Climate Change Reconsidered: The Report of the Nongovernmental International Panel on Climate Change*, The Heartland Institute, available online at: http://www.nipccreport.org/reports/2009/2009report.html.

Idso, Craig, Robert M. Carter, and S. Fred Singer, 2011, *Climate Change Reconsidered: 2011 Interim Report*, The Heartland Institute, available online at: http://www.nipccreport.org/reports/2011/2011report.html.

Idso, Craig D., Robert M. Carter, and S. Fred Singer, 2013, *Climate Change Reconsidered II: Physical Science*, The Heartland Institute, Chicago, IL, available online at: http://www.nipccreport.org/reports/ccr2a/ccr2physicalscience.html

591

Idso, Craig D., Robert M. Carter, and S. Fred Singer, 2014, *Climate Change Reconsidered II: Biological Impacts*, The Heartland Institute, Chicago, available online at:
http://www.nipccreport.org/reports/ccr2b/ccr2biologicalimpacts.html

Inhofe, James, 2012, *The Greatest Hoax: How the Global Warming Conspiracy Threatens Your Future*, WND Books.

International Bank for Reconstruction and Development (IBRD), 1992, *World Development Report 1992, Development and the Environment*, Oxford University Press

Izrael, Yury, 2005, Climate Change: Not a Global Threat, *RIANovosti*, June 23, available online at:
http://en.ria.ru/analysis/20050623/40748412.html

Izrael, Yu. A., 2012, Investigations of a Possibility to Reduce the Solar Radiation Flux by a Layer of Artificial Aerosols Aimed at Stabilization of the Global Climate at Its Present-day Level: Results of Field Experiments Carried Out in Russia, in Yu. A. Izrael, A. G. Ryaboshapko, and S. A. Gromov, editors, *Investigation of Possibilities of Climate Stabilization Using New Technologies, Proceedings of International Scientific Conference, Problems of Adaptation to Climate Change* (Moscow, 7–9 November 2011), Russian Academy of Sciences, Moscow, pp. 5-11.

Iya, 2012, Milankovitch Cycles and Glaciations, Most-Likely Blog, March 20, available online at:
http://most-likely.blogspot.com/2012/03/milankovitch-cycles-and-glaciations.html

Jonsson, Patrik, 2013, Gulf's $1.5 Trillion Oil Wildcat Play Marks Post-Spill Drilling 'Renaissance,' *Christian Science Monitor*, September 14, available online at:
http://www.alaskadispatch.com/article/20130914/gulf-s-15-trillion-oil-wildcat-play-marks-post-spill-drilling-renaissance

Josh, 2013, Josh Channels the Antarctic Boat People, December 31, available online at:

References

http://wattsupwiththat.com/2013/12/31/josh-channels-the-boat-people/

Kagan, Rebecca A., Tabitha C. Viner, Pepper W. Trail, and Edgar O. Espinoza, 2014, Avian Mortality at Solar Energy Facilities in Southern California: A Preliminary Analysis, available online at: http://www.kcet.org/news/rewire/Avian-mortality%20Report%20FINALclean.pdf

Kealey, Terence, 2013, The Case Against Public Science, August 5, available online at: http://www.cato-unbound.org/2013/08/05/terence-kealey/case-against-public-science

Keet, Corinne, et al., 2015, Neighborhood Poverty, Urban Residence, Race/Ethnicity, and Asthma: Rethinking the Inner-City Asthma Epidemic, *Journal of Allergy and Clinical Immunology*, 135(3), 655-62, March.

Keim, Jonathan, 2014, Inventing Ambiguities in the Clean Air Act, *National Review*, June 9, available online at: http://www.nationalreview.com/bench-memos/379753/inventing-ambiguities-clean-air-act-jonathan-keim.

Kerry, John, 2014, Remarks on Climate Change, U.S. Department of State, February 16, available online at: http://www.state.gov/secretary/remarks/2014/02/221704.htm

Kirkby, Jasper, et al., 2011, Role of Sulphuric Acid, Ammonia, and Galactic Cosmic Rays in Atmospheric Aerosol Nucleation, *Nature*, 476, 429-33, August 25.

Klaus, Vaclav, 2011, The Global Warming Doctrine Is Not a Science: Notes for Cambridge, May 10, available online at: http://www.klaus.cz/clanky/2830

Klimakatastrophe.wordpress.com, 2009, Solar Sewage Treatment Plant On Fehmarn - State of the Solar Modules after 20 Years, May 18, available online at:

https://klimakatastrophe.wordpress.com/2009/05/18/solar-klarwerk-auf-fehmarn-%E2%80%93-zustand-der-solarmodule-nach-20-jahren/

Koningstein, Ross, and David Fork, 2014, What It Would Really Take to Reverse Climate Change, *IEEE Spectrum*, November 18, available online at http://spectrum.ieee.org/energy/renewables/what-it-would-really-take-to-reverse-climate-change

Koren, Ilan, Guy Dagan, and Orit Altaratz, June 6, 2014, *From Aerosol—Limited to Invigoration of Warm Convective Clouds*, Science 344(6188), pp. 1143-1146

Knappenberger, Chip, 2010, IPCC 'Consensus'—Warning: Use at Your Own Risk, January 29, available online at:
http://www.masterresource.org/2010/01/ipcc-consensus-warning-use-at-your-own-risk

Knappenberger, Paul C., and Patrick J. Michaels, 2014, 0.02°C Temperature Rise Averted: The Vital Number Missing from the EPA's "By the Numbers" Fact Sheet, June 11, available online at:
http://www.cato.org/blog/002degc-temperature-rise-averted-vital-number-missing-epas-numbers-fact-sheet

Kramer, J. Bryan, 2013, If Things Continue as They Have Been, in Five Years, at the Latest, We Will Need to Acknowledge that Something Is Fundamentally Wrong with Our Models, June 20, available online at:
http://wattsupwiththat.com/2013/06/20/if-things-continue-as-they-have-been-in-five-years-at-the-latest-we-will-need-to-acknowledge-that-something-is-fundamentally-wrong-with-our-climate-models/.

Laframboise, Donna, 2010, The Great Peer-Review Fairy Tale, March 6, available online at:
http://nofrakkingconsensus.blogspot.com/2010/03/great-peer-review-fairy-tale.html

Laframboise, Donna, 2011, *The Delinquent Teenager Who Was Mistaken for the World's Top Climate Expert*, CreateSpace

References

Larsen, Eric, 2014, China's Growing Coal Use Is World's Growing Problem, January 27, available online at:
http://www.climatecentral.org/blogs/chinas-growing-coal-use-is-worlds-growing-problem-16999

Lavigne, Franck, *et al.*, 2013, *Source of the Great A.D. 1257 Mystery Eruption Unveiled*, Samales Volcano, Rinjani Volcanic Complex, Indonesia, *Proceedings of the National Academy*, abstract available online at:
http://www.pnas.org/content/early/2013/09/26/1307520110

Legates, D.R., Soon, W., Briggs, W.M., Monckton of Brenchley, 2013, Climate Consensus and Misinformation: A Rejoinder to Agnotology, Scientific Consensus, and the Teaching and Learning of Climate Change, *Science and Education*, available online at:
http://link.springer.com/article/10.1007%2Fs11191-013-9647-9

Levine, Steve, 2015, The Car of the Future May Run on Gasoline, *Wall Street Journal*, January 30, available online at:
http://www.wsj.com/articles/the-car-of-the-future-may-run-on-gas-1422646049?google_editors_picks=true

Lewis, Marlo, 2013, Climate Change 'Deniers' Not Welcome at Interior—Secy. Jewell, July 31, available online at:
http://www.globalwarming.org/2013/07/31/climate-change-deniers-not-welcome-at-interior-secy-jewell/

Lewis, Marlo, 2013a, Rep. Pompeo Questions EPA Administrator McCarthy on Obama Climate Plan, September 18, available online at:
http://www.globalwarming.org/2013/09/18/rep-pompeo-questions-epa-administrator-mccarthy-on-obama-climate-plan

Lewis, Marlo, 2014, Broken Record: EPA Administrator Gina McCarthy Plays the Asthma Card, available online at:
http://www.globalwarming.org/2014/06/02/broken-record-epa-administrator-gina-mccarthy-plays-the-asthma-card.

Lewis, Marlo, 2014a, EPA's Carbon Rules: Same Old Way to Skin the Cat, June 3, available online at:

http://www.globalwarming.org/2014/06/03/epas-carbon-rules-same-old-way-to-skin-the-cat/#more-20295

Lewis, Marlo, 2014c, The Unbearable Lightness of UARG v. EPA, July 4, available online at:
http://www.globalwarming.org/2014/07/04/the-unbearable-lightness-of-uarg-v-epa/

Lewis, Marlo, 2014d, EPA's Carbon "Pollution" Rules: War on Coal by the Numbers, June 9, available online at:
http://www.globalwarming.org/2014/06/09/epas-carbon-pollution-rules-war-on-coal-by-the-numbers/#more-20419

Lewis, Marlo, et al., 2014, Clean Power Plan Comment, Submitted to EPA Docket ID No. EPA-HQ-OAR-2013-0602, December 1, available online at: http://www.scribd.com/doc/248928031/Marlo-Lewis-CLean-Power-Plan-Comment

Lewis, Marlo, 2015, Some Free Market Talking Points on the Keystone XL Pipeline Amendments, January 13, available online at:
http://www.globalwarming.org/2015/01/13/some-free-market-talking-points-on-the-keystone-xl-pipeline-amendments/#more-22593

Lewis, Marlo, 2015a, Is Climate Policy Sustainable: Sobering Slides on the EU's 60-by50 Climate Treaty Proposal, March 4, available online at:
http://www.globalwarming.org/2015/03/04/sobering-slides-on-the-eus-60-by-50-climate-treaty-proposal/

Lewis, Marlo, William Yeatman, and David Bier, 2012, The Illusory Benefits of the Utility MACT, Competitive Enterprise Institute, Issue Analysis 2012 No.5, Washington, DC, available online at https://cei.org/sites/default/files/Marlo%20Lewis,%20William%20Yeatman,%20and%20David%20Bier%20-%20All%20Pain%20and%20No%20Gain.pdf

Lewis, Nicholas, and Marcel Crok, 2014, Oversensitive: How the IPCC Hid the Good News on Global Warming, The Global Warming Foundation, Report 12, available online at:

References

http://www.thegwpf.org/content/uploads/2014/02/Oversensitive-How-The-IPCC-hid-the-Good-News-on-Global-Warming.pdf

Lerner, Josh, 2009, *Boulevard of Broken Dreams*, Princeton University Press.

Lilley, Peter, 2012, *What is Wrong with Stern? The Failing of the Stern Review of the Economics of Climate Change*, Report 9, The Global Warming Policy Foundation, available online at:
http://www.thegwpf.org/wp-content/uploads/2012/09/Lilley-Stern_Rebuttal.pdf.

Lloyd, Philip, 2013, *Who Are the Denialists?*, November 9, available online at: http://WattsupWithThat.com/2013/11/09/who-are-the-true-denialists/#more-97119

Lindzen, Richard S., and Yong-Sang Choi, 2011, On the Observational Determination of Climate Sensitivity and Its Implications, *Asia-Pacific J. Atmos. Sci.*, 47(4), 377-390, available online at: http://www-eaps.mit.edu/faculty/lindzen/236-Lindzen-Choi-2011.pdf

Lindzen, Richard, 2012, Climate Science: Is It Currently Designed to Answer Questions?, revised September 21, p.4, available online at: http://arxiv.org/ftp/arxiv/papers/0809/0809.3762.pdf,

Lindzen, Richard, 2012a, *Climate Science: Is It Designed to Answer Questions?*, Euresis Journal, Vol. 2, pp. 161-93, available online at: http://www.euresisjournal.org/default.asp?pagina=414&act=2&id=41

Lindzen, Richard S., Fall, 2013, Science in the Public Square: Global Climate Alarmism and Historical Precedents, *Journal of American Physicians and Surgeons,* 18(3), pp. 70-3, available online at: http://jpands.org/vol18no3/lindzen.pdf.

Lomborg, Bjorn, 2013, Windfarms, February 14, available online at: http://climaterealists.com/index.php?id=11150&utm_source=feedburner&utm_medium=feed&utm_campaign=Feed%3A+ClimaterealistsNewsBlog+%28ClimateRealists+News+Blog%29

Lomborg, Bjorn, 2013a, *A Golden Rice Opportunity*, Project Syndicate, February 15, available online at:
http://www.project-syndicate.org/commentary/the-costs-of-opposing-gm-foods-by-bj-rn-lomborg

Lomborg, Bjorn, 2014, Lomborg: Obama Energy Policy Hurts African Poor, *USA Today*, February 8, available online at:
http://www.usatoday.com/story/opinion/2014/02/08/bjorn-lomborg-africa-energy/5284631

Llovel, W., J.K. Willis, F.W. Landerer, and I. Fukumori, 2014, Deep Ocean Concentration to Sea Level and Energy Budget Not Detectable over the Past Decade, *Nature Climate Change*, 4, 1031-35

Luck, Micheal, Patrick T.T. Mayer, and Emma Stewart, 2010, *Cruise Tourism in Polar Regions: Promoting Environmental and Social Sustainability*, Google eBook, May 23.

Luckey, T. D., 1991, *Radiation Hormesis*, CRC Press.

Lund, Susan, James Manyika, Scott Nyquist, Lenny Mendonca, and Sreenivas Ramaswamy, 2013, *Game Changers: Five Opportunities for US Growth and Renewal*, McKinsey Global Institute, July, available online at:
www.McKinsey.com/insights/americas/us_game_changers

Luning, Sebastian, and Fritz Vahreholt, 2014, *Will the Solar Doldrums of the Coming Decades Lead to Cooling? A Look at the Latest Scientific Publications*, May 10, as translated from http://www.kaltessonne.de/?p=18141, available online at:
http://notrickszone.com/2014/05/10/flurry-of-scientists-recent-peer-reviewed-papers-warning-of-approaching-little-ice-age/

MacRae, Allan M.R., undated, Carbon Dioxide Is Not the Major Cause of Global Warming, [appears to be about 2007 or 2008], available online at:
http://icecap.us/images/uploads/CO2vsTMacRae.pdf

Macaes, Bruno, 2014, Send a Message to Putin with a Trans-Atlantic Energy Pact, *Wall Street Journal*, April 23, p. A15, available online at:

References

http://online.wsj.com/news/articles/SB1000142405270230481090457
9512100086900662

Macrae, Paul, 2011, Alarmist Climate Science and the Principle of Exclusion, available online at:
http://wattsupwiththat.com/2011/06/08/alarmist-climate-science-and-the-principle-of-exclusion.

Manning, Rick, 2014, Podesta's Challenge Puts EPA Funding on Hot Seat, *The Hill*, May 6, available online at:
http://thehill.com/blogs/pundits-blog/energy-environment/205347-podestas-challenge-puts-epa-funding-on-hot-seat

Marohasy, Jennifer, 2009, The Work of Ferenc Miskolczi (Part 1), May 2, available online at: http://jennifermarohasy.com/2009/05/the-work-of-ferenc-miskolczi-part-1/

Marsh, Gerald E., 2010, Interglacials, Milankovitch Cycles, and Carbon Dioxide, available online at:
http://arxiv.org/ftp/arxiv/papers/1002/1002.0597.pdf

Martinson, Erica, 2012, CO_2 Rules: Now You See'em, Now You Don't, *Politico*, April 16, available online at:
http://www.politico.com/news/stories/0412/75216.html

Martinson, Erica, 2013, Chris Horner, Master of FOIA, Bedevils the White House, *Politico Pro*, June 28, available online at:
http://www.politico.com/story/2013/06/chris-horner-foia-epa-white-house-93264.html

Mattson, Mark P., and Edward J. Calabrese (editors), 2009, *Hormesis: A Revolution in Biology, Toxicology and Medicine/Edition*, Springer-Vertag, New York

McKitrick, Ross, and Elmira Aliakbari, 2014, *Energy Abundance and Economic Growth: International and Canadian Evidence*, Fraser Institute, May, available online at:

http://www.fraserinstitute.org/uploadedFiles/fraser-
ca/Content/research-news/research/publications/energy-abundance-
and-economic-growth.pdf

McMahon, Jeff, 2014, U.S. Already Teaches Climate Change in Schools,
Forbes, August 9, available online at:
http://www.forbes.com/sites/jeffmcmahon/2014/08/09/mccarthy-
proposal-to-teach-climate-change-in-schools-should-surprise-no-one/

Milloy, Steve, 2011, EPA's Air Quality Overkill: Costly New Air-Quality
Standards Based on Suspect Statistics, *Washington Times*, July 28, available
online at:
http://www.washingtontimes.com/news/2011/jul/28/epas-air-
quality-overkill/?page=all

Milloy, Steve, 2014, EPA's Claim That Its Coal Plant CO_2 Rules Will
Save Lives Is False, July, available online at:
http://junksciencecom.files.wordpress.com/2014/07/epa_s-health-
claims-for-its-coal-plant-co2-rules-are-false1.pdf

Mills, Mark P., 2012, *Unleashing the North American Energy Colossus:
Hydrocarbons Can Fuel Growth and Prosperity*, Manhattan Institute, July,
available online at http://www.manhattan-institute.org/pdf/pgi_01.pdf

Miskolczi, Ferenc Mark, 2014, The Greenhouse Effect and the Infrared
Radiative Structure of the Earth's Atmosphere, *Development in Earth
Science*, 2, 31-52, available online at:
http://www.seipub.org/des/Download.aspx?ID=21810

Mohan, Vishwa, 2014, India Invokes 'Right to Grow" to Tell Rich
Nations of Its Stand on Future Climate Change Negotiations, *Times of
India*, June 17, available online at:
http://timesofindia.indiatimes.com/home/environment/global-
warming/India-invokes-right-to-grow-to-tell-rich-nations-of-its-stand-
on-future-climate-change-negotiations/articleshow/36724848.cms

Monckton, Christopher, Willie W.-H. Soon, David R. Legates, and
William R. Briggs, 2015, Why Models Run Hot: Results from an

References

Irreducibly Simple Climate Model, *Science Bulletin*, 60(1), 122-35, available online at:
http://heartland.org/sites/default/files/why_models_run_hot_results_f rom_an_irreducibly_simple_climate_model.pdf

Montford, Andrew, 2013, *Consensus? What Consensus?*, The Global Warming Foundation, Note 5, available online at:
http://thegwpf.org/content/uploads/2013/09/Montford-Consensus.pdf

Montford, Andrew, 2015. *Unintended Consequences of Climate Policy*, Global Warming Foundation, Report 16, London, available online at:
http://www.thegwpf.org/content/uploads/2015/01/Unintended-Consequences.pdf

Moore, Patrick, 2014, Statement before the Senate Environment and Public Works Committee, Subcommittee on Oversight, Natural Resource Adaptation: Protecting Ecosystems and Economies, February 25, available online at:
http://www.epw.senate.gov/public/index.cfm?FuseAction=Files.View &FileStore_id=415b9cde-e664-4628-8fb5-ae3951197d03

Morano, Marc, 2013, Dem Sen. Boxer Blames Tornadoes on Global Warming—Plugs Her Carbon Bill to Fix Bad Weather: 'This is Climate Change. We Were Warmed about Extreme Weather...We Need to Protect Our People—Carbon Could Cost Us the Planet,' Climate Depot, available online at:
http://www.climatedepot.com/2013/05/21/dem-sen-boxer-blames-tornadoes-on-global-warming-plugs-her-carbon-tax-bill-to-fix-bad-weather-this-is-climate-change-we-were-warned-about-extreme-weather-we-need-to-protect-our/

Morano, Marc, 2014, Scientists and Studies Predict 'Imminent Global COOLING' Ahead—Drop in Global Temps 'Almost a Slam Dunk, June 29, available online at:
http://www.climatedepot.com/2014/06/29/scientists-and-studies-predict-imminent-global-cooling-ahead-drop-in-global-temps-almost-a-slam-dunk/

Moss, Todd, Roger Pielke, Jr., and Morgan Bazilian, 2014, *Balancing Energy Access and Environmental Goals in Development Finance: The Case of the OPIC Carbon Cap*, Center for Global Development Policy Paper 038, April, available online at:
http://www.cgdev.org/sites/default/files/balancing-energy-access-case-opic-carbon-cap_0.pdf

Nakamura, David, and Steven Mufson, 2014, China, US Agree to Limit Greenhouse Emissions, *Washington Post*, November 12, available online at http://www.washingtonpost.com/business/economy/china-us-agree-to-limit-greenhouse-gases/2014/11/11/9c768504-69e6-11e4-9fb4-a622dac742a2_story.html

National Research Council, 2015, *Climate Intervention: Reflecting Sunlight to Cool the Earth*, February, National Academies Press.

Neubacher, Alexander, 2012, Solar Subsidy Sinkhole: Re-Evaluating Germany's Blind Faith in the Sun, *Spiegel*, available online at:
http://www.spiegel.de/international/germany/solar-subsidy-sinkhole-re-evaluating-germany-s-blind-faith-in-the-sun-a-809439.html

Nova, Joanne, 2010, The Big Picture: 65 Million Years of Temperature Swings, February, available online at:
http://joannenova.com.au/2010/02/the-big-picture-65-million-years-of-temperature-swings/

Nova, Joanne, 2012, Government Burns $70 Billion a Year Subsidizing Renewables, and Wild Claims of "Fossil Fuel Subsidies," September 18, available at:
http://joannenova.com.au/2012/09/government-burn-70-billion-a-year-subsidizing-renewables-and-wild-claims-of-fossil-fuel-subsidies-debunked/

Nova, Joanne, 2014, Almost Everything The Media Tells You About Skeptics Is Wrong: They're Engineers And Hard Scientists. They Like Physics Too, available online at:
http://joannenova.com.au/2014/02/almost-everything-the-media-tells-you-about-skeptics-is-wrong-theyre-mostly-engineers-and-hard-scientists-they-like-physics-too/

References

Nova, Joanne, 2014a, Cook Scores 97% for Incompetence on a Meaningless Consensus, June 6, available online at: http://joannenova.com.au/2014/06/cook-scores-97-for-incompetence-on-a-meaningless-consensus

Nova, Joanne, 2014b, $7B Paid in Carbon Tax to Reduce CO_2 by 0.3% and Cool Us by Zero Degrees, February 6, available online at: http://joannenova.com.au/2014/02/7b-paid-in-carbon-tax-to-reduce-co2-by-0-3-and-cool-us-by-zero-degrees/

Nova, Joanne, 2015, Is the Sun Driving Ozone and Changing the Climate? January, available online at: http://joannenova.com.au/2015/01/is-the-sun-driving-ozone-and-changing-the-climate/

Oakley, Paul A., 2006, Radiation Hormesis: Low Doses Stimulate the Immune System and Do Not Cause Cancer-They Prevent It, *American Journal of Clinical Chiropractic*, 16(1), available online at http://www.idealspine.com/pages/ajcc_jan_06_radiation_hormesis.htm.

Obama, Barack, 2009, Transparency and Open Government, available online at: http://www.whitehouse.gov/the_press_office/TransparencyandOpenGovernment/

Obama, Barack, 2009, Scientific Integrity, March, available online at: http://www.whitehouse.gov/the-press-office/memorandum-heads-executive-departments-and-agencies-3-9-09

Obama, Barack, 2013, Remarks by the President on Climate Change, Georgetown University, June 25, available online at: http://www.whitehouse.gov/the-press-office/2013/06/25/remarks-president-climate-change

Orssengo, G., 2010, Predictions of Global Mean Temperatures & IPCC Projections, April, available online at: http://wattsupwiththat.files.wordpress.com/2010/04/predictions-of-gmt.pdf

Pangburn, Dan, 2013, The Sun Explains 95% of Climate Change over the Past 400 Years; CO_2 Had No Significant Influence, The Hockey Schtick, available at
http://hockeyschtick.blogspot.se/2013/11/the-sun-explains-95-of-climate-change.html.

Papagianis, Christopher, 2012, A simple plan to relieve airport congestion, March 26, Reuters, available online at:
http://blogs.reuters.com/great-debate/2012/03/26/a-simple-plan-to-relieve-airport-congestion.

Pearson, Byron E., 2000, We Have Almost Forgotten How to Hope: The Hualapai, the Navajo, and the Fight for the Central Arizona Project, 1944-1968, *Western Historical Quarterly,* 31, Autumn, pp. 297-316. Available online at:
http://connection.ebscohost.com/c/articles/3472541/we-have-almost-forgotten-how-hope-hualapai-navajo-fight-central-arizona-project-1944-1968

Pearson, Byron E., 2002, *Still the Wild River Runs: Congress, the Sierra Club, and the Fight to Save Grand Canyon*, University of Arizona Press, Tucson.

Peiser, Benny, 2014, Testimony to Committee on Environment and Public Works, Hearing on the Super Pollutants Act of 2014 (S.2911), December 2, available online at http://www.globalwarming.org/wp-content/uploads/2014/12/Peiser-Senate-Testimony.pdf

Perry, Mark J., 2013, *America's Shale Revolution Is Providing Significant Economic and Environmental Benefits; CO_2 per Capita Lowest since 1964*, August 28, American Enterprise Institute, available online at:
http://www.aei-ideas.org/2013/08/americas-shale-revolution-is-providing-significant-economic-and-environmental-benefits-co2-per-capita-lowest-since-1964/

Perry, Mark J., 2014, An Amazing Chart of an Amazing Job-creating State; We Owe a Debt of Gratitude to 'Saudi Texas' and the Shale Boom, November 21, available online at:
http://www.aei.org/publication/822253/

References

Pielke, Jr., Roger, 2012, Drought and Climate Change, September 24, available online at:
http://rogerpielkejr.blogspot.com/2012/09/drought-and-climate-change.html,

Pielke, Jr., Roger, 2013, A Factual Look at the Relationship of Climate and Weather, Statement to the US House of Representatives, Committee on Science, Space, and Technology, Subcommittee on Environment, December 11, available online at:
http://science.house.gov/sites/republicans.science.house.gov/files/documents/HHRG-113-SY18-WState-RPielke-20131211.pdf

Polzin, Paul, and Bill Whitsitt, 2014, The U.S. Energy Boom Lifts Low-Income Workers Too, *Wall Street Journal*, April 5, p. A13

PopularTechnology.net, 2013, 97% Study Falsely Classifies Scientists' Papers, According to the Scientists that Published Them, May 31, available online at:
http://www.populartechnology.net/2013/05/97-study-falsely-classifies-scientists.html

PopularTechnology.net, 2014, 97 Articles Refuting the "97% Consensus," December 19, available online at:
http://www.populartechnology.net/2014/12/97-articles-refuting-97-consensus.html

Porter, Eduardo, 2015, A Call to Look Past Sustainable Development, *New York Times*, April 15, available online at:
http://www.nytimes.com/2015/04/15/business/an-environmentalist-call-to-look-past-sustainable-development.html?gwh=BD7996C30D7C9F4295922A70F19CD0F5&gwt=pay

Potts, Brian H., March 25, 2014, The EPA Doesn't Have the Legal Authority to Adopt Its New Power Plant Climate Rules, *Forbes*, available online at:
http://www.forbes.com/sites/realspin/2014/03/25/the-epa-doesnt-have-the-legal-authority-to-adopt-its-new-power-plant-climate-rules/.

605

Rappaport, Alan, 2014, On Climate Change, Voters Are as Partisan as Ever, *New York Times*, November 12, available online at http://www.nytimes.com/politics/first-draft/2014/11/12/?entry=5556&_php=true&_type=blogs&partner=rss&emc=rss

Rawls, Alec, 2012, Omitted Variable Fraud: Vast Evidence for Solar Climate Driver Rates One Oblique Sentence in AR5, WattsUpWithThat, February 22, available at:
http://wattsupwiththat.com/2012/02/22/omitted-variable-fraud-vast-evidence-for-solar-climate-driver-rates-one-oblique-sentence-in-ar5/

Renewable Energy Foundation, 2012, Wear and Tear Hit Wind Farm Output and Economic Lifetime, December 19, available online at:
http://www.ref.org.uk/press-releases/281-wearnandntearnhitsnwindnfarmnoutputnandneconomicnlifetime

Revkin, Andrew C., 2012, *The Next Ice Age and the Anthropocene*, January 8, available online at:
http://dotearth.blogs.nytimes.com/2012/01/08/the-next-ice-age-and-the-anthropocene

Ridley, Matt, 2014, The Scarcity Fallacy, *Wall Street Journal*, April 26-27, p. C1.

Ridley, Mathew, 2014a, Scientists Must Not Put Policy before Proof, *The Times*, December 8, available online at:
http://www.thetimes.co.uk/tto/opinion/columnists/article4290516.ece

Ridley, Matt, 2015, Fossil Fuels Will Save the World (Really), *Wall Street Journal*, March 13, available online at:
http://www.wsj.com/articles/fossil-fuels-will-save-the-world-really-1426282420

Roberts, Donald, and Richard Tren, 2010, *The Excellent Powder: DDT's Political and Scientific History*, Dog Ear Publishing.

References

Robinson, Arthur, 2010, *How Government Corrupts Science*, Science and Public Policy Institute, March 3, available online at:
http://scienceandpublicpolicy.org/images/stories/papers/commentaries/how_govt_corrupts_sci.pdf

Robinson, Wills, 2014, Climate Change Scientist Claims He Has Been Forced from New Job in "McCarthy"-Style Witch-Hunt by Academics across the World, UK *Daily Mail*, May 15, available online at:
http://www.dailymail.co.uk/news/article-2629171/Climate-change-scientist-claims-forced-new-job-McCarthy-style-witch-hunt-academics-world.html

Roe, Gerard, 2006, In Defense of Milankovitch, *Geophysical Research Letters*, 33, L24703, available online at:
http://earthweb.ess.washington.edu/roe/Publications/MilanDefense_GRL.pdf

Rose, David, 2014, EXPOSED: How a Shadowy Network Funded by Foreign Millions Is Making Our Household Energy Bills Soar – for a Low Carbon Britain, *Daily Mail*, October 25, available online at:
http://www.dailymail.co.uk/news/article-2807849/EXPOSED-shadowy-network-funded-foreign-millions-making-household-energy-bills-soar-low-carbon-Britain.html

Rogers, Norman, 2013, The Climate-Industrial Complex, *American Thinker*, November 9, available online at:
http://www.americanthinker.com/2013/09/the_climate-industrial_complex.html

Rose, David, 2014, BBC's Six-Year Cover-Up of Secret 'Green Propaganda' Training for Top Executives, *Daily Mail*, January 11, available online at:
http://www.dailymail.co.uk/news/article-2537886/BBCs-six-year-cover-secret-green-propaganda-training-executives.html

Ruddiman, William F., 2005, *Plows, Plagues and Petroleum*, Princeton University Press

Rust, James, 2015, Stifling Climate Research and Opinion: Another Desperado Mistake, March 5, available online at:
https://www.masterresource.org/debate-issues/nyt-stifled-research/

Sagan, Carl, and George Mullen, 1972, Earth and Mars: Evolution of Atmospheres and Surface Temperatures, *Science*, 177(4043), July 7

Sensenbrunner, Representative Jim, 2012, presentation at Heartland ICCC7 conference, May 22, available online at http://climateconferences.heartland.org/jim-sensenbrenner-iccc7

Sappenfield, Mark, 2014, Obama Mocks Climate Change Skeptics: Wise Move?, *Christian Science Monitor*, June 15, available online at:
http://www.csmonitor.com/USA/DC-Decoder/Decoder-Buzz/2014/0615/Obama-mocks-climate-change-skeptics-Wise-move

Scafetta, Nicola, 2012, Multi-scale Harmonic Model for Solar and Climate Cyclical Variation Throughout the Holocene Based on Jupiter–Saturn Tidal Frequencies Plus the 11-year Solar Dynamo Cycle, *Journal of Atmospheric and Solar-Terrestrial Physics*, 80, 296-311, May, available online at: http://arxiv.org/abs/1203.4143

Schröder, Catalina, 2012, Energy Revolution Hiccups: Grid Instability Has Industry Scrambling for Solutions, *Spiegel*, available online at:
http://www.spiegel.de/international/germany/instability-in-power-grid-comes-at-high-cost-for-german-industry-a-850419.html

Secchi, Silvia, and Bruce Babcock, 2007, *Impact of High Corn Prices on Conservation Reserve Program Acreage*, available online at:
http://www.card.iastate.edu/iowa_ag_review/spring_07/article2.aspx

Sen, Amartya, 2014, Stop Obsessing About Global Warming: Environmentalists Are Ignoring Poor Countries' Needs, *New Republic*, August 22, available online at:
http://www.newrepublic.com/article/118969/environmentalists-obsess-about-global-warming-ignore-poor-countries

References

Seddon, Duncan, 2013, Do Wind Farms/Gas Turbines Save Energy?, *Energy Generation*, October-December, pp. 23-5, available online at: http://powertrans.com.au/UserFiles//file//PDFs//EG-4-13.pdf

Sharman, Amelia, 2013, *Mapping the Climate Sceptical Blogosphere*, Centre for Climate Change Economics and Policy, Working Paper No. 144, Grantham Research Institute on Climate Change and the Environment, August, available online at:
http://www.lse.ac.uk/GranthamInstitute/wp-content/uploads/2014/02/Mapping-the-climate-sceptical-blogosphere.pdf

Singer, Fred, 2013a, IPCC's Bogus Evidence for Global Warming, *American Thinker*, November 12, available online at:
http://www.americanthinker.com/2013/11/ipcc_s_bogus_evidence_for_global_warming.html

Singer, Fred, 2015, Let's Rethink New Methane Policy, Letter to the Editor, *Wall Street Journal*, January 15, available online at:
http://www.wsj.com/articles/lets-rethink-new-methane-policy-letters-to-the-editor-1421363688

Simmons, Daniel R., 2013, Testimony, House Committee on Oversight and Government Reform, February 14, available online at:
http://www.instituteforenergyresearch.org/wp-content/uploads/2013/02/Simmons-testimony-FINAL.pdf

Siri, William E, 1979, *Reflections on the Sierra Club, the Environment, and Mountaineering, 1950s-1970s*, Regional Oral History Office, The Bancroft Library, University of California at Berkeley, California, available online at:
http://www.archive.org/stream/reflectsierraclub00siririch/reflectsierraclub00siririch_djvu.txt

Smith, Anne E., October 4, 2011, Testimony before the Subcommittee on Energy and the Environment Committee on Science, Space, and Technology, US House of Representatives on Quality Science for Quality Air, available online at: http://www.nera.com/nera-files/PUB_Smith_QualityAir_testimony_1011.pdf

Smith, Anne E., 2011a, *An Evaluation of the Pm$_{2.5}$ Health Benefits Estimates in Regulatory Impact Analyses for Recent Air Regulations*, December 1, NERA Economic Consulting, Washington, DC, available online at:
http://www.nera.com/nera-files/PUB_RIA_Critique_Final_Report_1211.pdf.

Smith, Anne E., 2012, Testimony before the Subcommittee on Energy and Power, Energy and Commerce Committee, US House of Representatives, *Hearing on The American Energy Initiative—A Focus on What EPA's Utility MACT Rule Will Cost US Consumers*, February 8, available online at:
http://democrats.energycommerce.house.gov/sites/default/files/docu ments/Testimony-Smith-EE-American-Energy-Initiative-2012-2-8.pdf

Smith, Anne E., 2012a, Testimony before the Subcommittee on Energy and Power, Energy and Commerce Committee, US House of Representatives, *Hearing on The American Energy Initiative—A Focus on EPA's New Proposal to Tighten National Standards for Fine Particulate Matter in the Ambient Air*, June 28, available online at:
http://www.nera.com/publications/archive/2012/nera-testimony-on-the-american-energy-initiative--a-focus-on-e.html

Soon, Willie, and Paul Driessen, 2011, The Myth of Killer Mercury, *Wall Street Journal*, May 25, available online at:
http://online.wsj.com/article/SB1000142405274870342120457632942 0414284558.html?mod=googlenews_wsj

Spencer, Roy W., 2014, *Time to Push Back Against the Global Warming Nazis*, February 20, available online at:
http://www.drroyspencer.com/2014/02/time-to-push-back-against-the-global-warming-nazis/

Stern, N., 2006, *Stern Review on the Economics of Climate Change*; HM Treasury: London, UK. Available online at:
http://webarchive.nationalarchives.gov.uk/+/http://www.hm-treasury.gov.uk/ sternreview_index.htm.

References

Steward, H. Leighton, 2014, *Obama Misled on Climate Change Facts*, February 6, available online at:
http://politix.topix.com/story/10271-obama-misled-on-climate-change-facts.

Strassel, Kimberly A., 2009, *The EPA Silences a Climate Critic*, *Wall Street Journal*, July 3, available online at:
http://online.wsj.com/article/SB124657655235589119.html

Svensmark, Henrik, 1998, Influence of Cosmic Rays on Earth's Climate, *Physical Review Letters*, 81(22):5027-30; October 15, available online at:
http://hep.physics.indiana.edu/~rickv/quarknet/article2.pdf.

Tanton, Tom, 2014, *Reforming Net Metering: Providing a Bright and Equitable Future*, March, American Legislative Exchange, Washington, DC

Taylor, James, 2014, 'Dark Money' Funds To Promote Global Warming Alarmism Dwarf Warming 'Denier' Research, January 2, available online at: http://www.forbes.com/sites/jamestaylor/2014/01/02/dark-money-funds-to-promote-global-warming-alarmism-dwarf-warming-denier-research/

Taylor, James, 2014a, *Colorado Consumers Paying Steep Price for Renewable Power Mandates*, available online at:
http://environmentblog.ncpa.org/colorado-consumers-paying-steep-price-for-renewable-power-mandates

ThePointman, 2012, *The Sun Is Setting on Solar Power, the Money's Gone, and Nobody's Asking Any Questions*, April 13, available online at:
http://thepointman.wordpress.com/2012/04/13/the-sun-is-setting-on-solar-power-the-moneys-gone-and-nobodys-asking-any-questions/

ThePointman, 2013, Why Would Anyone Believe a Single Word Coming Out of Their Mouth, August, 2, available online at:
http://thepointman.wordpress.com/2013/08/02/why-would-anyone-believe-a-single-word-coming-out-of-their-mouth/

Tierney, John, 1996, Recycling Is Garbage: Rinsing Out Tuna Cans and Tying Up Newspapers May Make You Feel Virtuous, but Recycling Could

Be America's Most Wasteful Activity, *New York Times*, June 30, available online at:
http://web.williams.edu/HistSci/curriculum/101/garbage.html

Tisdale, Bob, 2012, *Who Turned on the Heat? The Unsuspected Global Warming Culprit—El Nino-Southern Oscillation*, available online at:
http://bobtisdale.wordpress.com

Tisdale, Bob, 2013, On Cowtan and Way (2013) "Coverage Bias in the HadCRUT4 Temperature Series and Its Impact on Recent Temperature Trends," available online at:
https://bobtisdale.wordpress.com/2013/11/19/on-cowtan-and-ray-2013-coverage-bias-in-the-hadcrut4-temperature-series-and-its-impact-on-recent-temperature-trends/

Tobin, Mitch, 2013, Gallup Finds Rising Hostility Toward Environmentalists, May 30, Ecowest.com, available online at:
http://www.ecowest.org/2013/05/30/gallup-finds-rising-hostility-toward-environmentalists

Tol, Richard S. J., 2009, The Economics of Climate Change, *Journal of Economic Perspectives*, 23(2), Spring, pp.29-51, available online at:
http://www.econ.yale.edu/~nordhaus/homepage/documents/Tol_impacts_JEP_2009.pdf

Tol, Richard, 2014, Testimony before the US House Committee on Science and Technology, May 29, available online at:
http://science.house.gov/sites/republicans.science.house.gov/files/documents/HHRG-113-SY-WState-RTol-20140529_0.pdf

Tol, Richard, S. J., 2014a, Quantifying the Consensus on Anthropogenic Global Warming in the Literature: A re-analysis, *Energy Policy*, 73, October, 701-5.

Tribe, Laurence H., and Peabody Energy Corporation, 2014, Comments on EPA Carbon Emission Guidelines for Existing Stationary Sources: Electric Utility Generating Units, EPA Docket ID No. EPA-HQ-OAR-2013-0602, December 1, available online at

References

http://www.masseygail.com/pdf/Tribe-
Peabody_111(d)_Comments_(filed).pdf

Tribe, Laurence H., The Clean Power Plan Is Unconstitutional, 2014a,
Wall Street Journal, December 22, available online at:
http://www.wsj.com/articles/laurence-tribe-the-epas-clean-power-
plan-is-unconstitutional-
1419293203?mod=WSJ_LifeStyle_LatestHeadlines#livefyre-comment

United Nations Framework Convention for Climate Change, 2013,
*Warsaw International Mechanism for Loss and Damage Associated with Climate
Change Impacts*, available online at:
http://unfccc.int/files/meetings/warsaw_nov_2013/in-
session/application/pdf/fccc.cp.2013.l.15.pdf

United Nations Fund for the Development of West Irian, 1968, *A Design
for Development in West Irian: Report Prepared for the Government of Indonesia by
a Survey Mission Acting as Consultant for the Administrator of the United Nations
Development Programme*.

United Nations Intergovernmental Panel on Climate Change, 2001, *Third
Assessment Report (AR3)*, available online at:
http://www.ipcc.ch/ipccreports/tar/wg1/505.htm

United Nations Intergovernmental Panel on Climate Change, 2007,
Fourth Assessment Report (AR4), Cambridge University Press; available
online at: http://www.ipcc.ch

United Nations Intergovernmental Panel on Climate Change, 2012:
*Managing the Risks of Extreme Events and Disasters to Advance Climate Change
Adaptation. A Special Report of Working Groups I and II of the Intergovernmental
Panel on Climate Change*, C.B Field *et al.* (eds.). Cambridge University
Press, Cambridge, UK, and New York, NY.

United Nations Intergovernmental Panel on Climate Change, 2013, *Fifth
Assessment Report (AR5),Working Group 1 Draft Report*, Cambridge University
Press; available online at: http://www.ipcc.ch

United Nations World Commission on Environment and Development, 1987, *Our Common Future*, Oxford University Press, available online at: http://www.un-documents.net/our-common-future.pdf

US Chamber of Commerce, 2015, *In Their Own Words: A Guide to States' Concerns Regarding the Environmental Protection Agency's Proposed Greenhouse Gas Regulations for Existing Power Plants*, available online at: http://www.energyxxi.org/sites/default/files/FINAL%20EPA%20CPP%20Report%20FINAL.pdf

US Global Change Research Program, 2014 National Climate Assessment, available online at: http://nca2014.globalchange.gov.

US Energy Information Administration, 2011, Review of Emerging Resources, U.S. Shale Gas and Shale Oil Plays, July, available online at: http://www.eia.gov/analysis/studies/usshalegas/pdf/usshaleplays.pdf

US Energy Information Administration, 2013, *Updated Capital Cost Estimates for Utility Scale Electricity Generating Plants*, April, available online at: http://www.eia.gov/forecasts/capitalcost/pdf/updated_capcost.pdf

US Energy Information Administration, 2014, *Coal-Fired Power Plant Operators Consider Emissions Compliance Strategy*, March 28, available online at: http://www.eia.gov/todayinenergy/detail.cfm?id=15611

US Environmental Protection Agency, 1994, *Comments on Proposed NOAA/DOI Regulations on Natural Resource Damage Assessment*, October, available online at: http://yosemite.epa.gov/ee/epa/eerm.nsf/vwAN/EE-0309-1.pdf/$file/EE-0309-1.pdf

US Environmental Protection Agency, 2009, *Endangerment and Cause or Contribute Findings for Greenhouse Gases Under Section 202(a) of the Clean Air Act: EPA's Response to Public Comments*, available at: http://www.epa.gov/climatechange/endangerment/downloads/RTC Volume 1.pdf

References

US Environmental Protection Agency, 2009a, *Technical Support Document for Endangerment and Cause or Contribute Findings for Greenhouse Gases Under Section 202(a) of the Clean Air Act*, December 7, available online at: http://www.epa.gov/climatechange/Downloads/endangerment/Endangerment_TSD.pdf

US Environmental Protection Agency, 2009b, Endangerment and Cause or Contribute Findings for Greenhouse Gases under Section 202(a) of the Clean Air Act, December 7, available online at: http://www.epa.gov/climatechange/endangerment/

US Environmental Protection Agency, 2009c, Proposed Endangerment and Cause and Contribute Findings for Greenhouse Gases under Section 202(a) of the Clean Air Act, April 24, 74 *Federal Register* 18886, 18894.

US Environmental Protection Agency, 2011, Office of the Inspector General, *Procedural Review of EPA's Greenhouse Gases Endangerment Finding Data Quality Processes*, Report No. 11-P-0702, September 26, available online at: http://www.epa.gov/oig/reports/2011/20110926-11-P-0702.pdf

US Environmental Protection Agency, 2013, Office of the Inspector General, *Quick Reaction Report: EPA Must Take Steps to Implement Requirements of Its Scientific Integrity Policy*, Report No. 13-P-0364, August 28, available online at: http://www.epa.gov/oig/reports/2013/20130828-13-P-0364.pdf

US Environmental Protection Agency, 2014, Standards of Performance for Greenhouse Gas Emissions From New Stationary Sources: Electric Utility Generating Units, *Federal Register*, January 8, Vol. 79, Number 5, p. 1430, available online at: http://www.federalregister.gov/articles/2014/01/08/2013-28668/standards-of-performance-for-greenhouse-gas-emissions-from-new-stationary-sources-electric-utility.

US Environmental Protection Agency, 2014a, Carbon Pollution Emission Guidelines for Existing Stationary Sources: Electric Utility Generating Units, June 2, available online at:

http://www.globalwarming.org/wp-content/uploads/2014/06/EPA-Prepublication-Proposal-June-2-2014.pdf

US Geological Survey, 2009, Natural Oil and Gas Seeps in California, available online at: http://walrus.wr.usgs.gov/seeps/where.html.

US Geological Survey, 2011, Volcanic Gases and Climate Change Overview, available online at:
http://volcanoes.usgs.gov/hazards/gas/climate.php

US Government Accountability Office, 2013, *Climate Change: State Should Further Improve Its Reporting on Financial Support to Developing Countries to Meet Future Requirements and Guidelines*, Report GAO-13-829, September, available online at: http://www.gao.gov/assets/660/657985.pdf

US House of Representatives, 2009, H.R. 2454, American Clean Energy and Security Act of 2009, 111th Congress, available online at http://www.gpo.gov/fdsys/pkg/BILLS-111hr2454eh/pdf/BILLS-111hr2454eh.pdf

US House of Representatives, 2013, *H.R. 367, Regulations from the Executive in Need of Scrutiny Act of 2013*, 113th Congress, available online at https://www.congress.gov/bill/113th-congress/house-bill/367

US Senate, 2012, Committee on Environment and Public Works, Minority Staff, *A Look Ahead to EPA Regulations for 2013*, October, available online at: http://www.inhofe.senate.gov/download/?id=28b57b78-30ba-4d78-bf32-2259797e513f&download=1

US Senate, 2014, Committee on Environment and Public Works, Minority Staff Report, *EPA's Playbook Unveiled: A Story of Fraud, Deceit, and Secret Science*, March 19, available online at:
http://www.epw.senate.gov/public/index.cfm?FuseAction=Files.View&FileStore_id=b90f742e-b797-4a82-a0a3-e6848467832a

US Senate, 2014a, Committee on Environment and Public Works, Minority Staff Report, *How a Club of Billionaires and Their Foundations Control the Environmental Movement and Obama's EPA*, July 30, available online at:

References

http://www.epw.senate.gov/public/index.cfm?FuseAction=Files.View&FileStore_id=8af3d005-1337-4bc3-bcd6-be947c523439

US Senate, 2014b, Committee on Environment and Public Works, Minority Staff Report, *Critical Thinking on Climate Change, Evidence to Consider before Taking Regulatory Action and Implementing Economic Policies*, September 4, available online at:
http://www.epw.senate.gov/public/index.cfm?FuseAction=Files.View&FileStore_id=3f33b3c9-a28b-4f6c-a663-50c7d02fda24

US Senate, 2014c, Committee on Environment and Public Works, Minority Staff Report, *Setting the Record Straight: Hydraulic Fracturing and America's Energy Revolution*, available online at:
http://www.epw.senate.gov/public/index.cfm?FuseAction=Files.View&FileStore_id=beed1c2e-1210-48f1-8367-f967fab49c38

Vartabedian, Ralph, and Peter Pae, 2003, A Barrier that Could Have Been, *Los Angeles Times*, September 9, available online at:
http://articles.latimes.com/2005/sep/09/nation/na-surge9

Viscusi, W. Kip, Wesley A. Magat, Alan Carlin, and Mark Dreyfus, 1994, Environmentally Responsible Energy Pricing, *Energy Journal*, Vol. 15, No. 2, April, pp. 23-42, available online at:
http://law.vanderbilt.edu/files/archive/137_Environmentally_Responsible_Energy_Pricing.pdf.

Walden, Andrew, 2010, Wind Energy's Ghosts, *American Thinker*, February 15, available online at:
http://www.americanthinker.com/articles/2010/02/wind_energys_ghosts_1.html

Wall Street Journal Editorial Board, 2014, Lima's Magic Beanstalk, December 19, available online at: http://www.wsj.com/articles/limas-magic-climate-beanstalk-1418947069

Wall Street Journal Editorial Board, 2015, Meth Heads in the White House, January 4, available online at: http://www.wsj.com/articles/meth-heads-in-the-white-house-1420412043

Walsh, Brian, 2013, The Medieval Volcanic Eruption that Triggered a Year without Summer, *Time*, available online at http://science.time.com/2013/10/02/the-medieval-volcanic-eurption-that-triggered-a-year-without-summer/

Wang, et al., 2013, Residential Radon and Lung Cancer Risk in a High-exposure Area of Gansu Province, China, *American Journal of Epidemiology*, 155(6), pp. 554-64.

Watts, Anthony, 2010, The scandal deepens—IPCC AR4 riddled with Non Peer Reviewed WWF papers, available online at:
http://wattsupwiththat.com/2010/01/24/the-scandal-deepens-ipcc-ar4-riddled-with-non-peer-reviewed-wwf-papers/

Watts, Anthony, Evan Jones, Stephen McIntyre, and John R. Christy, 2012, An Area and Distance Weighted Analysis of the Impacts of Station Exposure on the U.S. Historical Climatology Network Temperatures and Temperature Trends, available online at:
http://wattsupwiththat.com/2012/07/29/press-release-2

Watts, Anthony, 2013, US Meets Kyoto Protocol Goal—without Ever Embracing It, April 5, available online at:
http://wattsupwiththat.com/2013/04/05/usa-meets-kyoto-protocol-without-ever-embracing-it/

Watts, Anthony, 2013a, Pielke, Jr. Agrees—'Extreme Weather to Climate Connection' Is a Dead Issue, October 3, available online at:
http://wattsupwiththat.com/2013/10/03/pielke-jr-agrees-extreme-weather-to-climate-connection-is-a-dead-issue/

Watts, Anthony, 2014, 5.7 Million Dollar NSF Grant to Columbia University for Climate 'Voice Mails from the Future, May 26, available online at: http://wattsupwiththat.com/2014/05/26/5-7-million-nsf-grant-to-columbia-university-for-climate-voice-mails-from-the-future/

Watts, Anthony, 2014a, The Color of the Climate Social Movement Zombie Says 'Red," September 24, available online at:
http://wattsupwiththat.com/2014/09/24/the-color-of-the-climate-social-movement-zombie-says-red/

References

Watts, Anthony, 2014b, Touchy Feely Science—One Chart Suggests There's a 'pHraud' in Omitting Ocean Acidification Data in Congressional Testimony, December 23, available online at: http://wattsupwiththat.com/2014/12/23/touchy-feely-science-one-chart-suggests-theres-a-phraud-in-omitting-ocean-acidification-data-in-congressional-testimony/

Wayland, Robert, 2013, EPA Responses to Initial OMB/Interagency Comments, April 19, US Environmental Protection Agency, available online at: http://www.globalwarming.org/wp-content/uploads/2014/04/OMB-Comments-and-EPA-response-on-carbon-pollution-rule-April-19-2014.pdf.

Wetzel, Daniel, 2013, Germany's Renewables Fiasco: Doldrums and Clouds Bring Green Electricity Production to a Halt, *Die Welt*. December 24, English translation available from The Global Warming Foundation, December 25 and available online at: http://www.thegwpf.com/renewables-fiasco-doldrums-clouds-bring-green-electricity-production-halt/

White, Joseph P., 2014, Detroit Gallops Ahead: New Cars Have More Horsepower, Weigh More and Go Farther, *Wall Street Journal*, January 16, available online at: http://online.wsj.com/news/articles/SB10001424052702304419104579324801579241562

White House, 2014, US-China Joint Announcement on Climate Change, November 12, available online at: http://www.whitehouse.gov/the-press-office/2014/11/11/us-china-joint-announcement-climate-change

Wilde, Stephen, 2012, Evidence that Oceans Not Man Control CO_2 Emissions, April 26, ClimateRealists.com, available online at: http://climaterealists.com/index.php?id=9508&linkbox=true&position=4

WND, 2015, New Inquisition: Punish Climate Change 'Deniers,' March, available online at http://www.wnd.com/2015/03/new-inquisition-punish-climate-change-deniers/

Wohlstetter, Albert, 1959, The Delicate Balance of Terror, *Foreign Affairs*, January

Wojick, David E., 2001, The UN IPCC's Artful Bias, July 13, available online at: http://www.john-daly.com/guests/un_ipcc.htm

Yeatman, William, 2011, The EPA Cannot Be Trusted to Keep the Lights On, August 19, available online at:
http://www.globalwarming.org/2011/08/19/the-epa-cannot-be-trusted-to-keep-the-lights-on/.

Yeatman, William, 2013, *The U.S. Environmental Protection Agency's Assault on State Governments, American Legislative Exchange Council*, available online at: http://alec.org/docs/EPA_Assault_State_Sovereignty

Yeatman, William, 2014, Don't Ask about Climate Rule's Impact on Reliability, Because EPA Administrator Gina McCarthy Is "Tired" of Hearing about It, June 3,available online at:
http://www.globalwarming.org/2014/06/03/dont-ask-about-climate-rules-impact-on-reliability-because-epa-administrator-gina-mccarthy-is-tired-of-hearing-about-it

Yeatman, William, 2014a, Can FERC Chair Nominee Norman Bay Be Trusted to Keep the Lights On?, June 11, available online at:
http://www.globalwarming.org/2014/06/11/can-ferc-chairman-nominee-norman-bay-be-trusted-to-keep-the-lights-on/#more-20460

Yeatman, William, 2014b, Primer: The Ongoing Controversy over Whether Clean Air Act §111(d) Authorizes EPA's Clean Power Plan, July 2, available online at:
http://www.globalwarming.org/2014/07/02/primer-the-ongoing-controversy-over-whether-clean-air-act-%C2%A7111d-authorizes-epas-clean-power-plan/

References

Yeatman, William, 2014c, Graph: Obama's EPA Is Killing Cooperative Federalism, January 9, available online at:
http://www.globalwarming.org/2014/01/09/graph-obamas-epa-is-killing-cooperative-federalism/#more-18389

Yeatman, William, 2014d, Greens' Pending Suit over Aviation GHGs Demonstrates Dangers of EPA's Climate Change Regulatory Regime, August 7, available online at:
http://www.globalwarming.org/2014/08/07/greens-pending-suit-over-aviation-ghgs-demonstrates-dangers-of-epas-climate-change-regulatory-regime-2/#more-21120

Zdanowicz, Christian, *et al.*, 2012, Summer Melt Rates on Penny Ice Cap, Baffin Island: Past and Recent Trends And Implications For Regional Climate, *Journal of Geophysical Research*, Vol. 117(F2), June

Index

CPSIA information can be obtained
at www.ICGtesting.com
Printed in the USA
LVOW08s1157261216
518698LV00005B/574/P